Diether Neubert Robert J. Kavlock
Hans-Joachim Merker Jane Klein (Eds.)

Risk Assessment of Prenatally-Induced Adverse Health Effects

With 106 Figures and 100 Tables

Springer-Verlag
Berlin Heidelberg New York
London Paris Tokyo
Hong Kong Barcelona
Budapest

Professor Dr. Diether Neubert
Institut für Toxikologie und Embryopharmakologie
Universitätsklinikum Rudolf Virchow
Freie Universität Berlin
Garystrasse 5, W-1000 Berlin 33, FRG

Dr. Robert J. Kavlock, Director
Developmental Toxicology Division
US Environmental Protection Agency
Health Effects Research Laboratory (MD-71)
Research Triangle Park, NC 27711, USA

Professor Dr. Hans-Joachim Merker
Institut für Anatomie
Freie Universität Berlin
Königin-Luise-Strasse 15, W-1000 Berlin 33, FRG

Jane Klein
Institut für Toxikologie und Embryopharmakologie
Freie Universität Berlin
Garystrasse 5, W-1000 Berlin 33, FRG

Library of Congress Cataloging-in-Publication Data
Risk assessment of prenatally-induced adverse health effects / Diether Neubert ... [et al.], hrsg. "Compiles the presentations and discussion of a symposium organized by the Institute of Toxicology and Embryopharmacology, Free University of Berlin, on May 21-23 1991, in Berlin-Dahlem". – Pref. Includes bibliographical references and index.
ISBN-13: 978-3-642-77755-4 e-ISBN-13: 978-3-642-77753-0
DOI: 10.1007/ 978-3-642-77753-0

1. Reproductive toxicology–Congresses. 2. Developmental toxicology–Congresses. 3. Human reproduction–Effect of chemicals on–Congresses. 4. Health risk assessment–Congresses. I. Neubert, D. (Dieter) RA 1224.2.R44 1992 618.3'2–dc20

Softcover reprint of the hardcover 1st edition 1992

Typesetting: Camera ready by editors
27/3145 - 5 4 3 2 1 0 - Printed on acid-free paper

Preface

This publication compiles the presentations and discussion of a symposium organized by the *Institute of Toxicology and Embryopharmacology*, Free University of Berlin on May 21-23, 1991, in Berlin-Dahlem. Participants from seven countries from university institutions, industry and regulatory agencies joined in these discussions.

This compilation has been supplemented with some contributions which were not presented at the symposium but which fit into the framework of the topic of the meeting, therefore, allowing us to include new information. Some authors updated their original manuscripts, thus making this a current presentation of the data. Where available, a short protocol of the podium discussions have also been included at the end of each contribution.

Our institute has regularly organized symposia dealing with topics from the field of *Reproductive Biology and Toxicology* for the last 20 years. This the 9th symposium of this kind.

The Institute of Toxicology and Embryopharmacology at the Free University of Berlin is unique in its size (about 120 coworkers, including 35 scientists) and research orientation, and for more than 20 years has concentrated on problems of reproductive biology and toxicology.

The Institute of Toxicology is the center of a Special Research Association ("Sonderforschungsbereich 174") devoted to: *Risk Assessment of Prenatally-Induced Lesions*. In Germany, "Special Research Associations" are local concentrations of 10 to 25 intensively cross-linked groups concentrating their research on a defined topic. They represent the highest level of funding by the

Deutsche Forschungsgemeinschaft (DFG), controlled by an extensive peer-review procedure. Predominantly located within the *Universitätsklinikum Rudolf Virchow* of the Free University of Berlin, problems of the above mentioned research area are presently studied jointly by eight clinical and twelve experimental groups, also collaborating with one group from the Bundesgesundheitsamt (Federal Health Office).

Our institute has also been selected as *WHO Collaborative Centre* for developmental toxicology.

We want to thank the Free University of Berlin and the Deutsche Forschungsgemeinschaft for the financial help in organizing this symposium. The financial support received from Bayer AG, Wuppertal and the Verband der Chemischen Industrie is also greatly appreciated.

Of course, the success of such a meeting results from the joint effort of many (not to say all) members of the institute. While the contribution of each single member of the department is highly appreciated, the main load of the organization always rests on a few. In this case our special thanks go to Doris Webb for the general organization of the meeting and her help during the work on this book, and to Peter Abt for taking care of all the financial details.

We hope that this publication will aid in recognizing and perhaps solving the many problems connected with risk assessment in the area of reproductive and developmental toxicology.

Berlin, May 1992

Diether Neubert
Robert J. Kavlock
Hans-Joachim Merker
Jane Klein

Contents

List of First Authors and Discussants

Hashem Abdul-Khaliq
Universitätsklinikum Rudolf Virchow
Kinderklinik und Poliklinik
Abteilung für Neonatologie
Kaiserin Auguste Viktoria Haus
Heubnerweg 6
W-1000 Berlin 19

Dr. Elizabeth Anderson
President
Clement International Corporation
9300 Lee Highway
Fairfax, Virginia
USA 22031

Prof. Dr. Rolf Baß
Institut für Arzneimittel
des Bundesgesundheitsamtes
Seestraße 10
W-1000 Berlin 65

Prof. Dr. Dr. H.M. Bolt
Institut für Arbeitsphysiologie
Ardeystraße 67
D-4600 Dortmund 1

Dr. Jochen Buschmann
ITA - Frauenhofer-Institut
für Toxikologie und
Aerosolforschung
Nikolai-Fuchs-Straße 1
D-3000 Hannover 61

Prof. Dr. Dr. Jorge Cervós-Navarro
Institut für Neuropathologie
der Freien Universität Berlin
Hindenburgdamm 30
W-1000 Berlin 45

Dr. Ibrahim Chahoud
Institut für Toxikologie
und Embryopharmakologie
der Freien Universität Berlin
Garystr. 5
W-1000 Berlin 33

Dr. Gerald F. Chernoff
California Dept. of Health Services
Reproductive and Cancer Hazard
Assessment Section
Health Hazard Assessment Division
714 P Street, Room 476
Sacramento, California
USA 95814

Prof. Dr. Jean R. Claude
Faculte des Sciences Pharmaceutiques
6, avenue de l'Observatoire
75006 PARIS
France

Prof. Dr. Joachim W. Dudenhausen
Universitätsklinikum Rudolf Virchow
Abt. Geburtsmedizin
Pulsstraße 4
W-1000 Berlin 19

Prof. Dr. med. Harald Frohberg
Direktor
E. Merck
Frankfurter Straße 250
D-6100 Darmstadt 1

Dr. Janine Goujard
INSERM Unité 149
123, Boulevard de Port Royal
F-75014 PARIS

Priv.-Doz. Dr. Renate Graf
Institut für Anatomie der FUB
FB 01, WE 01
Königin-Luise-Str. 15
W-1000 Berlin 33

Priv.-Doz. Dr. Annette Grüters
Kinderklinik der Freien Universität
Heubnerweg 6
W - 1000 Berlin 19

Priv.-Doz. Dr. Wolfgang Heger
Umweltbundesamt
Mauer Str. 52
O-1080 Berlin

Dr. Ruth Hildebrandt
Frauenklinik und Poliklinik (WE 4)
Hindenburgdamm 30
W-1000 Berlin 45

Dr. Barbara Heinrich-Hirsch
Max von Pettenkofer-Institut
Unter den Eichen 82 - 84
W-1000 Berlin 33

Prof. Dr. med. Hans Helge
Kinderklinik der Freien Universität
Heubnerweg 6
W-1000 Berlin 19

Dr. Robert J. Kavlock
Director,
Developmental Toxicology
Division MD-71
US Environmental Protection Agency
Research Triangle Park Laboratories
Research Triangle Park, North
Carolina
USA 27711

Dr. Stephan Klug
Institut für Toxikologie und
Embryopharmakologie
Freie Universität Berlin
Garystraße 5
W-1000 Berlin 33

Dr. Elisabeth Koch
Institut für Toxikologie und
Embryopharmakologie
Freie Universität Berlin
Garystraße 5
W - 1000 Berlin 33

Dr. Ralf Krowke
Institut für Toxikologie und
Embryopharmakologie
Freie Universität Berlin
Garystraße 5
W-1000 Berlin 33

Dr. Wolfgang Lingk
Max von Pettenkofer-Institut
Unter den Eichen 82 - 84
W-1000 Berlin 33

Dr. Ludwig Machemer
Bayer AG
Fachbereich Toxikologie
Pharma Research Centre
Wuppertal-Elberfeld

Dr. Jeanne Manson
Merck Sharp & Dohme
Research Laboratory
Dept. of Safety Assessment
West Point, PA
USA 19486

Prof. Dr. Reinhard Meister
Technische Fachhochschule Berlin
Luxemburger Str. 10
W-1000 Berlin 65

Prof. Dr. Hans-Joachim Merker
Institut für Anatomie der FUB
FB 01, WE 01
Königin-Luise-Str. 15
W-1000 Berlin 33

Prof. Dr. Dr. hc. Heinz Nau
Institut für Toxikologie
und Embryopharmakologie,
der Freien Universität Berlin
Garystr. 5
W-1000 Berlin 33

Prof. Dr. Diether Neubert
Institut für Toxikologie und
Embryopharmakologie
Freie Universität Berlin
Garystraße 5
W-1000 Berlin 33

Dr. Reinhard Neubert
Kinderklinik der Freien Universität
Heubnerweg 6
W-1000 Berlin 19

Prof. Dr. Michael Obladen
Universitätsklinikum Rudolf Virchow
Kinderklinik und Poliklinik
Abteilung für Neonatologie
Kaiserin Auguste Viktoria Haus
Heubnerweg 6
W-1000 Berlin 19

Anthony K. Palmer
Huntingdon Research Centre Ltd.
P.O. Box 2
Huntingdon, Cambridgeshire
PE18 6ES England

Prof. Dr. Francisco R. Paumgartten
Ministério da Saúde
Fundaçao Oswaldo Cruz (FIOCRUZ)
Instituto Nacional de Controle
de Qualidade em Saúde - INCQS
Departamento de Farmacologia e
Toxicologia
Av. Brasil, 4.365 - Manguinhos
CEP 21045 - RIO DE JANEIRO, RJ
Brazil

Prof. Dr. Paul Peters
National Institute for Public Health
and Environmental Protection
P.O. Box 1
3720 BA Bilthoven
The Netherlands

Priv.-Doz. Dr. Thomas Platzek
Max von Pettenkofer-Institut
des Bundesgesundheitsamtes
Thielallee 88 - 92
W-1000 Berlin 33

Priv.-Doz. Dr. Gabriela Rune
Institut für Anatomie der FUB
FB 01, WE 01
Königin-Luise-Str. 15
W-1000 Berlin 33

Dr. Andrew C. Scallet
Division of Reproductive and
Developmental Toxicology
National Center for Toxicological
Research
NCTR Drive, HFT-132
Jefferson, Arkansas 72079-9502
USA

Dr. Horst Schleusener
Medizinisches Klinik und
Poliklinik
Hindenburgdamm 30
W-1000 Berlin 45

Dr. Thomas Schulz-Schalge
Institut für Toxikologie und
Embryopharmakologie
Freie Universität Berlin
Garystraße 5
W-1000 Berlin 33

Dr. William J. Scott
Department of Pediatrics
The Children's Hospital Research
Foundation
College of Medicine
University of Cincinnati
Cincinnati, Ohio 45229
USA

Dr. Hugo Segerer
Universitätsklinikum Rudolf Virchow
Kinderklinik und Poliklinik
Abteilung für Neonatologie
Kaiserin Auguste Viktoria Haus
Heubnerweg 6
W-1000 Berlin 19

Dr. R. Woodrow Setzer
Mail Drive 68
Genetic Toxicology Division
Environmental Protection Agency
Research Triangle Park, North Carolina
USA 27711

Prof. Dr. Arpad Somogyi
Max von Pettenkofer-Institut
Unter den Eichen 82 - 84
W-1000 Berlin 33

Priv.-Doz. Dr. Ralf Stahlmann
Institut für Toxikologie und Embryopharmakologie
Freie Universität Berlin
Garystraße 5
W-1000 Berlin 33

Dr. Helmut Sterz
Institut de Recherches Internationales (IRIS)
14 rue de Bezons
92414 Courbevoie Cédex
France

Prof. Dr. med. Ulrich Wahn
Kinderklinik der Freien Universität
Heubnerweg 6
W-1000 Berlin 19

Dr. Judi Weissinger
Center for Drug Evaluation and Drug Research
Food and Drug Administration
5600 Fisher's Lane, 13 B - 28
Rockville, Maryland
USA 20857

Dr. Ashley Wickramaratne
Central Toxicology Laboratory, ICI Plc
Alderley Park
Cheshire SK10 4TJ
England

Dr. Keisuke Yamashita
University of Hiroshima
Department of Anatomy
School of Medicine
1-2-3, Kasumi, Minami-ku,
Hiroshima 734
Japan

Prof. Dr. Mineo Yasuda
Department of Anatomy
Hiroshima University
School of Medicine
1-2-3 Kasumi, Minami-ku
Hiroshima 734
Japan

Introduction

Diether Neubert and Robert J. Kavlock

Risk assessment of prenatally-induced adverse health effects is still a considerable problem, and it will continue to be so for the next decade.

The reasons for this are both:

- *scientific*, i.e. the basis for our understanding of developmental processes is still very rudimentary. However, developmental biology is entering a rapidly growing phase, with knowledge of the molecular events during mammalian pattern formation increasing steadily. The probability that this knowledge will begin to impact risk assessment within the next decade seems to be rather high, and
- *pragmatic*, i.e. no clear-cut strategy has been agreed upon nor followed in the various institutions involved in such a risk assessment.

The aim of this symposium was to discuss scientific prerequisites for a risk assessment in reproductive toxicity, and especially to discuss the possibilities for a *quantitative* risk assessment of prenatally-induced toxic effects. In some respects, the general principles of assessing toxic effects must be resolved, therefore some of the aspects covered are not only confined to reproductive toxicity.

Aspects to be considered in such a risk assessment include:

- endpoints to be evaluated,
- terminology,
- problem of using excessive doses,
- problem of "maternal toxicity",
- qualitative vs. quantitative aspects of toxicity,
- dose-response relationships,

- pharmacokinetic aspects,
- species differences, and extrapolations between species,
- the use of mathematical models,
- role of observations in man,
- significance of *in vitro* data,
- significance of prenatally-induced dysfunctions.

Of course, these are not all of the problems existing in this field, and every aspect could not be covered during these discussions. Several of the problems have been reviewed by us before (Neubert et al. 1980; Neubert et al. 1987).

Endpoints to be Evaluated

In the emotional opinion of the public, and also from a scientific point of view, not all of the possible endpoints of reproductive toxicity are given the same weight. Most parents would generally consider gross-structural abnormalities ("malformations") to represent the greatest threat. However, an increased risk for spontaneous abortion (the majority of which are unrecognized) may be more insidious, and the chronic consequences of behavioural and cognitive deficits as well as defects of the hormonal and immune system in children may actually presently be underestimated.

Although in the course of a scientific evaluation all possible deviations from normal reproduction and development should be taken into account (including many other toxicological and kinetic aspects), again a considerable controversy exists in defining what weight should be given to the individual types of manifestations deviating from normal (or from controls). For this two main reasons may be offered:

(a) the significance of certain deviations from the norm for the situation possibly existing in man may be largely *unknown* or controversial, or

(b) some embryofeto-toxic effects may be induced "*unspecifically*", i.e. by any agent at a high enough dose.

ad (a): While the significance of the occurrence of many drastic gross-structural "malformations" (e.g. amelia, transposition of the great vessels, etc.) does not pose major problems in risk assessment, this is not so in the case of an increased frequency of several minor structural abnormalities or especially of morphological variations (also occurring normally at a certain rate). Prenatally-induced dysfunctions have hardly begun to be considered, up till now.

There has never been a general agreement as to what should be considered a "malformation". However, almost all investigators would agree that it should be:

- *congenital* (i.e. induced prenatally, and preferentially recognizable at birth), and
- *persistent* (i.e. not reversible during early postnatal life).

In two earlier papers Lorke (1963, 1977) defined such defects with respect to the *skeleton* as representing the *absence, bending* or *shortening, fusion* or *clefts* of a bone anlage.

According to this definition a dilated pelvis induced during infancy would not represent a "congenital malformation", nor would this term apply for "wavy ribs" or delayed ossifications which are reversible postnatally and could only be described as "transient congenital structural defects".

Another non-scientific but pragmatic aspect concerns the terms "major" or "minor" abnormality (or "malformation"), because of the consequences to be drawn in a risk assessment. Again, the definition is easy in the case of very pronounced effects (e.g. amelia or phocomelia), but otherwise the terminology is rather a matter of the definition used in the evaluating laboratory, and no international agreement has been attempted so far. In the case of human defects again each evaluating study group has to state their definition of individual defects (*see*, e.g., Heinonen et al. 1977).

A special point of controversy may arise when a substantial increase in a defined type of "variation" is observed in a study. While many investigators do not give too much weight to such deviations from normal development if they occur in the absence of clear-cut abnormalities, there have been some agencies suggesting that such increased rates of variations may be judged as an indication of a "teratogenic" potential. The problem becomes more complex when at higher doses clear-cut "major" abnormalities are observed. The question then arises as to what should be considered a NOAEL (no-observed-adverse-effect-level), and whether the doses only inducing the increased rate of variations should be considered as "clean". Again, no international agreement has been reached up till now on such a situation, and every investigator has to use his/her own judgement.

Another aspect concerns the terms "retardation" or "abnormality" with respect to skeletal deviations from the norm. Per definition a retardation would be of little concern, since it should be reversible, but this reversibility has by no means been proven for all of the so-called retardations. The only scientific way out is to demonstrate that the defect called "retardation" is in fact reversible, which may be achieved by studying later (e.g. postnatal) stages of development. In a number of cases such investigations have been performed, sometimes ending with the surprise that the defect was not readily reversible. Such defects are often located at the vertebral column (e.g. bipartite vertebrae, dumb-bell shaped vertebrae, etc.). Also in these cases no international agreement has been achieved on the nomenclature and the consequences for risk assessment. Furthermore, a retarded development or "dystrophy" may appear "minor" in experimental studies, but it may create considerable problems in human neonatology.

ad (b): The second considerable problem concerns the question as to which types of abnormalities may be inducible by "unspecific" means, e.g. by "maternal toxicity". Such manifestations would then be demonstrable with any agent at a high enough dose.

This is in fact one of the major problems for the assessment of an embryofetotoxic potential or potency with relevance for man. Especially if we try to consider *all* possible deviations from normal development; *there is no doubt that some "abnormal" manifestations can be induced with any agent*, provided the doses are high enough (as in all toxicological evaluations it may not be practical to apply such a [huge] dose in the case of some agents). Manifestations to be expected subsequent to sufficiently high doses of any agent (often in the presence of toxic manifestations in the dam) may include (Neubert et al. 1980):

- prenatal mortality,
- prenatal growth retardation,
- increased rates of certain "variations",
- possibly some minor or major gross-structural abnormalities.

Since this fact is well known to any experienced investigator in this field of toxicology, the assessment of a possible prenatal toxicity will turn out to be a *quantitative* problem. Some further aspects of this problem will be discussed later.

Terminology

Terminology is often used to facilitate understanding of complex facts. Unfortunately, the extended use (and misuse) of such terms contributes more to confusion and not to clarification. Often such simplified terminology has been used by regulatory agencies for classification and labelling, as well as for information for the public. *Presently such efforts are prone to contribute rather to misinformation, since complicated facts are oversimplified to an unacceptable extent.*

Scientific knowledge within the last decades has led to the understanding that the causes for many adverse effects in toxicology are manifold, as are the final manifestations. For this reason simplified terms such as "cancerogenic" or "teratogenic" have largely lost all scientific as well as pragmatic meaning, since they may be justified for a few strongly acting agents, but do not properly describe the majority of effects observed in experimental studies.

This situation is regrettable for the public and especially for journalists with a minimum specific knowledge who then find it is easy to paint black-and-white pictures: cancerogenic or teratogenic or non-cancerogenic or non-teratogenic, suggesting that a substance is capable of inducing such effects in almost every individual. The result is the easy to publish and well read horror stories that we are all surrounded by thousands of life-threatening substances in a completely poisoned environment.

Such a scenario only inadequately mirrors reality, and it intrinsicly carries the danger that our senses are detracted and no longer focussed on the real problems surrounding us, i.e. the agents which pose real danger.

From a scientific point of view and with respect to risk assessment we have not achieved any clarification when a substance is qualitatively designated as "cancerogenic" or "teratogenic". A scientist would have to ask: just what exactly is meant with these labels? Therefore, it is only helpful to provide extensive information, e.g.:

> *"... an increased cleft palate frequency (32%) has been induced in XY-mice at doses causing 20% weight loss in the dams; no difference to the frequency compared with controls was seen with 2/3 of this dose, nor was there an increase in prenatal mortality seen under these conditions; there was no clear-cut adverse effect seen in rats and rabbits at the maximally tolerated doses..."*.

It seems highly questionable whether this observation should be considered alarming with respect to a hazard for man. Simply labelling such substances as "teratogenic" would be neither informative, comparable nor "fair" when compared with, e.g., defects observed in monkeys with thalidomide, a classical "teratogenic" substance for man.

The confusion is perfect when the term "teratogenic" (well-defined as prenatal induction of permanent structural abnormalities) is misused as "behavioural teratology", indicating possible postnatal manifestations in CNS functions induced via prenatal exposure. In an analogous way prenatally-induced permanent arterial hypertension should then be "hypertentional teratology", the induction of persistent hormonal changes would be "hormonal teratology" and the prenatal triggering of permanent lesions of the immune system "immunologic teratology". What confusion of the mind! Since "teratogenic" is still defined as referring to "gross-structural" and "persistent", most of the "behavioural changes" observed in experimental animals do not deserve this term anyway, because they are transient. Why don't we say what we mean, e.g. prenatally-induced dysfunctions, or prenatally-induced (reversible) learning disability, e.g. in rats.

More serious is the lack of consistency in the description of defects in reports and in the literature. A certain alteration of cartilage is sometimes, e.g., designated as "chondrodystrophy", although in the medical field this name is reserved for a special type of defect, certainly not related to the experimental abnormality observed. Similarly, other defects are quite differently designated without giving a detailed description of the defect. In this area it would be extremely helpful if regulatory agencies could help to establish a catalogue in which the terminology is combined with a typical picture of this type of defect. Using the same nomenclature among scientists would greatly facilitate communication, and finally also risk assessment. Attempts in this direction have been initiated by the Bundesgesundheitsamt (Federal Health Office).

Problem of Using Excessive Doses

A strategy in toxicological evaluations of the last decades has been to compensate for small numbers of experimental animals by using excessive doses. Although pragmatic and easy to perform, there is increasing concern that this strategy may be basically wrong and unsuitable for risk assessment relevant to man, and that the results obtained from such studies may easily give rise to misinterpretation.

Such a strategy is used to a lesser extent for medicinal substances, because here the dose range used in experimental studies often does not deviate that much from the anticipated human exposure, especially when serum concentrations are compared rather than doses. However, especially when evaluating environmental chemicals doses used in testing and human exposure often diverge by three to four orders of magnitude. One may argue that the exposure may also happen to three or more orders of magnitude individuals, and that it is public health policy to protect as many people on the tolerance distribution as possible. However, inevitably linked with the generation of data obtained after administration of excessive doses, is the necessity to extrapolate the results to much smaller exposures, with all the limitations intrinsic to such a procedure. It would also imply that there *is* an effect at extremely low doses, which is doubted by many experts in the field because of mechanistic considerations, and if there were still an effect the probability would rapidly decline and the cases would be hidden within the background level. An, even worse, alternative would be the labelling of substances according to the outcome of such studies with overdoses as "cancerogenic" or "teratogenic", i.e. the qualitative characterization of a toxic potential. *However, postgraduate students soon learn that toxicity is a quantitative phenomenon! It cannot be connected to a substance but only to a dose.*

While much was learned by this, apparently misleading, strategy, we have to be open-minded enough to look for alternatives. Data on toxicity have to bear relevance to the possible exposure of man, and the faint idea that something might happen at extreme exposures is of little help for everyday discussions and risk assessment.

Problem of "Maternal Toxicity"

One of the crucial problems in testing for possible reproductive or developmental toxicity is the fact that in mammals prenatal toxicity must always be assessed within the maternal organism. For this reason, the reproductive toxicity observed on the embryo, fetus or offspring must always be related to possible concurrent toxic effects in the mother.

It is generally agreed that prenatal toxicity is especially relevant if it occurs at doses below those affecting the maternal organism. This automatically leads to the problems that:

- prenatal toxicity should best be assessed in the absence of maternally toxic effects, and
- if maternally toxic effects have been observed at a given dose range, then the contribution of this to a possible adverse effect on the conceptus must also be assessed.

While it would be easy to satisfy the first prerequisite, this is in fact presently complicated by the recommendation of many regulatory agencies that the highest dose used in the studies should induce clear-cut signs of maternal toxicity. The intention of such a recommendation is apparently to ensure that the highest dose is not chosen to be far below a possible adult toxicity. However, this could be achieved much easier with dose-finding studies on maternal toxicity, and subsequently by choosing the highest dose in the main study just below the dose inducing maternal toxicity. In this way the inevitable difficulty of interpreting the results of a scientifically inadequate study, just satisfying regulatory demands, could largely be avoided (Neubert 1992).

The biggest difficulty is to define what "maternally toxic" means. Since the general term "toxic" cannot be defined, the same must hold for "maternal toxicity". It cannot be expected that all the different types of toxic alterations inducible within the maternal organism, leading to quite variable metabolic and functional manifestations, will have the same impact on the embryo or fetus. Furthermore, many pharmacological effects, which cannot simply be called "toxic", may interfere with certain facets of prenatal development. The presently used criterion of a reduction in body weight may be convenient and pragmatic, but it certainly is a far too simple approach. Severe defects in hormonal balance, in the immune system, or certain organ toxicities may also occur without concomitant weight loss.

Almost nothing is known on the possible influence of *specific* and defined maternally toxic effects on reproduction and development. With several pesticides and drugs, e.g., an increased frequency of cleft palates has been found in rats (per definition a clear-cut "teratogenic" effect) at maternally toxic doses only. Up till now, it is almost impossible to judge whether this was the result of a direct effect of the high dose of this substance on embryonic development, or whether this may be the result of, more or less specific, drastic functional or metabolic changes induced in the maternal organism.

The co-presence of maternal and developmental toxicity cannot *a priori* be used to diminish the significance of embryofeto-toxic effects. On the other hand, it will almost never be possible through animal experimentation to exclude the possibility of embryofeto-toxic effects occurring in humans in the presence of severely toxic symptoms in the mother (e.g. after accidental releases).

Several attempts have been made to press the relationship of maternal and embryofetal toxicity into a single or a set of number(s). While some of the more complex attempts may be helpful to illustrate the complex situation:

mother/conceptus in a condensed form, the value of such ratios should not be overestimated. They are misleading and scientifically worthless when a single ratio is calculated from different species or from data of various laboratories, since the basic laws of pharmacology and toxicology are violated in such an attempt. In such cases several prerequisites cannot be met:

- only data on acute or repeated-dose toxicity obtained in the same animals or in the same strain under the exact laboratory conditions are directly comparable.
- LD_{50} values are well-known to vary up to 10-fold when assessed in different laboratories. The variability in the lower range (e.g. LD_{10}) is bound to be even less accurate, even in the same study.
- NOEL or LOEL values per definition cannot represent exact doses (since "zero" cannot be defined in a dose-response curve), but they are rough estimates of upper or lower limits; according to the dose range tested these dose estimates are more or less accurate (also note the statistical problems).
- It is no longer acceptable in pharmacology or toxicology to compare effects between species on the basis of doses, since it is well-known that the kinetics differ considerably. Therefore, any attempt in this direction is bound to fail.

For these reasons, and also for statistical reasons (Setzer and Rogers 1989), single so-called "A/D ratios" cannot be calculated with any scientific background, and they should no longer (and should not have been from the beginning) be considered in risk assessment.

Qualitative vs. Quantitative Aspects of Toxicity

It is without any doubt that a few substances with well-defined toxic potentials may induce special effects at low and relevant doses at a high frequency in many species. From the action of these substances terms such as "cancerogenic" or "teratogenic" have been created, and it was deduced that some substances may trigger such an effect and others not. There certainly are substances for which it seems obvious that such effects cannot be induced, even at high doses.

If the world would consist of such black-and-white substances and black-and-white effects, a qualitative labelling would make sense. However, the world consists of all shades of gray, and, as was pointed out before, for the great majority of substances such a simple classification is not possible or meaningful. *A complex state of affairs most often cannot be squeezed into a simple "yes" or "no".*

It has already been stated that toxicities, including reproductive toxicity, represent *quantitative* phenomena, and that qualitative labelling is largely meaningless with respect to risk assessment.

One of the aims of this symposium was to clarify whether sufficient scientific information is available to agree on certain strategies of a quantitative risk assessment in the area of reproductive and developmental toxicology. We feel that this information is presently not available and it would be helpful to define the lines of research to reach the level of knowledge required.

Risk may be defined as *"incidence or probability of the occurrence of an adverse effect, in dependence of the exposure or dose"* (WHO 1978). This then boils down to establishing dose-response relationships.

Dose-Response Relationships

As all other toxic effects, also reproductive and developmental toxicity occurs dose-dependently, and such relationships can be revealed in experimental studies.

For establishing a dose-response curve generally several experimental doses are required; the more doses the better the relationship can be defined. There are a number of aspects that limit this approach. These are again scientific as well as pragmatic:

- *scientifically*, a dose-response may be limited in reproductive or developmental toxicity by the fact that certain outcomes may obscure other effects: a polydactyly cannot be assessed at doses leading to a high incidence of amelia, the frequency of any gross-structural abnormality cannot be assessed accurately at simultaneously occurring high embryomortality, developmental retardation may shift the susceptible stage beyond the treatment period, etc.

- *Pragmatically*, the design of the study (three doses plus one control group, as recommended by most guidelines) would not give sufficient information to establish a reliable dose-response curve. Generally, the aim is to find a "no-observed-adverse-effect-level" (NOAEL). This certainly is a severe draw-back in the present study design if a quantitative risk assessment is the goal.

Prenatal toxicity is characterized by a multiplicity of possible outcomes. It should be remembered that the dose-response curves for the different outcomes are by no means parallel. For this reason quite different situations may occur at different dose levels (Neubert et al. 1980).

In the case of gross-structural abnormalities, the dose-response curves are usually very steep (Neubert et al. 1987), and again the curves for various abnormalities may not run parallel. This may complicate attempts to extrapolate abnormality rates down to lower exposure levels. Very few substances exhibit a very shallow dose-response relationship, the herbicide nitrofen may fit this criterion in rats. In the monkey, the teratogenic dose-range for thalidomide and it's derivatives extends over several orders of magnitude (Merker et al. 1988; Heger et al. 1988). This, like many other characteristics, also make thalidomide an exception to the rule.

Pharmacokinetic Aspects

Although the importance of kinetic data for toxicological evaluations is well recognized today, many studies on reproductive and developmental toxicity, especially when performed by smaller companies, suffer from the lack of the availability of such data before the studies are initiated. For this reason an optimal dosing schedule may not be possible. This inadequacy is especially relevant when later a very short elimination half-life is found for this substance, suggesting the need for several applications per day or the possible use of minipumps. For these reasons many studies have been considered inadequate and had to be repeated (a waste of experimental animals!).

Another frequent inadequacy comes from the choice of the second species: in general the rabbit is used for this purpose in studies on embryotoxicity (segment II). Usually for this species no kinetic data are established, and since the rabbit is not used for other studies in routine toxicology, such data may have to be obtained just for this one study.

Kinetic aspects should also apply to environmental chemicals. In fact, this type of information is often available (at least in non-pregnant animals) and is woefully underutilized. Merely obtaining maternal blood time-concentrations would probably resolve many ambiguities about shape of dose-response curves and interspecies differences in response. Industry and regulatory agencies should begin to pay greater attention to integrating toxicological and kinetic information.

Since problems of kinetics in prenatal toxicology have been discussed as early as 1978 (Neubert 1978) and in two symposia (Neubert et al. 1978; Nau and Scott 1987), this shall not be further elaborated here.

Species Differences

In prenatal toxicology species and strain differences are rather common, and therefore may represent a problem in risk assessment. Frequently teratogenic effects are only seen in one of the species used (either the rat or the rabbit), and the reasons for this difference in outcome are usually unknown.

There are two main reasons for species and strain differences in prenatal toxicology:

- *kinetic* reasons, i.e. mostly that maternal kinetics differs in the two species used, and
- differences in *embryonic development*, and apparently in the susceptibility of the corresponding gestational stages.

It is often claimed that embryonic development proceeds according to the same principles in all warm-blooded animals, and, e.g., the chick embryo has frequently been used and is still used as a convenient experimental model in developmental biology. It also has, over the last decades, been recommended as a model for revealing "teratogenic" effects of substances. While many basic processes may proceed rather similarly during embryogenesis in the chick and in man, for toxicological evaluations the *differences* between species may be more important. Probably many such differences exist, which may explain the pronounced differences in the response to chemicals between the chick and the mammalian embryo. Furthermore, in contrast to mammals, virtually any agent tested is capable of inducing abnormal development in the chick embryo if tested at a high enough dose. This would necessitate a quantitative extrapolation of data found in the chick embryo to the situation possible existing in man, and no intelligent suggestion has been made as to how to proceed in such a risk assessment. For these reasons the chick embryo is not appropriate as an experimental model for revealing the teratogenic potency of chemicals, despite other claims.

Several other aspects of species and strain differences have been discussed before (Neubert and Chahoud 1985).

Extrapolations Between Species

Extrapolations of data between species, and especially the extrapolation from an experimental species to the situation possibly existing in man, is the main challenge in toxicology. Since, with the exception of medicinal products, data in humans are scarce or will never become available (hopefully not, if adverse effects of suspicious substances should be avoided) the only acceptable risk assessment aiming at a primary prevention must be based on experimental data.

There are two extremes which should be avoided during a risk assessment:

- a possible hazard for man should not be *overlooked*, and
- the *use* of a valuable substance should not be *prevented* by an overconservative interpretation of the data, including the use of hypersensitive test systems.

Any formal approach will be unsatisfactory since the situation is known to be different for almost every substance to be evaluated. Since the presently used whole-animal model systems represent rather empiric tests and "black-boxes", any kind of basic information on the cellular and molecular biology of morphogenetic differentiations may improve our understanding of crucial developmental processes. For this reason, similar to the situation in cancerogenesis (cf. e.g. Neubert 1992), information on mechanisms of developmental processes may help in the interpretation of substance-induced adverse effects. Presently, such information is lacking for the majority of substance-induced abnormal developments.

Use of Mathematical Models

Especially in the USA the use of mathematical models has been applied to extrapolations of animal data from high dose levels in the experimental range to dose levels below any possible experimental verification. This approach has predominantly been applied for assessing cancerogenic risk, and no extensive attempts have been published in the area of reproductive or developmental toxicology.

Since experimental data are extrapolated to a range below any verification (otherwise no extrapolation would be necessary), under conditions of an unknown dose-response relationship, *the result must a priori always be speculative*. This is acceptable if such estimates are used for a conservative hazard assessment; through the addition of many conservative assumptions the error may be on the safe side. However, *it is completely unacceptable if the results of such a conservative probability calculation are turned around to a positive statement, e.g. so-and-so many additional people will suffer malformations or die from cancer.* Unfortunately, just this is frequently stated by poorly trained "scientists" or irresponsible and ill-informed journalists.

As with all extrapolations the possible error will become greater when decreasing the probable incidence to be assessed. In the usual extrapolation down to a frequency of 1 case in a population of 10^6 (starting with a frequency of the event in $<10^2$ animals tested) this uncertainty must be drastic. For this reason, for reproductive toxicity a more rational extrapolation to incidences of 1 in 10^2 to 10^3 has been suggested (*see:* Platzek et al., this book).

Very much on these lines are recent considerations within the US-EPA to introduce mathematical modelling to developmental toxicity through use of the "*benchmark*" dose concept. These empirical models generally operate within the observable response range and do not attempt to extrapolate directly to low dose levels. For this reason such attempts have marked advantages (as well as a few disadvantages) over the NOAEL approach. This strategy has a similar background as the suggestion of calculating an ED_{10} (Platzek et al. 1983, 1987). The "benchmark" dose concept will be discussed in the paper by Kavlock and Kimmel in this book.

Up till now, biologically-based models have not been used for risk assessment in reproductive toxicity. One should remember that such models (similar to biomonitoring) during the evaluation of dose-response relationships predominantly provide a measure of "*dose*" (cellular or molecular) rather than a reasonable estimate for "*effect*". However, attempts to use *effect-based* models using biomarkers of early response in the embryo to establish links between delivered dose, biochemical responses, tissue response, and finally, manifestations of developmental toxicity at term appear worthwhile, especially when attempting to establish "safe" dose levels. The future may see greater emphasis put on such approaches, especially for agents with widespread human exposure or for which regulation is likely to cause large socio-economic impacts. One should keep in mind that with modern technology the detection of concentrations in tissue as well as molecular reactions (e.g. agent-receptor interactions) may only depend on the sensitivity of the method used, and is as such almost unlimited. Not all of such measures give relevant information on possible toxic effects on tissues or the organism, and one intrinsic property of developing tissues and systems is that deviations from the norm are readily reversible.

Significance of "Safety Factors"

In order to "extrapolate" non-cancer toxicity data to the situation probably existing in man, frequently "safety factors" or "uncertainty factors" are used. This approach is pragmatic but rests on poor scientific ground. The intention is to suggest low dose levels or ranges which are likely to be without adverse effects (i.e. below some "threshold"). The rationale behind it rests on the fact that the frequency of a pharmacologic or toxic effect decreases when the dose is reduced. Therefore, at very low exposures the effect may be expected not to be induced anymore, or the probability of its occurrence will become vanishingly small.

For most toxic effects another aspect which must be considered is the background of the spontaneous occurrence of such adverse effects (in our case: e.g. gross-structural abnormalities). Substance-related effects are only detectable if they do not significantly change this background. This poses a crucial question for any attempts to "extrapolate" to "acceptable safe" doses. Is it **(a)** sufficient to extrapolate the possible effect of a given chemical to an incidence well within the background level, or is it **(b)** necessary to extrapolate to extremely low incidences? To answer this question it doesn't make any difference whether a mathematical model (e.g. extrapolation to 1 to 10^3 or to 10^6), a NOAEL-safety factor or a benchmark approach is used.

Since single abnormalities do not occur at less than 1 : 10^3 in humans, an extrapolation to this level may be felt to be sufficient; certainly it will never be possible to prove a substance-related effect at such a level. The opposite argument relates to the concern that we are not exposed to a single substance only, and therefore a higher margin of safety may be warranted. This latter argument is certainly conservative, but scientifically not very good, since it boils down to the problem of effects of combinations. It is an oversimplification to assume that only additive effects are to be expected when a large variety of substances act on a given target. This discussion shows that the level of effects to be extrapolated to cannot be decided scientifically. It is a political or pragmatic decision.

When employing such factors a number of uncertainties must be taken into account. These include:

- different susceptibilities between species (which are not known for man), and
- the fact that, e.g., the NOAEL was established in a relatively small number of experimental animals.

Role of Observations in Man

There is no doubt that well-conducted and properly interpreted studies in man represent the most relevant information for assessing a possible risk of prenatally-induced lesions.

However, the significance of such observations in man is mostly overestimated by laymen and often also by physicians. There are a number of reasons for this unfavourable situation. For instance:

- it is almost impossible to *exclude* a risk, and multiple studies with very large numbers of mother/child-pairs (several thousand) are necessary. This is very time consuming and costly.
- Clues for possible adverse effects are usually found in multiple *case reports*. Although no causal relationship may be drawn from such reports, if such suspicions are serious it may be unethical to initiate additional large scale studies; the agent should rather be precautiously removed from the market.
- No *primary prevention* is possible through observations in man. The intention of preclinical studies is to prevent humans from being exposed to potentially dangerous agents.
- Most of the newer medicinal products are *seldom prescribed* to women of childbearing age. Therefore, it is often difficult to obtain enough cases for a satisfactory risk assessment. This holds for substances such as: H2-blockers, calcium channel antagonists, many newer chemotherapeutic agents, etc.

- With respect to "*environmental chemicals*" verification of an exposure (and especially its quantification) is usually extremely difficult, but it is possible for some persistent chemicals (e.g. lead, TCDD, PCBs). Efforts to accumulate human exposure information is another unmet need.
- Many *outcomes* are difficult or impossible to assess (abortion, dysfunctions, etc.).
- Induction of special adverse effects may be confined to an exposure at a *defined* developmental stage. However, in order to evaluate a sufficient number of cases, mother/child-pairs with different exposures are often "lumped together" (e.g. first trimester of pregnancy), thus greatly reducing the predictive power of the data.

For the reasons mentioned, sufficient information cannot be expected for the majority of substances on the basis of observations in man. Animal studies are still indispensable for this purpose.

Sometimes, studies in humans resulting in "null" effects, if adequately conducted, may at least place an upper bound on what the risk could have been, and in that sense, may provide more information than is obvious at first analysis. In addition, so-called "meta-analyses" can sometimes pull out information from several studies which alone are inconclusive.

Significance of In Vitro Data

A vivid discussion has begun on the possibility of supplementing (and possibly in the future replacing) whole-animal studies with *in vitro* studies, also in the area of prenatal toxicology.

In this area considerable progress has been made in recent years, and special aspects will be discussed in the paper by Klug and Neubert (this book). Several aspects of *in vitro* systems for assessing abnormal prenatal development have been extensively reviewed and discussed before (Barrach and Neubert 1980; Neubert 1981, 1982, 1985; Neubert and Barrach 1983; Neubert et al. 1985; 1986).

It is important to recognize that presently no *in vitro* method is available to completely replace *in vivo* tests in the area of reproductive and developmental toxicology. Furthermore, certainly no *single* test will ever be able to give information on all of the possible adverse effects in the complex area of prenatally-induced lesions (including morphological and functional abnormalities).

However, for certain, defined problems in developmental toxicology *in vitro* methods are extremely helpful, or they may even be the methods of choice. When dealing with chemical congeners, an *in vitro* system which detects or mimics the correct mechanism of action may have a strong potential. This approach is beginning to be pursued in several pharmaceutical companies.

Some of these aspects are also discussed by Klug and Neubert in this book.

Significance of Prenatally-Induced Dysfunctions

The assessment of possible prenatally-induced functional abnormalities (dysfunctions) which most often manifest themselves (late) postnatally has been largely ignored (in favour of morphological manifestations) in experimental studies as well as in studies in man over the last decades.

There are only two types of functional aspects that have been more or less routinely considered in routine studies: effects on the fertility of the next generation (Japanese modification of the segment II test), and some aspects of "behavioural" abnormalities.

Assessment of the risk of prenatally-induced functional abnormalities may be the greatest challenge of the next years. In this respect, especially the vast advances made in the field of *immunological research* pose the question whether (irreversible) impairments of the pre- and perinatal development of the numerous functions of the immune system may represent a risk for man inducible by medicinal products or environmental substances. No answer to this question is possible, up till now, and the methodology for testing has not been standardized or agreed upon.

It may be worth mentioning that postnatal deficits may be transient but still be of serious concern. This may especially be the case for any robust, repeated effect on postnatal physiology, regardless of it's presumed reversibility. This concerns, e.g., the well-known effect of "imprinting", i.e. transient hormonal changes which may irreversibly impair functions manifesting themselves late in postnatal life. Again, an intelligent case-by-case evaluation is called for, and not every slight and transient effect, e.g. on behavioural aspects, is significant for a risk assessment with respect to man.

Summary and Outlook

In the young field of reproductive and developmental toxicology, not much older than thirty years, many achievements have been made possible and our understanding of normal and abnormal processes have been broadened. Guidelines have been established to aid the routine testing required to reveal possible adverse effects in this area.

Nevertheless, experienced investigators in this field may continue to have the impression that we have hardly scratched the surface of the problems.

There seem to be three major areas in which our knowledge is completely unsatisfactory, and in which much more research and information is needed:

- the evaluation of the significance of the various types of structural defects for a risk assessment with relevance for man,
- the problem of quantification of various types of abnormal developments and extrapolations to the situation relevant for man, and
- the assessment of prenatally-induced dysfunctions (predominantly on the immune and the hormone systems) in experimental and clinical studies.

We hope that the presentations made at this symposium will help to solve, or at least to recognize, these problems.

References

Heger W, Klug S, Schmahl H-J, Nau H, Merker H-J, Neubert D (1988) Embryotoxic effects of thalidomide derivatives in the non-human primate *Callithrix jacchus*. 3. Teratogenic potency of the EM12 enantiomers. Arch Toxicol 62: 205-208

Lorke D (1963) Zur Methodik der Untersuchung embryotoxischer und teratogener Wirkungen an der Ratte. Naunyn-Schmiedeberg's Arch Exp Path Pharmakol 246: 147-151

Lorke D (1977) Evaluation of the skeleton. In: D Neubert, H-J Merker, TE Kwasigroch, eds, Methods in Prenatal Toxicology, Georg Thieme Publ, Stuttgart, pp 145-152

Merker H-J, Heger W, Sames K, Stürje H, Neubert D (1988) Embryotoxic effects of thalidomide derivatives in the non-human primate, *Callithrix jacchus*. 1. Effects of 3-(1,3-dihydro-1-oxo-2H-isoindol-2-yl)2,6-dioxopiperidine (EM12) on skeletal development. Arch Toxicol 61: 165-179

Nau H, Scott WJ Jr (eds) (1987) Pharmacokinetics in Teratogenesis, Vol I and II, CRC Press, Boca Raton, Florida

Neubert D (1978) Bedeutung pharmakokinetischer Parameter für die Teratologie. In: B. Schnieders, G. Stille, P. Grosdanoff, eds, Embryotoxikologische Probleme der Arzneimittelforschung. AMI-Berichte 1/78, Reimer, Berlin, pp 83-88

Neubert D (1981) On the predictability of developmental toxicity - especially prenatal toxicity - on the basis of culture experiments. In: D. Neubert, H.-J. Merker, eds, Culture Techniques, Applicability for Studies on Prenatal Differentiation and Toxicity. Walter de Gruyter, Berlin, New York, pp 567-582

Neubert D (1982) The use of culture techniques in studies on prenatal toxicity. Pharmacol Ther 18: 397-434

Neubert D (1985) Toxicity studies with cellular models of differentiation. Xenobiotica 15: 649-660

Neubert D (1992) Current efforts to use mechanistic information for risk assessment in cancerogenesis and it's relevance to man, and summary of the meeting. Symposium on Chemical Carcinogenesis. The relevance of mechanistic understanding in toxicological evaluation, April 29-30, 1991, BGA, in press

Neubert D (1992) Limitations to predicting risk in the evaluation of prenatal toxicity. Drug Information Association Workshop, Sept. 3-5, 1990, Montreux. DIA Journal, Pergamon Press, New York, Oxford, Seoul, Tokyo, in press

Neubert D, Barrach H-J (1983) Effect of environmental agents on embryonic development and the applicability of *in vitro* techniques for teratological testing. In: A. Kolber, T. Wong, L.D. Grant, R.S. DeWoskin, T.J. Highes, eds, In Vitro Toxicity Testing of Environmental Agents, Part B, Plenum Publ Corp, pp 147-172

Neubert D, Barrach HJ, Merker H-J (1980) Drug-induced damage to the fetus (molecular and multilateral approach to prenatal toxicology), In: E Grundmann, ed, Drug-induced Pathology, Springer-Verlag, Berlin, Heidelberg, New York, Current Topics Pathol 69: 241-331

Neubert D, Blankenburg G, Chahoud I, Franz G, Herken R, Kastner M, Klug S, Kröger J, Krowke R, Lewandowski C, Merker H-J, Schulz T, Stahlmann R (1986) Results of *in vivo* and *in vitro* studies for assessing prenatal toxicity. Envir Health Perspect 70: 89-103

Neubert D, Blankenburg G, Lewandowski C, Klug S (1985) Misinterpretations of results and creation of "artifacts" in studies on developmental toxicity using systems simpler than *in vivo* systems. In: Lash JW, Saxén L (eds), Developmental Mechanisms, Normal and Abnormal, Progress in Clinical and Biological Research. Vol 171, Alan R Liss Inc, New York, pp 241-266

Neubert D, Bochert G, Platzek T, Chahoud I, Fischer B, Meister R (1987) Dose-response relationships in prenatal toxicity. Congen Anom 27: 275-302

Neubert D, Chahoud I (1985) Significance of species and strain differences in pre- and perinatal toxicology. Acta Histochem 31: 23-35

Neubert D, Chahoud I, Platzek T, Meister R (1987) Principles and problems in assessing prenatal toxicology. Arch Toxicol 60: 238-245

Neubert D, Merker H-J, Nau H, Langman J, eds (1978) Role of Pharmacokinetics in Prenatal and Perinatal Toxicology. 3rd Symp on Prenatal Development, Berlin 1978, Georg Thieme Publ, Stuttgart

Platzek T, Bochert G, Rahm U (1983) Embryotoxicity induced by alkylating agents: Teratogenicity of acetoxymethyl-methylnitrosamine - dose-response relationship, application route dependency and phase specificity. Arch Toxicol 52: 45-69

Platzek T, Meister R, Chahoud I, Bochert G, Krowke R, Neubert D (1987) Studies on mechanisms and dose-response relationships of prenatal toxicity. In: F. Welsch (ed) Approaches to Elucidate Mechanisms in Teratogenesis, Hemisphere Publ Corp, Washington, New York, pp 59-81

Setzer RW, Rogers JM (1989) The relative powers of statistical tests for developmental toxicology data: maximum likelihood beta-binomial, maximum quasi-likelihood, and ANOVA based on the Freeman-Tukey binomial tranform. Teratology 39: 481

WHO (1978) Principles and methods for evaluating the toxicity of chemicals. Part I, Environ Health Criteria 6: 19

Basis for Testing in Reproductive Toxicity

Risk Assessment for Pharmaceutical Products

Jeanne M. Manson

Introduction

As described in previous talks, pharmaceutical products are tested in a lengthy and highly redundant series of tests to identify whether they produce reproductive toxicity in laboratory animals. Considerable progress has been made in improving the design of these studies, and draft guidelines have been promulgated which reduce the redundancy and increase the flexibility of reproductive toxicity tests for drug safety evaluation (Bass et al. 1991; Lumley and Walker, 1991). With improvement in the design of animal studies, it is now possible to turn our attention to the risk assessment process for pharmaceuticals, the process in which results from animal studies are used to predict whether adverse effects will occur in humans. The perspective I will take is that of a laboratory investigator who condenses the results from animal studies into the narrowly defined categories for use of a drug in pregnancy. This is only the first step in the risk assessment process, and subsequently clinical, legal and regulatory perspectives must be obtained on the potential uses and abuses of the drug during human pregnancy before the final label is produced.

In the USA, the FDA has devised categories to describe levels of potential risk to human pregnancy based on findings from animal studies. These will be described and examples given of drugs that fall within each of these categories. Finally, recommendations for improving the labeling procedures will be given, especially for those agents with identified human developmental toxicity.

Pregnancy Categorization of Drugs

Category A (Table 1): This category is to be used for those agents where controlled studies in human pregnancy do not reveal evidence of maternal or developmental toxicity. Animal studies are also negative, or possibly not conducted. In practice, category A has been assigned to only seven agents, which include thyroid hormone supplements (six agents) and an iron/vitamin C supplement (one agent (PDR 1992). It has not been used to date for any pharmaceutical agent despite the wide-scale use of a number of pharmaceutical agents during pregnancy for treatment of pregnancy-related disorders (i.e., pregnancy-induced hypertension, pre-term labor). These drugs are not indicated for use during pregnancy in the package insert and controlled studies on the safety and efficacy in pregnancy are carried out after the drug is marketed for other indications. Use during pregnancy, even if intentional and well-established to treat pregnancy-related disorders, remains off-label. For these drugs the use pattern is determined by long-term clinical experience and reports in the literature.

Table 1. *Category A*: Controlled studies in human pregnancy negative, animal studies negative (or not conducted)

---

* Intentional use in human pregnancy for treatment of pregnancy-related disorders (e.g., antihypertensives).
* Unintentional exposure to drugs with wide-scale use in the general population (e.g., antibiotics).
* Use during pregnancy remains off-label despite demonstration of safety/efficacy.

Use of this category may be prohibited by the difficulties in conducting controlled studies in human pregnancy. It should be questioned whether "controlled" studies are necessary for demonstration of safety in humans. Use during pregnancy not associated with adverse outcomes, as documented in post-marketing surveillance, and clinical case reports in the literature could be considered sufficient for assignment to category A. Priority should be given by regulatory agencies and the pharmaceutical industry to identifying agents that are safe for use during pregnancy to make beneficial therapeutic agents available to this segment of the population.

Category B (Table 2): Category B is used for those agents in which there are no controlled studies in human pregnancy, and animal studies do not indicate a potential for developmental toxicity. There are currently 199 agents in this category out of a total of 1033 agents in the Physician's Desk Reference (PDR 1992).

This category is also used for new drugs that have been evaluated in animal models which are submitted for an NDA. A major limitation to its use is that with the current emphasis on testing to maternally toxic doses in animal studies, maternal toxicity frequently gives rise to some form of developmental toxicity. Discerning whether the developmental toxicity was selective to the conceptus or secondary to systemic toxicity in the mother can be difficult if not impossible for some agents. A further definition of a "negative" animal study is necessary before this category will be useful. A possibility is that category B should include those agents where developmental toxicity was demonstrated to be secondary to systemic toxicity in the mother, and systemic toxicity during human pregnancy under conditions of therapeutic use is considered unlikely.

Table 2. *Category B*: No controlled studies in human pregnancy, animal studies negative

* Extent of human data, case reports vs. controlled studies.
* Include agents where developmental toxicity secondary to systemic toxicity in the mother.

Category C (Table 3): Category C is assigned when there are no controlled studies in human pregnancy and animal studies show an adverse effect. This is the most frequently used pregnancy category for new pharmaceutical products, and

682 out of 1033 entries in the PDR have been assigned to this category (PDR 1992). Any type of adverse finding from animal studies can trigger a category C label, and discrimination between developmental effects that are meaningful for human risk assessment from those that are of questionable relevance rarely occurs. Also, the critical period during human pregnancy when the adverse effect is likely to occur on the conceptus is not usually estimated and consequently exposure throughout pregnancy rather than specific trimesters is implicated. Insufficient information is too often given to the prescribing physician to make a determination of the risk/benefit for use during human pregnancy.

Table 3. *Category C:*

* No controlled studies in human pregnancy, animal studies show an adverse effect.
* Drugs administered during pregnancy only if the benefit to the mother outweighs potential risk to fetus.
* More information on the dose-response specific outcome and presumed critical period for humans needed.
* Potential risk to the human conceptus should not be inferred based on a lack of human studies.
* Probability of adverse effects on the human conceptus should be ranked based on characteristics of patient population and results from animal studies.

Two examples will be given to illustrate these points. One is with the analgesic, anti-inflammatory agent diflunisal (DOLOBI[R]) (Clark et al. 1984) (Table 4). There was no evidence of adverse effects on the rat conceptus at doses up to 10 times the human therapeutic dose. The usual effects on parturition and closure of the ductus arteriosus expected of a prostaglandin synthetase inhibitor were observed at doses up to four times the human therapeutic dose. In rabbits, however, at two times the therapeutic dose, maternotoxicity (hemolytic anaemia) and developmental toxicity (embryolethality, axial skeletal malformations, decreased fetal weight) were observed. Toxicokinetic measurements revealed substantial maternal plasma drug levels but minimal drug levels in both the rat and rabbit embryos, suggesting that accumulation of the drug in the conceptus was not a likely cause of developmental toxicity in the rabbit. Subsequent investigations indicated the drug produced severe hemolytic anaemia in the adult rabbit (pregnant and non-pregnant), but not in any other species tested, including humans. The maternal hemolytic anaemia was shown to be associated with production of developmental toxicity possibly through a mechanism involving maternal hypoxia.

Table 4. Effects of diflunisal in laboratory animals

* No evidence of developmental toxicity in rats up to 10 times the human therapeutic dose.
* At 2 times the therapeutic dose, maternal toxicity (hemolytic anaemia) and developmental toxicity were produced in the rabbit.
* Developmental toxicity due to maternal hypoxia secondary to hemolytic anaemia. Rabbits the only species in which hemolytic anaemia was produced by drug administration.

In the risk assessment process (Table 5), the developmental toxicity observed in rabbits was identified as not being relevant to humans. Use during the third trimester was not recommended based on the known effects of prostaglandin synthetase inhibitors in late pregnancy, and on this basis category C was assigned. Use during the first two trimesters was recommended only if the "potential benefit justified the potential risk to the fetus". This obligatory language was used even though animal studies did not reveal any risk to the conceptus during the first two trimesters relevant to humans. Potential risk to the conceptus should not be inferred solely because there are no controlled studies in human pregnancy. Rather, the probability of adverse effects to the human conceptus and their severity should be ranked based on characteristics of the patient population and results from animal studies. In no other area of preclinical safety assessment is potential risk inferred because of a lack of controlled studies in humans.

Another example of pharmaceutical agents initially assigned category C, and later given a category D, are the angiotensin converting enzyme inhibitors (ACE inhibitors). These are widely used antihypertensive agents that inhibit the enzyme which converts angiotensin I to the potent vasoconstrictor, angiotensin II. There are a number of ACE inhibitors on the market and I will use one, enalapril (VASOTEC[R]) as an example to describe the developmental toxicity produced by these agents.

Table 5. Effects of diflunisal in humans

*	No indications of hemolytic anaemia in humans. No controlled studies or case reports in human pregnancy.
*	Pregnancy category C - Rabbit developmental toxicity studies showed an adverse effect that does not appear to be relevant to humans. Use during the first two trimesters only if benefit justifies the potential risk to fetus. Third trimester use not recommended.

In general, rodents are not very sensitive to the developmental toxicity of ACE inhibitors (Table 6). Examinations made on term fetuses have not revealed significant developmental toxicity except slight depressions in fetal body weight coincident with depressions in maternal body weight at large multiples of the human therapeutic dose. Treatment during the 3rd trimester in the rat, however, is associated with depressed weight gain in neonates and histologic changes in the kidneys of weanlings. Histologically, these renal changes range from hypertrophy/hyperplasia of the juxtaglomerular apparatus, a known pharmacologic effect of ACE inhibitors, to frank renal tubular dysplasia. Effects on neonates, however, were only seen at large multiples of the human therapeutic dose (Robertson et al. 1986).

ACE inhibitors are highly toxic to rabbits. Maternal toxicity (death, pre-renal elevations BUN) and fetotoxicity (late fetal death) occur at and below the human therapeutic dose. Maternal toxicity in the rabbit is prevented by saline supplementation, causing volume expansion and increased renal perfusion, indicating a pharmacologic basis for the maternal toxicity. Saline supplementation does not reproducibly reduce the fetal toxicity in the rabbit, and the fetal deaths may be due to reduced placental perfusion or a direct effect on fetal renal function. The fetal deaths typically occur late in pregnancy in the rabbit (Minsker et al. 1990).

Table 6. Effects of ACE inhibitors in laboratory animals

* Rodents relatively insensitive to the developmental toxicity of ACE inhibitors.
* Rabbits are highly sensitive: maternal toxicity (death, pre-renal elevation BUN) and fetotoxicity (late fetal death) at the human therapeutic dose.
* Non-pregnant rabbits more susceptible to the antihypertensive effects and toxicity (death, pre-renal elevation BUN) than pregnant rabbits and other species.
* Not all ACE inhibitors cross the placenta in sheep.

ACE inhibitors cause comparable reductions in blood pressure in both pregnant and non-pregnant rabbits, as measured by the blockade of the acute pressor response to exogenous angiotensin I in normotensive rabbits. The rabbit, pregnant or non-pregnant, is more sensitive to the antihypertensive actions of ACE inhibitors than rats or dogs (Robertson, 1984). Toxicity studies have revealed that the non-pregnant rabbit is slightly more susceptible than the pregnant rabbit based on survival data and pre-renal elevations of BUN. It has been speculated that the increased plasma volume associated with late pregnancy in the rabbit has a renal sparing effect to toxicologic exposure with ACE inhibitors, providing some amelioration of the pre-renal elevation in BUN. Consequently, even though substantial developmental toxicity has been obtained in the rabbit at therapeutic doses of ACE inhibitors, it has been difficult to discern whether this was due to systemic toxicity in the rabbit *per se* or to selective effects on the rabbit conceptus.

When the ACE inhibitor captopril was administered i.v. to pregnant sheep, a transient (approximately one hour) decrease in maternal blood pressure occurred. Fetal blood pressure, however, decreased rapidly and remained depressed for up to three days, resulting in four stillbirths out of five deliveries (Broughton Pipkin et al. 1982). Another ACE inhibitor, enalapril, administered i.v. under similar conditions by the same investigators, caused decreases in maternal blood pressure but no treatment-related effects on fetal blood pressure or fetal outcome. This compound was shown not to cross the placenta in sheep (Broughton Pipkin and Wallace, 1986). Consequently, the sheep model has not provided consistent results with all ACE inhibitors.

Based on these findings, category C was originally assigned to the ACE inhibitor, enalapril, with a lengthy description of effects in the rabbit. Since their introduction to the market approximately 10 years ago, there have been a growing number of reports associating use of ACE inhibitors during pregnancy with adverse perinatal outcomes (Table 7). The latest report on this topic identified 85 pregnancies in women who had received an ACE inhibitor at some time in pregnancy (Hanssens et al. 1991). Approximately 1/3 of the cases involved use of the agent to treat pregnancy-induced hypertension, and in these cases treatment was typically initiated after 16 weeks of pregnancy. The remaining cases involved women who were treated for renovascular or essential hypertension, and treatment was typically initiated prior to pregnancy and continued to term or was discontinued at some time prior to the 16th week of pregnancy. A recent paper by Pryde et al. (1991) indicated that there were 16 adverse outcomes in a total of approximately 88 exposed pregnancies. The true rate of adverse fetal effects from ACE inhibitor use in human pregnancy cannot be determined from available information due to the expected under-reporting of exposed pregnancies with no adverse outcomes.

Table 7. Effects of ACE inhibitors in humans

* Approximately 16 cases of ACE inhibitor fetopathy: mid-pregnancy onset of oligohydramnios, IUGR, delivery of infants with prolonged hypotension and anuria.
* Pulmonary hypoplasia, renal tubular dysplasia, hypoplasia of the membranous bones of the skull and patent ductus arteriosis are common sequelae.
* Primary cause likely to be drug-related fetal hypotension and decreased fetal renal perfusion persisting to the neonatal period.

The most commonly reported adverse effect in ACE inhibitor-exposed pregnancies is middle to late trimester onset of oligohydramnios and intrauterine growth restriction, followed by delivery of an infant with prolonged hypotension and anuria. Pulmonary hypoplasia, renal tubular dysplasia, hypoplasia of the membranous bones of the skull and patent ductus arteriosis are common sequelae of the syndrome in newborns. The primary cause is likely to be drug-related fetal hypotension and decreased fetal renal perfusion which persist into the neonatal period. The increased sensitivity of the fetus relative to the mother for these effects may be related to the importance of the renin-angiotensin system under conditions of low renal perfusion pressure (Hall et al. 1977). The fetus and neonate have low renal perfusion pressure and angiotensin-II mediated efferent arteriolar resistance is likely to be more important for maintenance of glomerular filtration and production of urine than in individuals with full developed renal function.

First trimester exposure alone has not been associated with the syndrome of oligohydramnios/ anuria/neonatal hypotension and renal failure in humans (Hanssens et al. 1991; Pryde et al. 1991). Second and third trimester exposure is associated with fetal and neonatal injury. Consequently, women who become pregnant while on ACE inhibitor therapy have a good chance for a normal pregnancy if treatment is terminated as early as possible in the first trimester and an alternative antihypertensive therapy initiated. Use of ACE inhibitors for treatment of pregnancy-induced hypertension should clearly be discouraged as the mid to late pregnancy onset of this condition coincides with the critical period for ACE inhibitor fetopathy.

Assignment to pregnancy category D is appropriate for ACE inhibitors as a small percentage of patients can achieve control of blood pressure only with this class of antihypertensive agent (Table 8). The benefit to the mother may outweigh the risk to the fetus in some patients. The issue then becomes how to transmit information about the clearly defined human fetal risk to patients and prescribing physicians. ACE inhibitors have been marketed for over ten years, and their use in clinical practice is well established. When drugs are newly marketed, there is a high level of scrutiny by prescribing physicians for human safety and efficacy (Fig. 1). The number of adverse case reports submitted by physicians is generally the highest the first two years after a drug is marketed. After that, as experience is gained with the product and side effects in humans become well understood, fewer adverse case reports are submitted even when the number of prescriptions written is greatly increased. A similar trend as depicted in Figure 1 could be imagined for scrutiny of the package insert. As experience is gained with the product it is unlikely the prescribing physician will continuously re-read the label. How then should information be transmitted on an adverse side effect identified in humans long after the product is initially marketed? Revision of the label is clearly an important step, but it seems that additional measures must also be taken. Description of effects during human pregnancy in promotional material and direct communication with prescribing physicians may be most effective. Direct labeling of the product, as carried out with Accutane[R], should also be considered for agents with clearly identified risk during human pregnancy.

Table 8. *Category D*: Positive evidence of human fetal risk (case reports) but benefit to the mother may outweigh risk to the fetus

- * Second and third trimester exposure to ACE inhibitors associated with fetal risk: no demonstrated risk with first trimester exposure.
- * Patients who become pregnant while on ACE inhibitor therapy should be switched to an alternate antihypertensive as soon as possible.
- * Use of ACE inhibitors for treatment of pregnancy-induced hypertension should be strongly discouraged.

Category X (Table 9): This category is reserved for agents with evidence of human fetal risk where the risk of use during pregnancy clearly outweighs any possible benefit. Consequently, the drug is contraindicated in women who are or may become pregnant. There are 69 agents with category D and 76 agents with category X in the PDR (PDR 1992). The necessity for having two separate labels, D and X, for products with clear evidence of risk during human pregnancy can be questioned. The determination of whether the benefit to the mother may or may not be worth the risk to the fetus can be difficult to make, and is best left up to the prescribing physician to make on a case-by-case basis. One category should be sufficient for agents with clear evidence of human fetal risk, with sufficient information provided to the physician to make an informed judgement on risk vs. benefit.

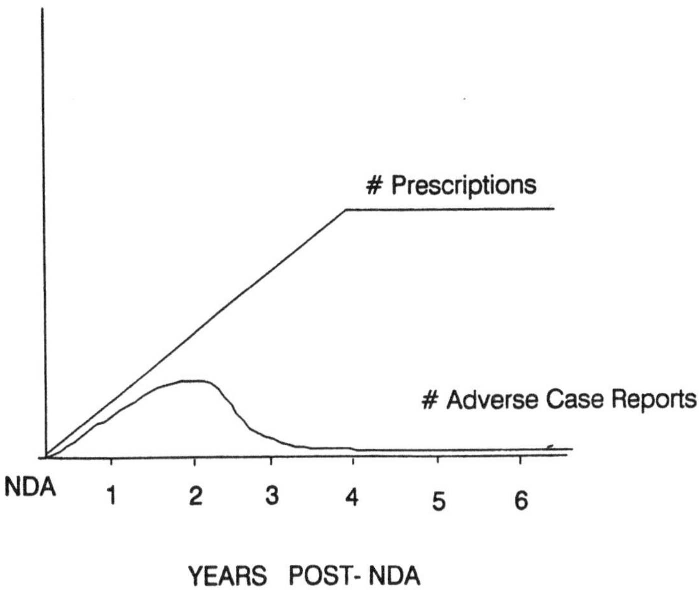

Fig. 1. Relationship between numbers of adverse case reports and prescriptions written post-NDA

Table 9. *Category X*: Positive evidence of human fetal risk; risk of use during pregnancy clearly outweighs any possible benefit

	Contraindicated in women who are or may become pregnant.
*	Only one category for human fetal risk is necessary.
*	Decision as to whether benefit is worth the risk should be made by the prescribing physician.

Conclusions

Unlike any other area in preclinical safety assessment, the tendency exists to obtain the most severe label possible for use of drugs in human pregnancy (Table 10). This is partially a reflection of abuses in the legal system where any adverse pregnancy outcome in a patient on drug therapy can be attributed to drug treatment, no matter how improbable the association. In this environment, legal risks to pharmaceutical companies are reduced when use of their products during pregnancy is strongly discouraged. Pharmaceutical companies are protected from law suits in these cases by strongly discouraging use of their products during pregnancy. This is also a reflection of the pregnancy categories themselves, where human risk is inferred based on the lack of controlled studies in humans. The reasons why the tendency to obtain the most severe label possible for use in pregnancy are understood, but the costs may not be fully realized. These are that there are few drugs available for treatment of common disorders related to pregnancy that can be associated with significant maternal and fetal morbidity, e.g., pregnancy-induced hypertension, pre-term labor, hyperthyroidism, cardiac arrhythmias, etc. Drugs should be considered not just for safety but also for efficacy in treatment of these conditions.

The pregnancy categorization of drugs can be considerably simplified to provide more useful information to physicians for risk/benefit decisions. Developmental toxicity secondary to systemic toxicity in the mother should not be considered indicative of potential human risk, but it is the responsibility of the investigator to *demonstrate* that effects on the conceptus were secondary to maternal systemic toxicity. Also, drugs placed in category C should be ranked in some fashion for their potential to cause human fetal risk. A drug causing 5% depression in fetal body weight, an adverse effect, should not be given the same weight as a drug causing malformation of the entire litter. Drugs causing adverse effects in animals but not studied in humans represent the most difficult situation and sufficient information should be given for the physician to make a risk/benefit decision. Finally only one category for human fetal risk is necessary, and the decision as to whether the benefit outweighs the risk should be made by the physician. For those few drugs with clear evidence of human fetal risk, direct labeling of the product should occur along with labeling in the package insert.

Table 10. Conclusions

- * Avoid tendency to obtain the most severe label possible for use during pregnancy.
- * Encourage specific indications for use during pregnancy when safety and efficacy data warrant.
- * Category B should be used when developmental toxicity is demonstrated to be secondary to systemic toxicity in the mother.
- * More information should be provided in Category C to allow the physician to make a risk/benefit decision.
- * Only one category for human fetal risk is necessary: risk/benefit decision should be made by the physician.
- * Direct labeling of the product should be considered for agents with clear evidence of human fetal risk.

References

Bass R, Ulbrich B, Hildebrandt A, Weissinger J, Doi 0, Baeder C, Fumero S, Harada Y, Lehman H, Manson J, Neubert D, Omori Y, Palmer A, Sullivan F, Takayama S, Tanimura T (1991) Draft guidelines on detection of toxicity to reproduction for medicinal products. Adverse Drug React Toxicol Rev 9(3): 127-141

Broughton Pipkin F, Symonds EM, Turner SR (1982) The effect of captopril (SQ 14, 225) upon mother and fetus in the chronically canulated ewe and in the pregnant rabbit. J Physiol (London) 323: 415-422

Broughton Pipkin F, Wallace CP (1986) The effect of enalapril (MK-421), an angiotensin converting enzyme inhibitor, on the conscious pregnant ewe and her fetus. Brit J Pharmacol 87: 533-542

Clark R, Robertson R, Minsker D, Cohen S, Tocco D, Allen H, James M, Bokelman D (1984) Diflunisal-induced maternal anaemia as a cause of teratogenicity in rabbits. Teratology 30: 319-332

Hall J, Guyton A, Jackson T, Coleman T, Lohmeier T, Trippodo N (1977) Control of glomerular filtration rate by renin angiotensin system. Am J Physiol 233: 366-372

Hanssens M, Keirse M, Vankelecom F, Van Assche F (1991) Fetal and neonatal effects of treatment with angiotensin converting enzyme inhibitors in pregnancy. Obstet Gynecol 78(1): 128-135

Lumley CE, Walker SR (1991) Current issues in reproductive and developmental toxicity: Can an international guideline be achieved. Quay Publishing, Lancaster, UK

Minsker DH, Bagdon W, MacDonald J, Robertson R, Bokelman D (1990) Maternotoxicity and fetotoxicity of an angiotensin converting enzyme inhibitor, enalapril, in rabbits. Fund Appl Toxicol 14: 461-470

Physicians' Desk Reference (PDR) (1992) ER Barnhart, New York

Pryde P, Nugent C, Sedman A, Barr M (1991) Ace inhibitor fetopathy. JAMA, in press

Robertson R (1984) Comparative ACE inhibitory activity of enalaprilat and captopril in anesthetized pregnant and nonpregnant rabbits. MSDRL memo May 29, 1984

Robertson RT, Minsker DH, Bokelman DL (1986) MK-421 (enalapril maleate): late gestation and lactation study in rats. Jpn Pharmacol Ther 14: 43-55

Discussion of the Presentation

Neubert: After listening to your presentation I wonder whether we shouldn't be primarily interested in pronounced toxic effects. One of the main topics of discussion today has been on borderline effects with questionable significance for man. As you said, it is difficult to detect additional exogenous effects caused by drugs or chemicals above the spontaneous background in man because of the normal fluctuations of these spontaneously occurring effects. Are we not overdoing toxicological evaluations by making tremendous efforts and using an enormous number of animals for possibly only little benefit?

Manson: I agree that too much emphasis is placed on borderline findings in animal studies. The primary responsibility for correcting this imbalance lies with laboratory investigators conducting studies who must be considered the most knowledgeable people to interpret the findings.

Stahlmann: You have rightly pointed out that for the use of a medicinal product during pregnancy the physician has the final responsibility. However, are we not passing the responsibility on to the weakest part of the chain, since the toxicological training and background knowledge of the specific effects of the drug in question is most likely lacking in the case of the practising physician? He generally receives the least information.

Manson: Your question has highlighted the most important issue in communication of human risk, particularly in the area of therapeutic drug use during human pregnancy. The prescribing physician, who is the most knowledgeable about the general health status of the patient, frequently has little training in toxicology. Other complicating issues arise with medical specialization, i.e., the obstetrician caring for a pregnant patient may not be aware of all the side effects of the cardiovascular drugs the patient is receiving, and likewise, the cardiovascular expert may not be aware of the importance of early pregnancy detection for some types of medications.

Despite the fact that the prescribing physician may be the weakest part of the chain in information transfer to patients, the responsibility for patient care must remain with the physician. We should be attempting to provide better information to the physician than the current pregnancy categorisations allow. These categories fail at both ends of the spectrum: they result in exaggerated warnings for agents with no demonstrated potential for human developmental toxicity and do not provide physicians with sufficient information to make risk/benefit decisions with agents that pose some risk for use during pregnancy. High priority should be placed on revising the pregnancy categories, or on abandoning them altogether. In the latter case risks to human pregnancy could be handled in the "Precaution" section of the label in the same manner as severe systemic toxicity to major organ systems.

Sterz: With regard to ACE-inhibitors, could you reproduce the adverse effects seen in man with the animal models used in segment II or III studies?

Manson: Conventional animal models have not been predictive of the effects of ACE-inhibitors on the human fetus. Doses of ACE-inhibitors in the therapeutic range cause late fetal death in rabbits but, as discussed, the rabbit per se is highly sensitive to ACE-inhibitors. Death and alterations in renal parameters occur under the same treatment conditions as fetal death.

It appears that the third trimester is the most sensitive period in rat development to ACE-inhibitors, and consequently the segment I and segment III studies have yielded the most important information on developmental toxicity. Effects are manifested in rat neonates as depressed weight gain and histologic changes in the kidneys. These effects occur only at very high doses and do not seem to be qualitatively similar to effects seen in humans.

There are no published studies in which non-human primates have been treated with ACE-inhibitors in the third trimester. It would be interesting to see if effects similar to those observed in humans could be produced in a primate model.

Neubert: Do you feel that the adverse effects you described in the human newborn with ACE-inhibitors are group characteristics? Would you extend your suggestion or even warning not to use ACE-inhibitors in the last trimesters to all members of this class of drugs?

Manson: There are published studies (*Hanssens et al., 1989*) in which an identical syndrome of effect (hypotension/oligohydramnios/anuria) was observed with three different ACE-inhibitors, captopril, enalapril and lisinopril. Consequently, it seems likely that this syndrome can be produced with all ACE-inhibitors, and can be considered a pharmaceutical class effect.

Neubert: I do not think it is any longer appropriate to present a dose used in experimental testing as multiple of the therapeutic dose. There are many and good examples that even a 100-fold therapeutic dose may lead to lower plasma concentrations in, e.g., rodents when compared with man. This is due to the fact that the elimination half-life is often much shorter in the rodent. For comparisons between species plasma concentrations should at least be used today, if target tissue concentrations are not available.

Manson: For the agent in question, diflunisal, complete toxicokinetic evaluations were carried out in rats and rabbits (*Clark et al., Teratology 30: 319-332, 1984*). Dose-dependent toxicokinetics were observed for maternal plasma in both species, and C_{max} values were 7.3 times higher in maternal rat plasma and 11 times higher in rabbit plasma at the highest dose tested in developmental toxicity studies than in human plasma at the therapeutic dose. The most important point is that only trace amounts of the drug were found in rat and rabbit embryos. These data were not mentioned because they did not prove to be useful in understanding the mechanism of developmental toxicity in the rabbit or in human risk assessment. All that can be said is that the drug was well absorbed and was systemically available in the test species. Insofar as there is no information on drug levels in the human conceptus, the findings of only trace drug levels in the rat and rabbit conceptus cannot be put into perspective for human risk assessment.

Neubert: You mentioned the example of a substance with an adverse effect found in animal tests, but then it was stated that this is irrelevant, since there is no indication for treatment with this substance during pregnancy. Why was so much effort spent in testing if the substance won't be used during pregnancy anyhow?

Manson: The point is that most drugs that are commonly used to treat pregnancy-related disorders are not specifically indicated for use during human pregnancy in the label. Controlled studies on the safety and efficacy in pregnancy are carried out after the drug is marketed for other indications. Use during pregnancy, even if intentional and well-established to treat pregnancy-related diseases, remains off-label. In these cases the use pattern in human pregnancy is determined by long-term clinical experience and reports in the literature.

A good example of this is the antihypertensive agent α-methyldopa (e.g. Aldomet[R]), which is the most commonly used drug for treatment of pregnancy-induced hypertension. There are controlled studies on safety and efficacy in pregnancy-induced hypertension in the literature, but the drug is not indicated in the label for use with this condition. The drug has a pregnancy category B because animal studies were negative and there is no human data with use during the first trimester.

Peters: I would like to challenge a remark that is made too often: "We cannot study effects of drugs (reproductive or developmental) in humans". I think that many women of childbearing age are exposed to (new)drugs and they first realize that they are pregnant when they are two or three weeks "overdue". This implicates an exposure until after closure of the neural walls, an important period of organogenesis. I think it is ethically imperative to perform post-marketing surveillance in that target group.

Manson: I agree that inadvertent exposure to many drugs occurs during the first trimester in humans, and that it would be highly beneficial to collect information on these exposed pregnancies. It is unlikely, however, that post-marketing surveillance programs set up by pharmaceutical companies would provide useful data on these exposures. The great majority of reports obtained during post-marketing surveillance deal with adverse pregnancy outcomes alone. Physicians and patients have no incentive to send in reports on exposed pregnancies with normal outcomes. Other data collection systems must be developed, possibly systems with prospective follow-up, to collect data on the entire population with exposure during pregnancy. Teratogen Information Services in the USA are currently providing valuable information of this kind because patients and physicians call in for information shortly after the exposure has occurred, allowing follow-up of pregnancy outcome.

Pritchard: Your suggested change to segment I dosing is based on pathology changes/organ weight effects in three month toxicity. However, we have the experience of affects on sperm but no pathological changes. Therefore, the pathology assessments alone currently practised may not be adequate.

Reproductive Toxicity Risk Assessment - Some Questions

Anthony K. Palmer

Introduction

The word "risk" is emphasised as I have decided to risk demonstrating my ignorance of the subject. I have also taken the risk of straying from customary styles of presenting scientific argument. However, on request I have changed the title from "Garbage in, Garbage out". The purpose of taking these personal risks is to attempt to stimulate examination of what we do from a different angle. If my attempt fails it should at least demonstrate that risk is relative since the contributions of other participants will appear more erudite.

I have not come in the usual role of a speaker who will be able to answer your questions. Instead I am seeking answers to questions that puzzle me. In consequence of this uncertainty, my efforts at risk assessment have been stillborn or have fallen with the first few steps. All of you must have faced the same questions, have you resolved them or merely glossed over them?

Questions

The questions I have are: What risks are we concerned with? Why are we concerned with them? Is our selection objective or subjective? When should we estimate risk? How do we estimate risk? Where do we obtain reliable information?

What Risks are Involved?

The most fundamental principle of toxicology is that:

"All substances are poisons, there is nothing that is not poisonous; only the dose determines that a substance does not poison" *(Paracelsus)*.

This principle shows that we cannot avoid risks entirely so it is a question of deciding which ones are acceptable and which are not. Restricting debate to toxicity in reproducing animals I think we should include any effect that might adversely affect reproduction directly or indirectly. I would also include general toxic effects that are enhanced in reproduction. For example, the increased sensitivity of pregnant females to iron compounds or non steroidal anti-inflammatory agents.

I believe that maternal toxicity is important in its own right but others dismiss it by implication. They dismiss effects on the conceptus as unimportant because they were due to maternal toxicity. Thereby they imply that maternal toxicity also is unimportant. Do you think that the mother of a malformed child would be consoled by the explanation that it was due to her own intoxication? In reality an effect realised on both the mother and conceptus presents a double jeopardy.

For effects more specifically associated with reproduction (Fig. 1) we need to consider those affecting mature adults. This is least complicated for males for which reduced male fertility is the main concern. In the case of the female we need to be concerned with effects on conception, maintenance of pregnancy, parturition and lactation. Curiously, however, many seem to consider adverse effects on the female only as a means to attenuate concern over effects on the developing organism. This is because greatest attention is given to effects on the developing organism (Fig. 2).

The developmental period is from conception through to sexual maturity. During this time a chemical may cause damage of a different nature or to a greater degree than that in the mature adult. Further, the consequences of this damage may not be apparent until later life as, for example, the consequences of exposure to (organic) lead, methyl mercury, and DES during the late foetal and postnatal periods of development.

Particularly considering the effects of lead on children I find it anomalous that testing for effects in infancy is still being ignored in discussions on guidelines. When such a relevant toxic effect is "staring us in the face" I find it difficult to understand why investigators should go to extreme lengths in an attempt to demonstrate that organic lead may induce physical defects in animals at high intravenous dosages so that they can apply the label "teratogen". This surely reflects a desire to apply quick, convenient labels rather than conduct thoughtful assessments of the reality of toxicological risk to humans.

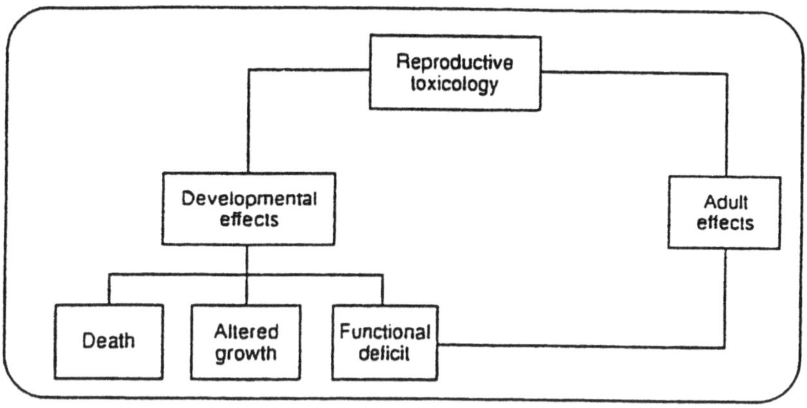

Fig. 1. Aspects of reproductive toxicity

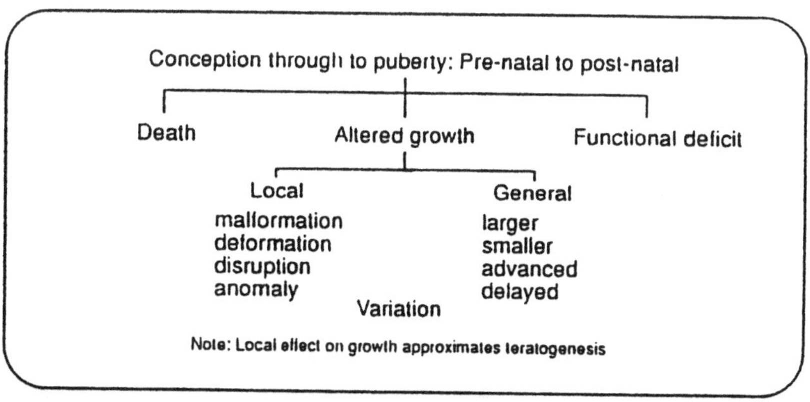

Fig. 2. Aspects of developmental toxicity

Consequent to this, in the hidden world of regulatory testing, it is not an uncommon experience to encounter panic, alarm and confusion over one or two abnormalities (of uncertain origin) observed at very high dosages whilst, at the same time, conclusive evidence of an effect on development at lower dosages is ignored or dismissed by applying a label such as "embryotoxicity" or, even worse, the meaningless term "foetotoxicity". The alternate desires, either to use or to avoid an attention catching label, have led to corruption of terminology causing further confusion.

Among all the many manifestations of developmental toxicity why is the greatest attention given a local effect on embryofoetal growth that may give rise to a physical defect? Described in these neutral, but realistic, terms (Fig. 2) it does not seem much to get excited about but, apply the alternate label teratogenesis, and it becomes the only risk of concern to a great majority.

Why are Some Risks Emphasised?

Why should people be excessively concerned with malformations relative to other effects? Why should abnormality be considered more important than the ultimate insult of death? Why should direct effects on the conceptus be considered more important than indirect ones? In reality, the mother and conceptus cannot be separated.

In numerical and biological terms the level of concern and the amount of effort expended in respect of the different manifestations of developmental toxicity would appear to be inversely proportional to their numerical probability of occurrence. Risk of malformation is limited to an extremely short period of time relative to the total developmental period and reproductive life span. The chance that exposure during this short time will result in malformation is even lower as mammals have developed extensive defence and repair mechanisms. This is why natural rates of malformation are so low (Fig. 3). At exposures relevant to humans, the proportion of chemicals that would induce any adverse effect on embryos is extremely low and the proportion of these that would induce malformations would be lower still.

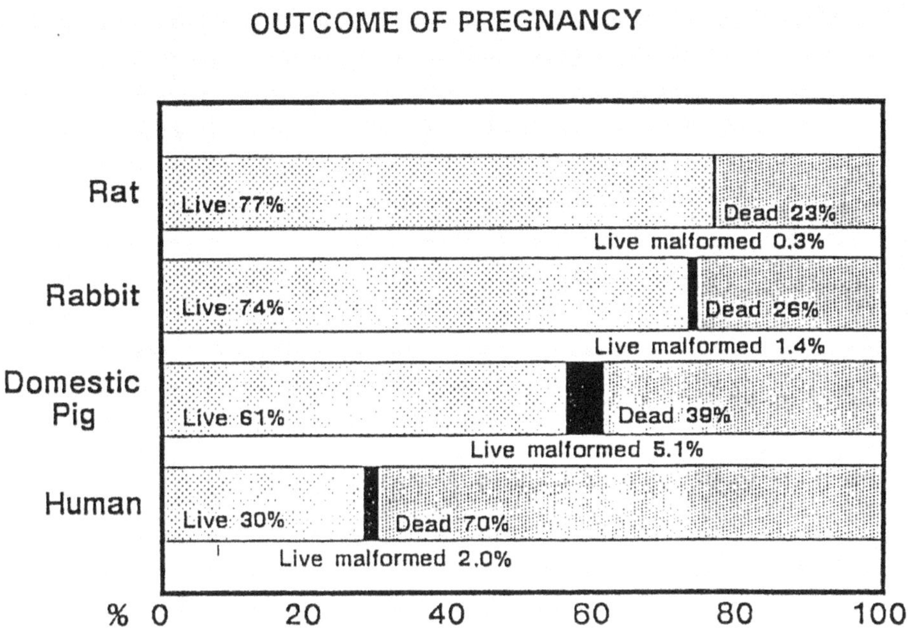

Fig. 3. The outcome of pregnancy in different species

It would seem then that it is the rarity of occurrence and the nature of the effect that combine to exaggerate concern rather than the likelihood of realisation. More concern is attached to an event that might occur in 1 per million but little or no attention is given to adverse effects that would occur in 1 per hundred. In other words it seems that scientists are as susceptible as anyone else to the primitive instincts of fear and dread. This leads to exaggerated perception of risk of rare events (Fig. 4).

It is *right* that we should take account of perceived risk but, as scientists, surely we should aim for a more equitable balance between what is real and what is imagined.

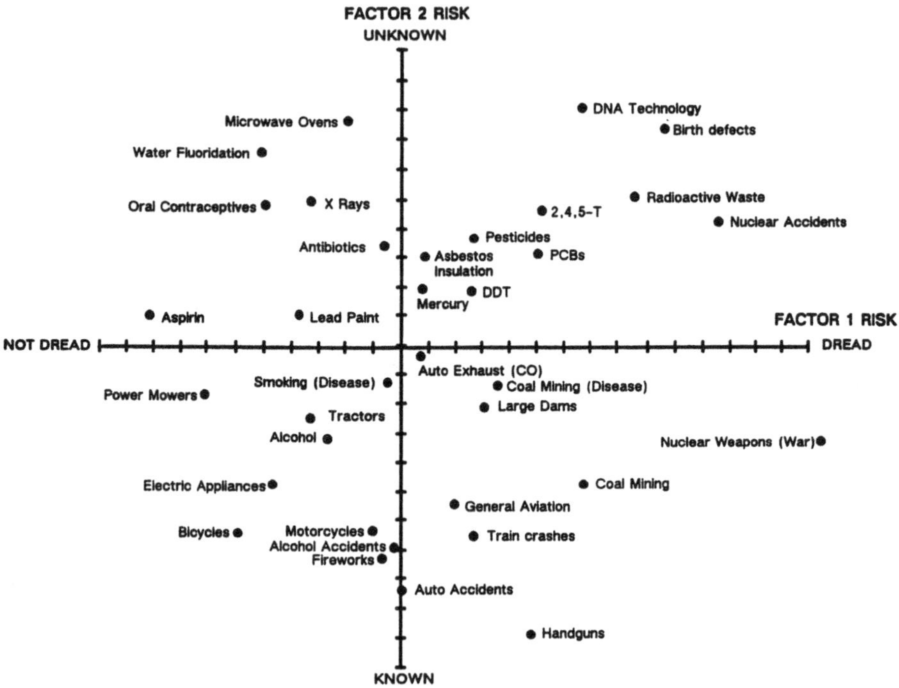

FACTOR 1 RISK		FACTOR 2 RISK	
ATTENUATED	EXAGGERATED	ATTENUATED	EXAGGERATED
Controllable	Uncontrollable	Observable	Unobservable
Not dread	Dread	Common	Rare
Not catastrophic	Catastrophic	Immediate effect	Delayed effect
Not fatal	Fatal	Scientifically known	Scientifically unknown
Voluntary	Involuntary	Known to exposed	Unknown to exposed
Current generation	Future generation	Old risk	New risk

Fig. 4. Perception of risk (courtesy of P. Slovic, Decision Research)

When to Assess Risks

The doubt concerning whether we are selecting the most appropriate risks to assess is linked with the question as to when we should do a risk assessment. The first necessity is to identify a (potential) hazard (Fig. 5) because if one cannot identify a hazard there is no point contemplating risk assessment. In respect of effects on reproduction and development, a risk assessment, worthy of the name, would not be required for the majority of compounds since the proportion that present a specific or selective hazard to reproduction is quite low.

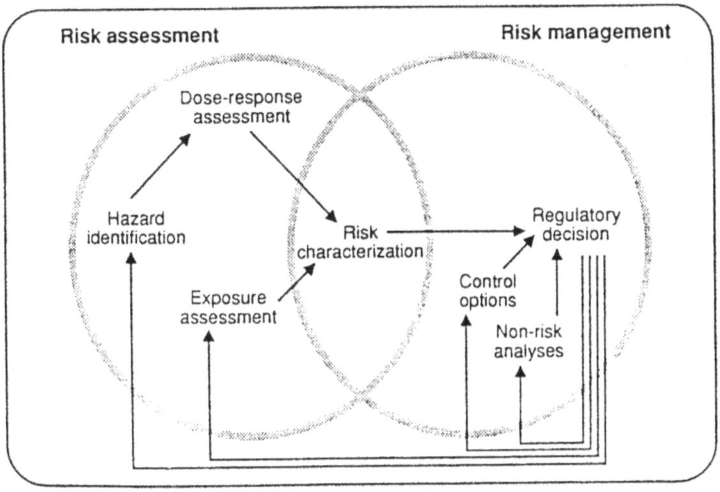

Fig. 5. Factors in risk assessment and management (EPA)

How to Assess Risk - The Safety Margin Approach

My next question is how to do a risk assessment on data from animal experiments. Since there are people here from regulatory agencies who use such procedures I will restrict myself to a brief, thumb-nail sketch of the two main methods. The first method is the safety margin approach, which I refer to as the bilateral pentadactyl limb system (Fig. 6). In this method, fudge factors, referred to in polite terms as uncertainty factors, are applied to the "no effect level" (NEL) determined in the animal experiment(s).

The first step is to apply, to the NEL, an uncertainty factor of 10, for intra species differences. This is multiplied by a further factor of 10, for inter species differences, to provide a factor of 100 times the NEL. For general toxicity it was (and sometimes still is) considered that a compound was safe if there was a hundredfold difference between the animal NOEL and the human dosage. After thalidomide, multiplication by a further factor of 10 was considered advisable to allow for differences between adults and embryos. Thus a compound was considered safe if there was a 1000 fold difference between the animal NEL and the human dosage. This often proved difficult to achieve so a factor of 500 became acceptable (presumably on the grounds that the embryo was smaller!).

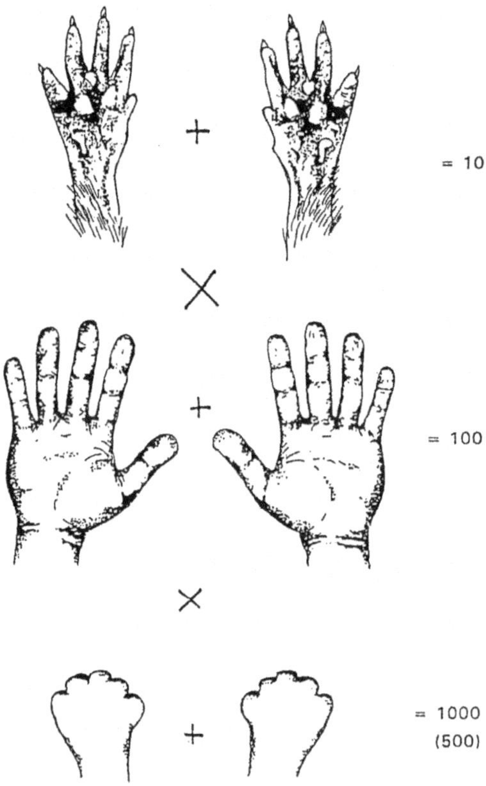

Fig. 6. The bilateral pentadactyl limb. Risk assessment using safety margins

The point to this childish description is to illustrate that scientific rationale and sophisticated argument may often camouflage inane reactions and guesswork. This is indicated by use of the psychological number of 10. This arbitrary base needs to be remembered as the safety margin approach to assessment may be the only one that can be applied for the great majority of chemicals.

The main weakness of the safety margin approach is the dependence on the somewhat intangible, experimental "no effect level" (NEL). There has been attempted refinement of the term, for example by using "no observed effect level" (NOEL) and even "no observed adverse effect level" (NOAEL) but such semantic machinations do not tackle the real problem. Being a negative, a no effect level cannot be proven, only presumed from failure to demonstrate an effect. Considering the many low frequency events (e.g. malformation, non pregnancy) that prevail in reproductive toxicity studies the chance of demonstrating a conclusive increase in low frequency events is extremely poor thereby making presumption of no effect extremely uncertain.

The uncertainty is especially exaggerated in respect of the emphasis on "teratogenicity" since the assumption that increasing dosage will produce an increasing manifestation of effects is less likely to be true than for general toxicity. It is not so much that the effect does not increase as the fact that an increase in effect is unrecognised. For example, it is not uncommon to encounter the conclusion that a compound was "embryotoxic (meaning embryolethal) but not teratogenic" thereby implying that it does not present a hazard. Not withstanding the fact that embryotoxicity and especially embryolethality are undesirable effects in their own *right* it would not be surprising that if the same dosages were applied on a single day rather than for several days the result would have been malformations rather than embryonic death.

Another reason for being wary of the safety margin approach to assessment is that, in general, the poorer the study the higher the NOEL. This encourages a manufacturer to do poor studies.

How to Assess Risk - Mathematical Modelling

The second method of risk assessment uses mathematical modelling. It can make better use of all the experimental groups in a study and the slope of the dose response. It need not be dependent on the unreliable NOEL nor need the cumulative addition of uncertainty factors. It reduces, but does not eliminate, prejudice. Also pseudo scientific phraseology can be replaced by pseudo statistical numeracy.

Basically the data from a study or series of studies are arranged to obtain a regression curve (Fig. 7). This can then be extrapolated to estimate the number of subjects that might be affected for a given dosage. However, when extrapolating beyond the observed range allowance must be made for deviation from linearity (Fig. 8), hence my personalised name of the broken umbrella system. Adding uncertainty factors to allow for deviation from linearity provides a wide range of estimates of human risk. The greater the difference between observed exposures and estimated human exposures the greater the variation in estimates.

Protagonists of mathematical modelling seem to be its worst enemy. They often become so involved with the mathematics that they lose sight of reality. They need to learn how to present results in terms that most people can visualise and that are within levels that could be detectable. It is no good demonstrating how clever their computers are at producing "n" decimal places or exponential values. Most people are unable to *put* in perspective ranges beyond 4 magnitudes, i.e. from 1 : 10 to 1 : 10,000. Prevalences less than 1 : 10,000 are virtually impossible to visualize.

Potentially, mathematical risk modelling has many advantages over the safety margin approach since some of the current doubts about the methodology relate to retention of concepts used in the safety margin approach. There is still a tendency to use the experimental NOEL as a starting point (the lowest observed effect level is better) and a tendency to add intuitive uncertainty factors. However, probably the main factor inhibiting the use of mathematical models is the paucity of appropriate studies and study data to use in the model.

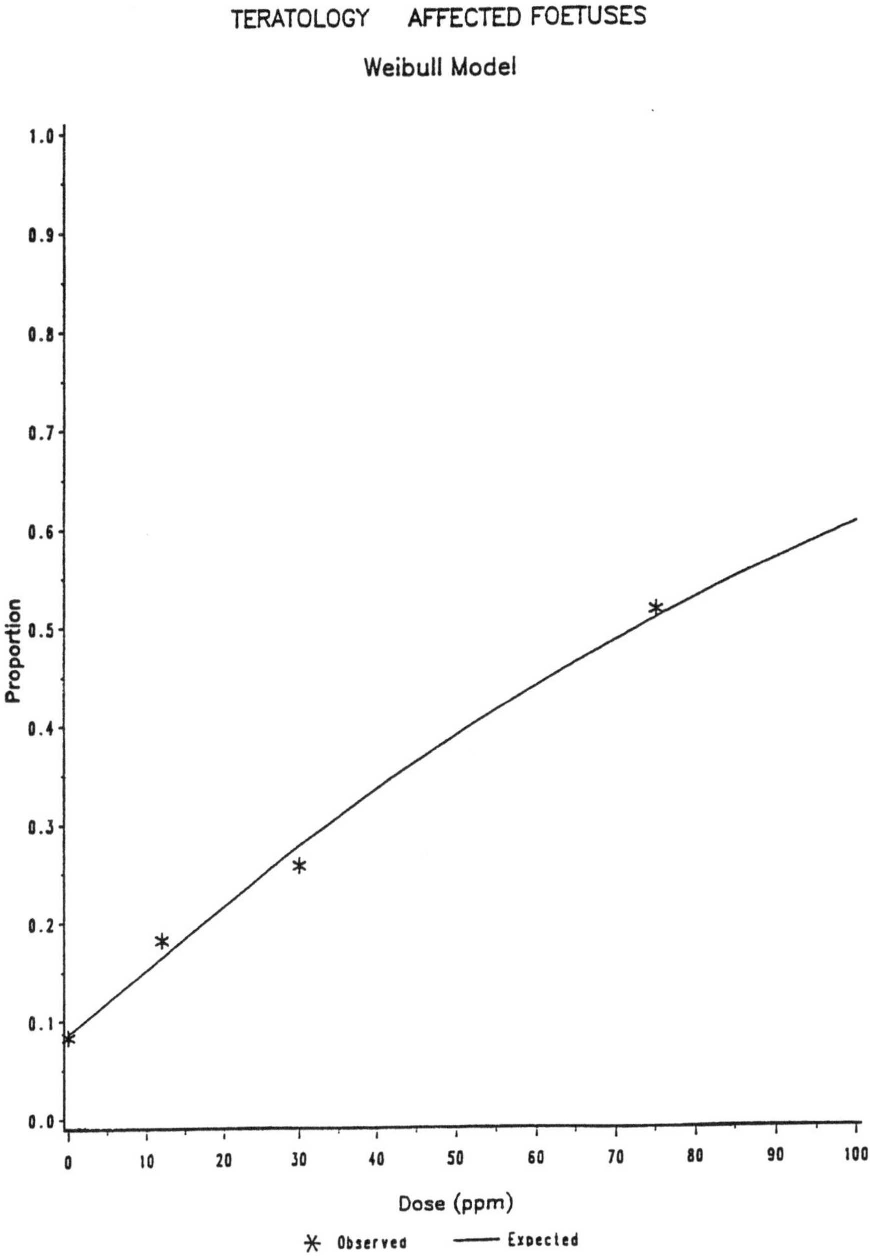

Fig. 7. Risk assessment using a mathematical model

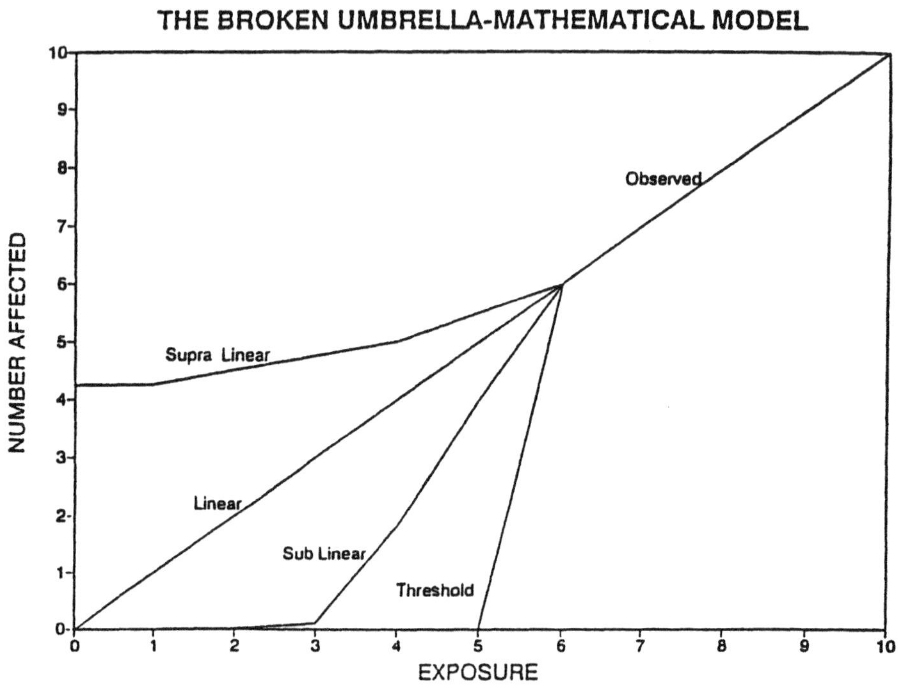

Fig. 8. Uncertainties of extrapolation. Extrapolation beyond observed values cannot assume that linearity of response will prevail. Adding uncertainty factors for different deviations from linearity would provide a wide spectrum of estimated risk at human exposures

Where to Obtain Data

For either method of risk assessment we need data that is relevant, reliable and consistent. Where do we get such data? Unfortunately, at this point we open Pandora's box and can become overwhelmed by a flood of inadequate studies and study presentations.

The first and often the only source of data are studies conducted to detect hazard (Fig. 5). I have some familiarity with such studies. I know what can be done and what cannot be done. I know what could be done and what is not done. I know that the criteria, methods and objectives for hazard detection are quite different from those that can be applied to hazard/risk characterisation or to academic research (Fig. 9). I know that whilst a hazard detection study might be just good enough to allow assessment by the safety margin approach, it would rarely have a structure appropriate for mathematical modelling. Applying the criteria and methodology for characterisation of risk, to the studies for hazard detection and vice versa can cause great confusion. It is a constant source of surprise and concern that so many investigators and regulatory agencies do not appreciate the differences between hazard detection (regulatory testing) and academic research.

Further deterioration of data reliability and consistency arises according to whether a study is reported in a scientific journal or in a report to a regulatory agency. Perhaps the least reliable source of data is in the open literature (Fig. 10). This is biased, censored and often inadequate. Editorial policy may enforce data presentation totally unsuitable for the variables involved. Distortions of presented data, however, pale in comparison with the data that is omitted. It is rarely possible to be able to reanalyse the results to check the author's conclusions.

The reports submitted to regulatory agencies generally contain individual data to allow alternate analysis. Opposite to the open literature, the major deficiency is the inclusion of irrelevant trivia and a profusion of manipulated data. Often, there is such a mass it is difficult to "see the wood for the trees". Perhaps the greatest impact on the data quality are flaws and fallacies of guidelines since these influence the planning, conduct and reporting of the study data. Many flaws in guidelines arise from the import of procedures from academic research and the open literature without appreciation of the deficiencies, bias and different objectives. Agencies do not appear to have validation procedures to ensure that these imports are appropriate for testing.

A third factor greatly influencing the reliability, relevance and consistency of the data from hazard detection studies is the enormous inter laboratory variation in skill level.

Aspect	Academia	Testing
Chemical activity	Used as a tool	Object of the study
Information	Extensive	Little or none
Aims	Specific	Wide spectrum
Methods	Free choice	Prescribed
Results	Predictable - positive -	Unpredictable - negative, equivocal -
Consequence - of error	Embarrassment	High expense or - human suffering

Fig. 9. Some common differences in reproductive toxicity

FACT OR FICTION?

-- A textbook "fact" can emerge from the close scrutiny of the scientific community and yet be falacious -

-- Editorial policy can considerably distort the findings of science. Censorship occasionally is seen to operate in a gross way.

-- They play safe with the accent on scientific "respectability" rather than on scientific advancement. They also represent establishment views.

-- Work - consistent with previous reports is most acceptable.

-- Novel ideas - are considered closely with a bias towards rejection.

??--?? Data are selected to fit a particular purpose - the opportunities here for distortion are immense.

(D.I. Williams (1977) The Social Evolution of Fact. Bull Br psychol Soc 30:241-243)

Fig. 10. Bias in the published literature

All told, we have an accumulation of confounding factors that make the reliability and consistency of the data available for risk assessment somewhat dubious. The following examples barely touch the surface of the mass of misleading data. The ultimate irony is that the worst cases are too unintelligible to use as examples and, in using examples, the impact of sheer volume of nonsense is lost.

The Zero Option

In the open literature the zero option is represented by the absence of relevant information. In reports to regulatory agencies it is represented by page after page of zeros. These can be presented as numbers or words, they can be inverted as 100% values rather than 0% values and they can be analysed and presented as mean and standard deviation of nothing to 4 or 5 places of decimals. Very often these provide a voluminous smoke screen to disguise inadequate study design, conduct and assessment. Regrettably it is often successful. The examples should be self explanatory.

Lost in the Forest

In the open literature the "lost in the forest" syndrome mostly is represented by lengthy dissertation, speculation and long reference listings. These often detract attention from the inadequacy of the study that has been performed. In reports to regulatory agencies it often takes the form of long listings of collected or manipulated data most of which are filed away without examination (e.g. individual weights of foetuses or neonates).

Most confusing and dangerous is the separation and fragmentation of related variables, especially variables with a natural low frequency such as abnormalities and embryonic deaths. Detailed examination of the parts often leads to neglect of the whole, especially in prenatal toxicity studies where an effect of treatment is often detected by an increase in the range of endpoints affected rather than an increase in a single endpoint.

The "lost in the forest" syndrome is often combined with the zero option such that relevant but low frequency events are difficult to detect amongst a sea of zeros.

Conclusion

Any method of risk assessment is only as good as the data available. Regarding toxicity studies in reproducing animals most of the data available is of poor quality. This is especially true for mathematical modelling since very few studies are designed to provide effects at several dosages (at least 3). The extent of the deficiencies in available data is impossible to demonstrate in a brief communication.

Discussion of the Presentation

Pritchard: You have given examples of both wedge- and L-shaped data. Whilst wedge-shaped may be preferred, the L-shape is often the result obtained. Are you suggesting that the L-shape is an inadequate assessment?

Palmer: No. The L-shape to the data is most frequently the consequence of setting dosage intervals that are too wide. With closer intervals the true wedge-shape would be seen. Sometimes the L-shape is true, a presumed reason being that up to a given dosage repair mechanisms can cope, leaving only small traces of effect, but once they are overwhelmed then lots of damage occurs.

Neubert: You mentioned "Karnofsky's law" stating that all substances are "teratogenic" at high enough doses. This statement was derived from studies in the chick embryo for which it may be correct. For mammals such a very general statement is certainly not true and there are good examples of substances given at doses which kill a high percentage of the mothers with no pronounced "teratogenic" effects on the surviving offspring. I am always surprised at how difficult it is to induce very severe gross-structural abnormalities (such as amelia) in experimental animals, and beside the effects of thalidomide in monkeys there are only few examples (mostly with alkylating agents or antimetabolites) for the induction of such drastic effects at a high frequency. However, inevitable adverse effects (increased mortality, reduced weight gain, minor abnormalities, etc.) will very often be induced with doses highly toxic to the mother.

Palmer: There are no rules that are infinite in biology. However, the basic principles of Paracelsus and Karnofsky are useful starting points and more useful than the perception that there is absolute safety.

Baß: In spite of all the bad examples shown in your talk (for understandable reasons), can you present the good example, taking into account GCO requirements as well as "scientific regulations" to present important data?

Palmer: Yes, if we have time for another workshop.

Biomarkers of Developmental Neurotoxicity

Andrew C. Scallet and William Slikker, Jr.

The Neurotoxicity Concept

Neurotoxicity may be defined as any adverse effect on the structure or function of the central and/or peripheral nervous system by a biological, chemical or physical agent. Neurotoxic effects may be permanent or reversible, produced by neuropharmacological or neurodegenerative properties of a neurotoxicant, or the result of direct or indirect actions on the nervous system (ICON 1990). Adverse effects include: unwanted side effects, effects due to overdosing, functional or structural compensatory responses to restore normal function; or any alteration from baseline which diminishes the ability to survive, reproduce or adapt to the environment. Some relevant effects can be measured directly by neurochemical, neurophysiological, and neuropathological techniques, whereas others must be inferred from observed behavior. Insults to the nervous system may take various forms and are often quite subtle (Anger 1986). Neurotoxicity may occur at any time in the life cycle from gestation through senescence. The developing nervous system may be particularly vulnerable to damage by certain chemicals, whereas adults may be more susceptible to other agents than the conceptus (Ali et al. 1986a; Ali et al. 1986b; Annau and Eccles 1986; Ecobichon et al. 1990; Lipscomb et al. 1989; Paule et al. 1986; Pearson and Dietrich 1985; Silbergeld 1986; Matthews and Scallet 1991). While the adult nervous system may also be acutely susceptible to new insults, the effects of earlier injuries may be revealed as it ages (Weiss 1990). Psychoactive substances may also indirectly impair health by inducing behaviors that decrease safety in the performance of numerous activities.

Exposure to Neurotoxicants

The number of potential neurotoxicants is difficult to assess because of limitations in the existing data base. Approximations from the current literature, however, suggest that 1) over 25% of chemicals for which threshold limit values have been set, had their threshold limit values determined all or in part on the basis of direct nervous system effects (Anger 1984), and 2) approximately 25% of chemicals frequently encountered in industry and of known toxicological significance have documented neurotoxic effects (OTA 1984). Based on these sources, the total estimated number of neurotoxicants that include drugs, food additives, cosmetics and pesticides would range from 1500-5000. If environmental and commercial agents are included, these numbers may well be tripled (McMillan 1987). Not only is the number of potential neurotoxicants large, but the cost of treatment of neurological disorders is staggering. The Congressional Office of Technology Assessment (OTA) estimates that care and treatment of neurological disorders and the accompanying loss of productivity can cost the U.S. as much as $300 billion a year (The Scientist 1990).

Accepting that there is the potential of exposure to a large number of neurotoxicants, a comprehensive, multidisciplinary approach that integrates neurochemical, neuropathological and behavioral assessments to elucidate effects and mechanisms in neurotoxicology has been developed at several institutions including the NCTR (Scallet et al. 1988; Schulze et al. 1988; Slikker et al. 1988; Slikker and Paule 1985). The requirement of this multidisciplinary approach is necessary because of the complexity and diverse functions of the nervous system.

Biomarkers and the Risk Assessment of Neurotoxicants

Risk assessment is the analytical process by which the nature and magnitude of risks are determined (OTA 1990). There are four steps or components of the risk assessment process: hazard identification, dose-response assessment, exposure assessment, and risk characterization (National Research Council 1983). Hazard identification entails a qualitative evaluation of toxicity produced by an agent and the relevancy of these data to the human situation. The dose-response assessment seeks to determine a quantitative relationship between exposure to an agent and the extent of toxicity. The third component, exposure assessment, is concerned with the numbers of individuals exposed and the magnitude and duration of the exposure. Finally, the combination of the results from the first three steps provides the risk characterization.

The nervous system's complexity not only provides a multitude of mechanisms by which toxicants can produce injury, but also provides a considerable challenge in the development of risk assessment strategies. Unlike risk assessment for carcinogens where tumor yield is often considered a universal endpoint, neurotoxicity may manifest itself in many ways. An interdisciplinary and comprehensive approach to assessment is therefore necessary for a full understanding of an agent's effects on the nervous system. Once those effects are defined, dose-response studies can serve to indicate which of the effects stem from a common cause. Finally, as with all risk assessments that rely on animal data, the extrapolation to the human situation must be accomplished. Such issues are currently being addressed directly by using the same endpoints and assessment procedures in both laboratory animals and humans (Paule et al. 1988a). Fortunately, as exemplified for lead (ATSDR 1988; U.S. EPA 1986), all the components necessary to conduct a risk assessment on a chemical affecting the nervous system can be described and exercised (Gaylor and Slikker 1990; Paule et al. 1988b; Sheehan et al. 1989). Therefore, risk assessments based solely on the basis of the neurotoxicities of chemicals are feasible.

The appropriate selection and use of biological markers or biomarkers is fundamental for the conduct of risk assessments for neurotoxicants. Biomarkers may be defined as indicators signaling events in a biological system, and are classified into three categories, those of exposure, effect and susceptibility (Committee on Biological Markers 1987). Exposure biomarkers may include either the quantitation of exogenous agents, or the complex of endogenous substances and exogenous agents within the system. Biomarkers of effect may be indicators of an endogenous component of the biological system or an altered state of the system that is recognized as an alteration or disease. A biomarker of susceptibility is an indicator that a biological system is especially vulnerable to toxic insult by an exogenous agent (Committee on Biological Markers 1987).

While the application of biomarkers to neurotoxicity risk assessment has not been widespread, examples of the three classes of biomarkers are available in the literature as shown in Table 1 and see Slikker 1991. In the majority of cases, of which this table lists only a few examples, the biomarkers have been used for hazard identification or to investigate the mechanism of action of a neurotoxicant. In only a few cases (e.g. Needleman 1987; O'Callaghan 1991; Gaylor and Slikker 1990; Slikker and Gaylor 1990) have attempts been made to validate the biomarker(s) by systematically examining specificity and sensitivity.

Mathematical Methods and Risk Assessment

With regard to dose-response assessment, the evaluation of neurotoxicants and other noncarcinogens is generally derived from observations of a no-observed-adverse-effect-level (NOAEL) or lowest-observed-adverse-effect-level (LOAEL) in people or animals exposed to a particular agent. The NOAEL is thought to approximate the theoretical threshold below which no effect is observed. In the final risk characterization step, these NOAEL or LOAEL values are usually divided by safety or uncertainty factors to produce a reference dose (RfD, Barnes and Dourson 1988). It is assumed that if human exposure is below the RfD, then little health risk exists.

There are several features of this RfD or safety factor approach which deserve consideration. First, the method assumes a theoretical threshold dose below which no biological effects of any type are observed. Not only is the determination of a threshold dose influenced by the sensitivity of the analytical methods employed, but the theoretical bases of a threshold dose may be questioned. If due to normal variation in cellular function an adverse effect can occur in untreated control subjects, then endogenous or exogenous factors may already be supplying a stimulus which is equivalent to a dose above the threshold dose. If exposure to an agent augments this stimulus, then an additional risk is expected and no threshold dose exists for that agent (Gaylor and Slikker 1990). Secondly, the magnitudes of the safety factors used to determine RfDs [interspecies extrapolation (10), intraspecies extrapolation (10) and acute vs chronic exposure (10) = 1000] are based more on best estimates than actual data (Sheehan et al. 1989;

McMillan 1987) and have been questioned for empirical reasons (Gaylor et al. 1990). Finally, the RfD approach uses a single observation, the NOAEL or LOAEL instead of a complete dose-response curve to calculate risk estimations. Chemical interactions with biological systems are often specific, stereoselective and saturable, such as enzyme-substrate binding leading to metabolism, transport and/or receptor-binding, any or all of which may be a requirement of an agent's effect or toxicity. Therefore, a chemical's dose-response curve may not be linear. The certainty of low dose extrapolation is markedly effected by the shape of the dose-response curve (Food and Drug Administration Advisory Committee on Protocols for Safety Evaluation 1971). Therefore, the appropriate use of a dose-response curve may enhance the certainty of risk estimations.

Risk Assessment of MDMA: A Numerical Example

To aid in the routine use of biomarkers in neurotoxicity, we exemplify here how biomarkers may be used in the case of risk assessment for the drug of abuse, methylenedioxymethamphetamine (MDMA), also known as "ecstasy". Gaylor and Slikker (1990) proposed a procedure for calculating health risk from exposure to MDMA. Risk is defined as the proportion of a population whose levels of a measure of toxicity, e.g., low levels of serotonin (5-HT), indicate an adverse level of the measure. In the absence of a well-established adverse level, an abnormal level may be used which is based upon the variation of levels observed among control animals.

Table 1. Biomarkers of neurotoxicity

Biomarkers of Exposure	Agents	Time of Exposure	Reference
Blood or Dentine Concentration	Lead	Pre- or Postnatal	Needleman, 1987
Cerebral Spinal Fluid Concentrations of Dopamine Metabolites	MPTP	Postnatal	Kopin & Markey, 1988
Cerebral Spinal Fluid Concentrations of a Serotonin Metabolite	MDMA	Postnatal	Ricaurte et al., 1988
Serum Esterase	Organo-phosphates	Postnatal	Levine et al., 1986
Biomarkers of Effect			
Neurotransmitters	Reserpine, MDMA	Pre- and Postnatal	Buelke-Sam et al., 1989 Gaylor & Slikker, 1990
Receptors	Reserpine	Prenatal	Ali et al., 1986b
Uptake Sites	MDMA	Postnatal	Battaglia et al., 1988
Enzymes Neurotransmitter Synthesis	Methamphetamine, MDMA	Postnatal	Bakhit & Gibb, 1981; Stone et al., 1986
Metabolic (ODC)	Methylmercury, TMT	Pre- and Postnatal	Slotkin et al., 1985; Ali et al., 1987
Neuropathology/ Neurohistology	TMT, MDMA	Postnatal	Chang et al., 1982; Scallet et al., 1988
Proteins (GFAP)	TMT, MPTP	Postnatal	O'Callaghan, 1991
Neurophysiological	Lead	Pre- and Postnatal	Johnson, 1980
Behavior	Cocaine, Diazepam	Pre- and Postnatal	Spear et al., 1989; Schulze et al., 1989
Biomarkers of Susceptibility			
Enzymes Delta-aminolevulinic acid dehydratase	Lead	Postnatal	Rogan et al., 1986
Glucose-6-posphate dehydrogenase	Lead	Postnatal	McIntire & Angle, 1972

The risk assessment process for neurotoxicants described by Gaylor and Slikker (1990) requires the use of a dose-response curve. A quadratic interpolation between doses of the logarithms of various measures of neurotoxicity was used. That procedure may be adequate within the experimental dose range but it is not recommended for extrapolation for low doses below the experimental dose range. Here, we consider a dose-response model which is biologically plausible in order to increase the validity of estimates of risk. The procedure will be demonstrated with examples of alterations in 5-HT neurotransmitter concentrations and concentrations of its metabolite 5-hydroxyindoleacetic acid (5-HIAA) in the hippocampus or frontal cortex of male rats exposed to MDMA. Doses corresponding to low levels of risk estimated from the proposed risk assessment procedure will be compared with doses obtained from the NOAEL or LOAEL divided by safety factors.

The rationale for the use of the saturation-type model is based on the biological effects produced by MDMA in the rat as reported in the literature. Our studies with ^{3}H-labeled MDMA indicate that peak concentrations of total radioactivity are evenly distributed in the rat brain after oral administration by 30 min and are eliminated by 24 hr (Ali et al. 1990). MDMA is not thought to be the proximate toxicant because direct injection of MDMA into regional brain areas fails to result in neuronal (Molliver et al. 1986) or neurochemical (Ali et al. 1990) alterations. Therefore, a metabolite of MDMA, probably other than methylenedioxyamphetamine (Schmidt et al. 1987; Gollamudi et al. 1989), is thought to be responsible for the neurotoxicity. Several metabolites have been described for MDMA (Lim and Foltz 1988). Although not all the enzymes necessary for the metabolism of MDMA have been characterized, we have demonstrated that some are stereoselective and likely to be saturable (Gollamudi et al. 1989). Therefore, the apparent requirement that MDMA be enzymatically bioactivated to produce toxicity provides at least one basis for a saturable dose-response model.

Another possible basis of a saturation-type biological model emanates from studies of the antagonism of MDMA effects by the 5-HT uptake inhibitor, citalopram (Schmidt et al. 1987). Although carrier-mediated transport of MDMA into the nerve terminal or carrier-mediated transport of monoamines out of the nerve terminal could explain the protective effects of citalopram on MDMA-induced alterations, either mechanism could provide a rationale for a saturable dose-effect relationship.

Currently, the exact mechanism whereby MDMA produces its neurotoxic effects is not known. It is known, however, that the stereoisomers of MDMA possess different behavioral, neurochemical and metabolic potentials (Anderson et al. 1978; Schmidt et al. 1987; Gollamudi et al. 1989). Even though a "receptor" site for MDMA has yet to be determined, these reported stereoselective interactions strongly suggest that MDMA does not produce its effects via nonspecific mechanisms. Therefore, the possibility of a further specific interaction of MDMA or metabolite with a saturable "receptor" also supports the need for a saturable biological model.

Based on these experimental and theoretical biological considerations, a saturation model (Slikker and Gaylor 1990) was used to describe the relationship between the regional brain concentrations of an endogenous neurotransmitter (expressed as % of the normal average) and dose of a neurotoxicant.

$$\% = \frac{1 + \text{ADose}}{1 + \text{BDose}} \times 100\%$$

At dose=0, the 5-HT concentration is 100% of the normal average. A and B are constants estimated from the data where A/B is the minimum % obtained at large doses and (A-B) 100% is the % reduction from the normal average per unit of dose for low doses approaching zero. A and B are estimated by nonlinear regression. The properties of the saturation curve are illustrated in Figure 1 for levels of 5-HT in the hippocampus of male rats. An equivalent procedure has been applied to quantitative analysis of dose-response data obtained from non-human primates (Gaylor and Slikker 1990).

Risk is defined as the proportion of a population whose levels of a measure of toxicity indicate an adverse level. In the absence of a well-established adverse level, an abnormal level may be used which is based upon the variation of levels observed among control animals. For example, a level of 5-HT may be considered abnormal if it is lower than the level observed at the 0.1 percentile or above the 99.9 percentile of the background levels observed in an untreated control population. That is, a level of a neurochemical may be considered abnormal if that level is less than that of 1 per 1000 individuals in a normal untreated population. In general, measurements are not available on 1000 or more individuals. Thus, the 0.1 percentile or 99.9 percentile cannot be observed directly. However, Gaylor and Slikker (1990) noted that many measures of neurotoxicity can be described by a log-normal distribution, i.e., the logarithms of the measures are described by a normal (Gaussian) distribution. In such cases, the levels corresponding to any percentile can be estimated.

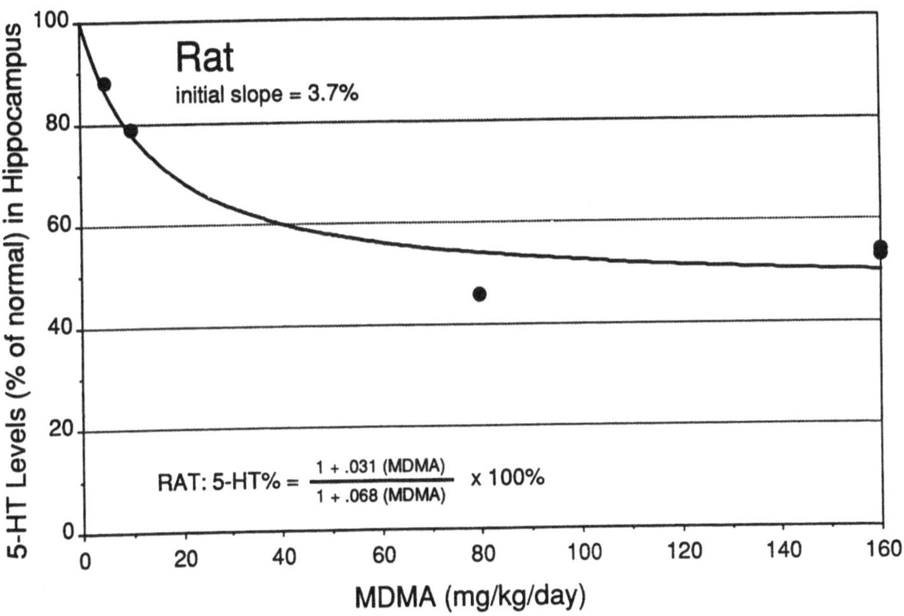

Fig. 1. Dose-response curve describing the hippocampal levels of 5-HT as % of normal controls one month after various doses of MDMA in the male Sprague Dawley rat. The MDMA was administered orally either once or twice a day for four consecutive days and the rats were sacrificed one month later to determine regional brain concentrations of 5-HT. The bold filled circles represent the experimental data and the solid line was determined by the saturation-model equation shown in the middle of the figure where (MDMA) = dose of MDMA in mg/kg/day. The correlation coefficient was $r > 0.98$

The numerical procedure for estimating risk as a function of dose is given by Gaylor and Slikker (1990). Here, the saturation curve described above is used for the dose-response relationship. The saturation model was fit to the geometric mean levels of 5-HT or 5-HIAA in the hippocampus or frontal cortex of male rats which had been administered MDMA (Slikker et al. 1988; Gollamudi et al. 1989; Slikker et al. 1989; Scallet et al. 1988). These rats were dosed once or twice per day for four consecutive days with various doses of MDMA and sacrificed one month later for regional brain neurochemical analyses. The Statistical Analysis System (SAS) Procedure NLIN, which is a nonlinear least squares regressions procedure, was used to fit the models (Table 2). The correlation coefficient between the observed and predicted values was high ($r > 0.98$). The saturation model was used to estimate the mean response as a function of dose. The levels of 5-HIAA in the frontal cortex showed the least change at high doses, i.e., the largest asymptote. At low doses the levels of 5-HT changed almost at twice the rate in the frontal cortex as in the hippocampus, as indicated by the initial slope (percent change per mg/kg/d of MDMA). Levels of 5-HIAA at low doses of MDMA, however, changed more rapidly in the hippocampus than in the frontal cortex. Since the standard deviations were similar in the four examples, the abnormal levels (expressed as a percent of the normal control level) corresponding to the 0.1 percentile are similar. Only 1 out of 1000 control animals would be predicted to have neurochemical levels below the abnormal level.

Since all four examples showed a dose-response without any apparent threshold dose, it is not clear if investigators would consider the lowest dose tested (5 mg/kg/d) the lowest-observed-effect-level or the highest no-observed-adverse-effect-level, as the mean effect at this dose was within one standard deviation of the mean control level. Thus, it is not clear whether a reference dose would be set at 5/1000 = 0.005 or 5/100 = 0.05 mg/kg/d of MDMA. In either case, the reference doses are well below doses that are estimated to produce additional risks above background of 0.0001. The probability (risk) of an abnormal animal in control animals was set to equal 0.001. Using the saturation model and the standard deviation, the doses estimated to produce additional risks of 1 in 10,000 and 1 in 1000 animals (See Table 2) were calculated using the technique described by Gaylor and Slikker (1990).

Table 2. Summary of selected neurochemical levels from exposures to MDMA

	5-HT Hippocampus	5-HT Frontal Cortex	5-HIAA Hippocampus	5-HIAA Frontal Cortex
Saturation Model[a]	$\frac{1+.031D}{1+.068} \times 100\%$ [b]	$\frac{1+.043D}{1+114D} \times 100\%$	$\frac{1+.069D}{1+.179D} \times 100\%$	$\frac{1+.078D}{1+.129D} \times 100\%$
Mean Minimum[c]	46%	38%	39%	60%
Initial Slope[d]	3.7%	7.1%	11.0%	5.1%
Standard Deviation[e]	0.318	0.312	0.354	0.250
Abnormal Level[f]	37%	38%	33%	46%
Reference Dose[g] (mg/kg/d)	0.005 or 0.05	0.005 or 0.05	0.005 or 0.05	0.005 or 0.05
Dose for (mg/kg/d) additional risk of 0.0001[h]	0.28	0.14	0.10	0.16
Dose for (mg/kg/d) additional risk of 0.001[h]	2.0	1.0	0.74	1.2

a) Percent of mean control level $= \frac{1 + AD}{1 + BD}$ 100%

b) D = MDMA dose in mg/kg/day given orally on four consecutive days

c) Mean (minimum) asymptote = $\frac{A}{B} \times 100\%$

d) (A-B) x 100% per mg/kg/day of MDMA; e) Standard deviation of $\log_e$ (neurochemical level); f) Risk = 0.001, i.e., 0.1 percentile of control levels; g) Lowest-observed-effect-level at 5 mg/kg/d ÷ 1000 or no-observed-effect-level at 5 mg/kg/d ÷ 100; h) Additional proportion abnormal estimated from expected saturation dose-response and standard deviation

Implications of the Models

Use of a dose-response model based upon plausible biological mechanisms provide more validity to prediction than purely empirical models. In the cases investigated here, saturation mechanisms were hypothesized and indeed saturation curves provided relatively good fits to the experimental results ($r > 0.98$). The saturation curve provides an estimate of the asymptote (minimum level) of a neurochemical at high doses. Also, the saturation model provides an estimate of the initial slope (percent change in the neurochemical level per mg/kg/d of MDMA) at low doses. For purposes of illustration, abnormal levels were defined as the 0.1 percentile. That is, only 1 out of 1000 animals would be expected to have neurochemical levels this low in normal control animals. Since these levels are described by a log-normal distribution (i.e., the logarithms of neurochemical levels are described by a normal/Gaussian distribution), other percentiles could readily be used for the abnormal level.

For these data it is not clear whether the lowest dose tested (5 mg/kg/d) would be considered a NOAEL or LOAEL. The subjectivity of setting reference doses with safety factors is one of the shortcomings of the RfD approach of risk assessment. On the other hand, use of a plausible dose-response curve in combination with the distribution of neurochemical levels about the dose-response curve provides a procedure to estimate the proportion of abnormal animals as a function of dose. Conversely, doses corresponding to specified levels of risk can be calculated. Hence, quantities with precise scientific definitions can be estimated and provide a rationale basis for decisions regarding risks from exposures to neurotoxicants.

References

ATSDR (Agency for Toxic Substances and Disease Registry) (1988) Nature and extent of lead exposure of children in the United States: A report to Congress. Atlanta, GA, U.S. Dept of Health and Human Services, U.S. Public Health Service

Ali SF, Buelke-Sam J, Newport GD, Slikker W Jr (1986a) Early neurobehavioral and neurochemical alterations in rats prenatally exposed to imipramine. Neurotoxicology 7(2): 365-380

Ali, SF, Buelke-Sam J, Slikker, W Jr (1986b) Prenatal reserpine exposure in rats decreases caudate nucleus dopamine receptor binding in female offspring. Toxicol Lett 31: 195-201

Ali SF, Newport GD, Slikker W Jr, Bondy SC (1987) Effects of trimethyltin on ornithine decarboxylase in various regions of the mouse brain. Toxicol Lett 36: 67-72.

Ali SF, Tandon P, Tilson HA, Lipe GW, Newport GD, Slikker W Jr (1990) Intracerebral and oral administration of methylenedioxymethamphetamine (MDMA): distribution and neurochemical alterations in rat brain. Presented at the XIth International Congress of Pharmacology, Amsterdam, July, 1990. Eur J Pharmacol 183: 450

Anderson GM, Braun G, Braun U, Nichols DE, Shulgin AT (1978) In: G Barnett, M Trisc and R Willette (eds) Quasar Research Monograph 22, National Institute on Drug Abuse, Washington, DC, pp 8

Anger WK (1984) Neurobehavioral testing of chemicals: Impact on recommended standards. Neurobehav Toxicol Teratol 6: 147-153

Anger WK (1986) Worker exposures In: Z Annau (ed) Neurobehavioral Toxicology, The Johns Hopkins University Press, Baltimore, pp 331-347

Annau Z, Eccles CU (1986) Prenatal exposure. In: Z Annau (ed) Neurobehavioral Toxicology, The Johns Hopkins University Press, Baltimore, pp 153-169

Barnes DG, Dourson M (1988) Reference dose (RfD): Description and use in health risk assessments. Reg Toxicol Pharmacol 8: 471-486

Bakhit C, Gibb JW (1981) Methamphetamine-induced depression of tryptophan hydroxylase: recovery following acute treatment. Eur J Pharmacol 76: 229-233

Battaglia G, Yen SY, Desouza EB (1988) MDMA-induced neurotoxicity: parameters of degeneration and recovery of brain serotonin neurons. Pharmacol Biochem Behav 29: 269-274

Buelke-Sam J, Ali SF, Kimmel GL, Slikker W Jr, Newport GD (1989) Postnatal function following prenatal reserpine exposure in rats. Neurobehavioral toxicity. Neurotoxicol Teratol 11: 515-522

Chang LW, Tiemeyer TM, Wenger GR, McMillan DE (1982) Neuropathology of mouse hippocampus in acute trimethyltin intoxication. Neurobehav Toxicol Teratol 4: 149-156

Committee on Biological Markers of the National Research Council (1987) Biological markers in environmental health research. Environ Health Perspect 74: 3-9

Ecobichon D Davies JE, Doull J, Ehrich M, Joy R, McMillan D, Macphall R, Reiter LW, Slikker W, Jr, Tilson H (1990) The effect of pesticides on human health. In: S Baker and C Wilkinson (eds) Advances in Modern Environmental Toxicology, Vol XVIII, Princeton Scientific Publishing Co, Inc, pp 131-199

Food and Drug Administration Advisory Committee on Protocols for Safety Evaluation (1971) Panel on carcinogenesis. Report on cancer testing in the safety evaulation of food additives and pesticides. Toxicol Appl Pharmacol 20: 419-438

Gaylor DW, Slikker W Jr (1990) Risk assessment for neurotoxic effects. Neurotoxicology 11: 211-218

Gaylor DW, Chen JJ, Sheehan DM (1990) Uncertainty in cancer risk estimates. Risk Analysis (submitted)

Gollamudi R, Ali SF, Lipe G, Newport G, Webb P, Lopez M, Leakey JEA, Kolta M, Slikker W, Jr (1989) Influences of inducers and inhibitors on the metabolism *in vitro* and neurochemical effects *in vivo* of MDMA. Neurotoxicology 10: 455-466

ICON, Interagency Committee on Neurotoxicity, Personal communication (1990)

Johnson BL (1980) Electrophysiological methods. In: P Spencer and H Schaumburg (eds) Neurotoxicity Testing in Experimental and Clinical Neurotoxicology, Williams and Wilkins, pp 726-742

Kopkin IJ, Markey SP (1988) MPTP toxicity: Implications for research in Parkinson's disease. Ann Rev Neurosci 11: 91-96

Levine MS, Fox NL, Thompson B, Taylor W, Darlington, AC, Van Der Hoeden J, Emmett EA, Rutten W (1986) Inhibition of esterase activity and an undercounting of circulating monocytes in a population of production workers. J Occup Med 28: 207-211

Lim HK, Foltz RL (1988) *In vivo* and *in vitro* metabolism of 3,4-(methylenedioxy) methamphetamine in the rat: identification of metabolites using an ion trap detector. Chem Res Toxicol 1: 370-378

Lipscomb JC, Paule MG, Slikker W Jr (1989) The disposition of ^{14}C-trimethyltin in the pregnant rat and fetus. Neurotox Teratol 11: 185-191

Matthews JC, Scallet AC (1991) Nutrition, neurotoxicants, and age-related neurodegeneration. Neurotoxicology 12: 547-558

McIntire MS, Angle CR (1972) Air lead: relation to lead in blood of black school children deficient in glucose-6-phosphate dehydrogenase. Science 177: 520-521

McMillan DE (1987) Risk assessment for neurobehavioral toxicity. Environ Health Perspect 76: 155-161

Molliver ME, O'Hearn E, Battaglia G, DeSouza EB (1986) Direct intracerebral administration of MDA and MDMA does not produce serotonin neurotoxicity. Soc Neurosci Abs 12: 336.3

National Research Council (1983) Risk assessment in the Federal Government: managing the process. Washington, DC, National Academy Press

Needleman HL (1987) Introduction: Biomarkers in neurodevelopmental toxicology. Environ Health Perspect 74: 149-152

OTA, (Office of Technology Assessment) U.S. Congress (1990) Neurotoxicity: identifying and controlling poisons of the nervous system. OTA-BA-436 Washington, DC, U.S. Government Printing Office (April 1990)

OTA (Office of Technology Assessment) (1984) Impact of neuroscience: A background paper. OTA-BP-BA-24 Washington, DC, U.S. Government Printing Office

O'Callaghan JP (1991) Assessment of neurotoxicity: use of glial fibrillary acidic protein as a biomarker. Biomed Environ Sci: in press

Paule MG, Reuhl K, Chen JJ, Ali SF, Slikker W Jr (1986) Developmental toxicology of trimethyltin in the rat. J Toxicol Appl Pharmacol 84(2): 412-417

Paule MG, Cranmer JM, Wilkins JD, Stern HP, Hoffman EL (1988a) Quantitation of complex brain function in children: Preliminary evaluation using a nonhuman primate behavioral test battery. Neurotoxicology 9(3): 367-378

Paule MG, Schulze GE, Slikker W Jr (1988b) Complex brain function in monkeys as a baseline for studying the effects of exogenous compounds. Neurotoxicology 9: 463-470

Pearson DT, Dietrich KN (1985) The behavioral toxicology and teratology of childhood: Models, methods and implications for intervention. Neurotoxicology 6(3): 165-182

Ricaurte GA, Delanney LE, Weiner SG, Irwin, Langston JW (1988) 5-Hydroxyindoleacetic acid in cerebrospinal fluid reflects serotonergic damage induced by 3,4-methylenedioxymethamphetamine in CNS of non-human primates. Brain Res 474: 359-363

Rogan WJ, Reigart Jr, Gladen BC (1986) Association of amino levulinate dehydratase and ferrochelatase inhibition in childhood lead exposure. J Pediatr 109: 60-64

Scallet AC, Lipe GW, Ali SF, Holson RR, Frith CH, Slikker W Jr (1988) Neuropathological evaluation by combined immunohistochemistry and degeneration-specific methods: application to methylenedioxymethamphetamine. Neurotoxicology 9: 529-538

Schmidt CJ, Levin JA, Lovenberg W (1987) *In vitro* and *in vivo* neurochemical effects of methylenedioxymethamphetamine on striatal monoaminergic systems in the rat brain. Biochem Pharmacol 36(5): 747-755

Schulze GE, McMillian DE, Bailey Jr, Scallet AC, Ali SF, Slikker W Jr, Paule MG (1988) Acute effects of delta-9-tetrahydrocannabinol (THC) in rhesus monkeys as measured by performance in a battery of cognitive function tests. J Pharmacol Exp Ther 245: 178-186

Schulze GE, Slikker W Jr, Paule MG (1989) Multiple behavioral effects of diazepam in rhesus monkeys. Pharmacol Biochem Behav 34: 29-35

Sheehan DM, Young JF, Slikker W Jr, Gaylor DW, Mattison DR (1989) Workshop on risk assessment in reproductive and developmental toxicology: addressing the assumptions and identifying the research needs. Reg Toxicol Pharmacol 10: 110-122

Silbergeld, EK (1986) Maternally mediated exposure of the fetus: *In utero* exposure to lead and other toxins. Neurotoxicology 7: 557-56

Slikker W Jr, Ali SF, Scallet AC, Frith CH, Newport GD, Bailey Jr, (1988) Neurochemical and neurohistological alterations in the rat and monkey produced by orally administered methylenedioxymethamphetamine (MDMA). Toxicol Appl Pharmacol 94: 448-457

Slikker W, Jr, Holson RR, Ali SF, Kolta MG, Paule MG, Scallet AC, McMillan DE, Bailey Jr, Hong JS, Scalzo FM (1989) Behavioral and neurochemical effects of orally administered MDMA in the rodent and nonhuman primate. Neurotoxicology 10: 529-542

Slikker W Jr, Paule MG (1985) Symposium overview: developmental neuropharmacology/neurotoxicology. Proc West Pharmacol Soc 28: 309-310

Slikker W Jr, Gaylor DW (1990) Biologically-based dose-response model for neurotoxicity risk assessment. Korean J Toxicol 6: 205-213

Slikker W Jr (1991) Biomarkers of neurotoxicity: an overview. Biomed Environ Sci 4: 192-196

Slotkin TA, Pachman S, Kavlock, RJ, Bartlome J (1985) Early biochemical detection of adverse effects of a neurobehavioral teratogen: influence of prenatal methylmercury exposure on ornithine decarboxylase in brain and other tissues of fetal and neonatal rat. Teratology 32: 195-202

Spear LP, Kirstein CL, Bell J, Yoottanssumpun V, Green Baum, R, O'Shea J, Hoffmann H, Spear NE (1989) Effects of prenatal cocaine exposure on behavior during the early postnatal period. Neurotoxicol Teratol 11: 57-63

Stone DM, Stahl DC, Hanson GR, Gibb JW (1986) The effects of 3,4-methylenedioxymethamphetamine (MDMA) and 3,4-methylenedioxyamphetamine (MDA) on monoaminergic systems in the rat brain. Eur J Pharmacol 128: 41-48

The Scientist (1990) Neurotoxicologists call for more research, Regulation P. 5, February 5, 1990

U.S. EPA (Environmental Protection Agency) (1986) Air quality criteria for leads. Vols. I-IV (Document No. EPA-600/8/028aF). Research Triangle Park, NC

Weiss B (1990) Risk assessment: the insidious nature of neurotoxicity and the aging brain. Neurotoxicology 11: 305-323

Discussion of the Presentation

Meister: Concentration-response data can be modelled and analyzed for continuous as well as for quantal variables. There is no need to introduce a dichotomy from effect measured on a continuous scale, as this has been proposed for the SM-levels.

Scallet: Our example (5 HT levels) happened to be for an effect biomarker that appears to be a normally distributed continuous or real number random variable. This by no means excludes the application of quantitative risk assessment procedures to other types of data, and indeed this has been done (references available and can be added).

Meister: Statistics should not be misused for the decoration of results with high evidence based on a very low number of animals.

Scallet: In fact, the original data in this case is probably more convincing and instructive than the means table slide taken as a part from a previously published paper of mine. However, I point out that this test (the student's t-test) was specifically developed as a small sample method and the required size of the difference between means *is* very large to ensure "significance". The point being made is only that the differences were so large, they show up *even with* only the few data. Not that this is a desirable sample size to use.

A Call for Increased Flexibility in Current Teratogenicity Testing

William J. Scott, Jr.

Introduction

The testing of new drugs and chemicals for the potential risk they might pose for the human population is a very difficult proposition. In the field of developmental toxicity this task is more difficult due to the added complexity of species differences in embryogenesis and the placental interface.

In 1966 the U.S. Food and Drug Administration made public a set of Guidelines for Reproduction Studies which included recommendations for the conduct of teratology studies now commonly referred to as segment 2. These recommendations utilized very well the scientific concepts of teratology available at that time. Twenty-five years later the same "guidelines" are still being used and with little significant change have been the basis for testing strategy in other countries as well as other governmental agencies within the U.S.

The premise of this presentation is that scientific knowledge accrued over the last twenty-five years should allow a reordering of the testing strategy and will attempt to demonstrate this premise using a single concept, duration of action.

The 1966 FDA "Guidelines" provided suggestions for many of the experimental parameters such as dose, days of gestation on which treatment should be given and route of administration. However, there was no indication of how often a drug should be given on each day of treatment. In my limited experience, the overwhelming majority of studies utilize once a day administration and the reason is most likely pragmatic rather than having a scientific basis. It is easier to dose animals once a day rather than having to catch and inject or intubate at multiple intervals on a single day. It could be argued that such an arrangement lessens the amount of stress to which the maternal animal is subjected thereby lessening a possible confounding influence on pregnancy outcome. Yet many clinical therapeutic regimes include multiple daily administrations in an attempt to maintain effective blood levels. Coupled with the fact that the rodent species often used in testing, i.e., mice and rats, frequently metabolize and/or clear therapeutic agents within minutes or a few hours leads to the conclusion that the testing strategy may fall far short of mimicking clinical use of many therapeutic agents. It should also be borne in mind that within a 24-hour period during early rodent organogenesis that a great deal of embryologic development takes place. Thus it is conceivable that a specific developmental event sensitive to perturbation by the agent under study may routinely occur during a part of the day when levels of the test agent are minimal or absent altogether. These admittedly unlikely series of events could lead to an erroneous estimation of teratogenic risk.

To continue this argument regarding duration of action as an important parameter in defining teratologic risk it would seem prudent to offer some data in support of the concept. There are available data indicating that clinically important teratogens such as ethanol and retinoids are more potent when administration regimes produce a prolonged exposure. I have chosen, however, to broach the topic of duration using experimental work performed on acetazolamide primarily due to the long history of work on this agent in our laboratory. Three levels of inquiry (teratogenic response, pharmacokinetics, mechanism of action) will attempt to show that duration of action is a critical feature in determining the fetal outcome.

Teratogenic Response

The most frequent malformation induced by acetazolamide is postaxial right forelimb ectrodactyly as first reported by Layton and Hallesy (1965). Subsequently it has become clear that malformations of other organ systems can be produced by acetazolamide (see Hirsch and Scott 1983 for review) but all comments in this manuscript will concern the forelimb malformations.

From the outset studies of acetazolamide teratogenesis have pointed to prolonged exposure as an integral requirement for expression of limb malformations although no single study has been expressly designed to investigate this idea. For examples of duration of action as an important factor we can point to the original description of acetazolamide teratogenesis by Layton and Hallesy (1965). This study utilized dietary administration to pregnant rats or mice which could be expected to cause a protracted exposure compared to a bolus dosing regime. The idea was taken a step further by Wilson et al. (1968) who compared subcutaneous versus intraperitoneal injection of acetazolamide on day 10 of rat gestation. Typical limb malformations were found in 23% of the offspring from mothers treated subcutaneously versus 4% from those treated intraperitoneally. In general subcutaneous administration prolongs exposure to a toxicant when compared to intraperitoneal administration thus contributing to the notion that duration of acetazolamide in the pregnant rodent is an important determinant of teratologic response.

Finally, a study conducted to determine the site of acetazolamide action (i.e., mother or embryo) also suggests duration to be important (Scott et al. 1990). Based on pharmacokinetic information provided by Wilson et al. (1968), multiple intrauterine injections of acetazolamide were made 7 hours apart. Three of eight surviving embryos (38%) had typical postaxial forelimb ectrodactyly when given 0.25 mg 4 times compared to 0/11 given the same dosage 3 times. Even when dosage was doubled to 0.5 mg/administration the frequency of affected survivors was only 11% (2/18) after three injections. All of these results plus others not described herein strongly suggest that acetazolamide must be present for a considerable time in order to produce typical limb deformity.

Pharmacokinetics

The transplacental distribution of acetazolamide has not been well studied but the available data indicate that after a teratogenic regime effective levels of drug can be measured for about 24 hours. Wilson et al. (1968) measured acetazolamide concentration in the plasma of non-pregnant rats who had received 500 mg/kg subcutaneously twice, 7 hours apart. Assuming that 2 μg/ml is sufficient for pharmacodynamic activity (i.e., inhibition of carbonic anhydrase) this treatment schedule achieved complete inhibition for almost 24 hours. Subsequently (Hirsch and Scott 1983) it was shown that the concentration of acetazolamide in mouse embryos exposed to a teratologic regime remains above this inhibitory concentration (2 μg/ml) for more than 20 hours. Thus the pharmacokinetic data available are consistent with the idea of a prolonged exposure as a requirement to reveal acetazolamide teratogenicity. A study utilizing constant infusion to produce prolonged but low concentrations of acetazolamide in maternal plasma and embryo would certainly help resolve this question.

Mechanism of Action

Our laboratory has suggested that the mechanism of acetazolamide-induced forelimb malformation is reduction of intracellular pH (pH_i) as a consequence of carbonic anhydrase inhibition (Scott et al. 1990). The information utilized to form this hypothesis and the results produced to test it cannot be discussed here. However, one set of results again points to duration as an important parameter. The above study utilized two inbred mouse strains, one very sensitive (C57BL/6) to ectrodactyly induction by acetazolamide and one very resistant (SWV). Measurement of embryonic pH_i showed a similar reduction in both strains but this reduction was prolonged for 8-12 hours in the sensitive strain but short-lived (ca. 3 h) in the resistant strain. These data indicate that the resistant strain embryos can recover more quickly from the insult than sensitive strain embryos and contribute further to the idea of duration as an important determinant of teratologic response.

Discussion

If the argument just put forward has merit, then the next logical step would be the proposal of alternate experimental designs to circumvent the problem. Attempts to create "guidelines" for dealing with duration of exposure need to incorporate information from the proposed clinical usage. This simple idea is, however, confounded by the fact that rodent and primate embryogenesis does not proceed at the same rate. Thus simple matching of exposure times may lead to an overestimate of potential clinical risk from studies conducted in rodents.

These sort of dilemmas point out the shortcomings in scientific information about the questions under study. Our knowledge is woefully deficient of rodent embryonic development in which the studies are done and in human embryonic development to which the results are extrapolated. These shortcomings coupled with incomplete pharmacokinetic and pharmacodynamic information of the agent under study diminish the usefulness of "guidelines" if they are rigidly adhered to.

Thus the message of this presentation is a call for greater flexibility in the design of hazard identification studies. This call for greater flexibility is meant to encompass all the parameters that can be varied in a typical FDA segment 2 study not just duration of action as discussed herein. The choice of species, dose, route of administration, and the period of gestational exposure should be based on the known and predicted attributes of the agent under study. Pragmatic considerations should be given less attention than they presently command.

The call for an expanded, individualistic approach to experimental design is not new. The original 1966 FDA "Guidelines" hinted that "modifications may be necessary to get meaningful experiments." The Canadian version of guidelines (1973) say it better. "Each chemical must be considered individually and based on sound *scientific* (italics added) judgment, together with a basic knowledge of the physical, pharmacological and toxicological properties of the chemical under study, suitable protocols should be developed for each specific compound." Some of the scientific rationale for a varied experimental design in teratology testing was discussed by Wilson (1975) and from that same publication Leon Golberg (pp. 53-55) argued strongly for a flexible rather than a rigid approach.

It is such a varied, individualistic and scientifically creative approach that I advocate if we are to improve the relevance of developmental toxicity testing. For too long we have relied on a single, narrowly focused experimental design strong on pragmatic considerations but weak scientifically.

References

Hirsch KS, Scott WJ (1983) Searching for the mechanism of acetazolamide teratogenesis. In: H. Kalter (ed), Issues and Reviews in Teratology, Plenum Press, New York, pp 309-347

Layton WM, Hallesy D (1965) Deformity of forelimb in rats; association with high doses of acetazolamide. Science 149: 306-308

Scott W, Duggan C, Schreiner C, Collins M (1990) Reduction of embryonic intracellular pH: a potential mechanism of acetazolamide-induced limb malformations. Toxicol Appl Pharmacol 103: 238-254

Wilson JG (1975) Critique of current methods for teratogenicity testing in animals and suggestions for their improvement. In: T. Shepard, J. Miller, M. Marois (eds), Methods for detection of environmental agents that produce congenital defects, North-Holland/American Elsevier, New York, pp 29-57

Wilson JG, Maren TH, Takano K, Ellison A (1968) Teratogenic action of carbonic anhydrase inhibitors in the rat. Teratology 1: 51-60

Discussion of the Presentation

Neubert: I appreciate that you took up the problem of duration of action of a substance, as a prerequisite for inducing a "teratogenic" effect. As you have shown it is not only the extent of a metabolic alteration, but also the duration of an impairment, which is critical. This is often not taken into account. Ralf Krowke made similar observations about 15 years ago with cytosine arabinoside and the duration of inhibition of DNA synthesis (*Handbook of Teratology 2: 117-151, 1977*). Also under these conditions it was not the extent of the inhibition of DNA synthesis, but the period for which it persisted which correlated with the induction of malformations.

I also appreciate that you do not call this AUC (Area Under the Curve), since this term is greatly misleading when used to designate an effect. It is a useful pharmacokinetic mathematical expression especially for characterizing bioavailability, but it has, of course, no direct relevance to, e.g., the degree of an enzyme inhibition or receptor occupancy, and thereby to the potency of a xenobiotic.

Specific and Non-Specific Developmental Effects

Ludwig Machemer, Ulrich Schmidt and Beate Holzum

Introduction

Embryo damage can arise in a number of different ways. Essentially two possibilities exist:

A. Noxae exert specifically developmental toxicity. In such cases the embryo often demonstrates greater sensitivity than the maternal organism, and embryotoxic effects show no correlation with maternal toxicity. The effects are often described as specific or primary embryotoxicity.

B. Embryo damage results from maternal damage, and a relationship between a female response and the offspring can be demonstrated. These effects are often described as non-specific, or secondary. However, the mechanism is not usually known, so that in principle a direct or specific damage to the embryo cannot be ruled out. All that can be stated in the majority of cases is that maternal and developmental toxicity are correlated, i.e., there is a correspondence between the sensitivity of the mother and that of the embryo.

It has long been known that virtually all chemical substances are capable of inducing embryo damage. Fraser (1963) and Karnofsky (1965) postulated that most substances damage the embryo if tested at the correct developmental stage, with an adequately high dose, and by the appropriate administration route.

For a separate risk evaluation of real developmental toxic substances their hazard potentials must first of all be identified in order to differentiate them from other, less critical substances. This differentiation is also a necessary first step before these substances or products containing them can be definitely labelled as embryotoxic.

There are various ways in which this differentiation can be carried out. One method is to classify substances on the basis of the dose-response relationship, as was attempted by Jusko (1972, 1973). Two classes were established, differing in their dose-response curve. In one case it was linear down to the low-dose range whereas the curve was non-linear in the other. The hazard potential of substances exhibiting a linear curve is higher than that of compounds with a non-linear curve, where a non-embryotoxic dose range exists.

Another and more recent method is to classify substances by the relationship between the maternal toxic dose range and the embryotoxic dose range (Schwetz and Tyl 1987; Hart et al. 1988; Johnson 1988; DFG 1990). This type of differentiation seems to be appropriate for toxicological and practical reasons, which probably accounts for its broad acceptance. However, the limitations of this approach should not be neglected (Setzer and Rogers 1991).

In order to come to conclusions regarding developmental toxic hazard potential it is therefore vital to identify and evaluate, as accurately as possible, both the maternal effects and the effects on the embryo.

Examples

As examples we present the results of embryotoxicity investigations in rats and rabbits using three substances, with particular emphasis on the association between the maternal toxic and embryotoxic effects.

The first example shows results obtained with fenamiphos, an *insecticidal organophosphorus ester* (Tables 1 and 2). As can be seen, *fenamiphos* induced distinct maternal toxicity up to lethality in both rats and rabbits without in any way recognizably impairing numerical, gravimetric, or morphological parameters in the descendants. In this example the maternal organism is therefore much more sensitive than the embryo. This can tentatively be said to apply to all the organophosphorus esters tested in our laboratory.

Table 1. Fenamiphos. Embryotoxicity study in rats. Correlation between maternal and developmental toxicity in rats obtained with fenamiphos

Dose (mg/kg p.o.)	Maternal Toxicity	Developmental Toxicity
0.25	--	--
0.85	--	--
3.0	Body weight gain ↓ Food consumption ↓ CHE ↓ (plasma, erythrocytes) Tremor, salivation, lacrimation, mortality	--

-- : No effect observed

Table 2. Fenamiphos. Embryotoxicity study in rabbits. Correlation between maternal and developmental toxicity in rabbits obtained with fenamiphos

Dose (mg/kg p.o.)	Maternal Toxicity	Developmental Toxicity
0.1	--	--
0.5	--	--
2.5	Body weight gain ↓ Food consumption ↓ Ataxia, salivation, dyspnoea, diarrhoea, mortality	--

-- : No effect observed

Figures 1 and 2 show the relationship between the maternal toxic and embryotoxic effects of *12 organophosphorus esters* in the rat and rabbit. The substances are coded, as it is the substance class, not the individual compound, that is important.

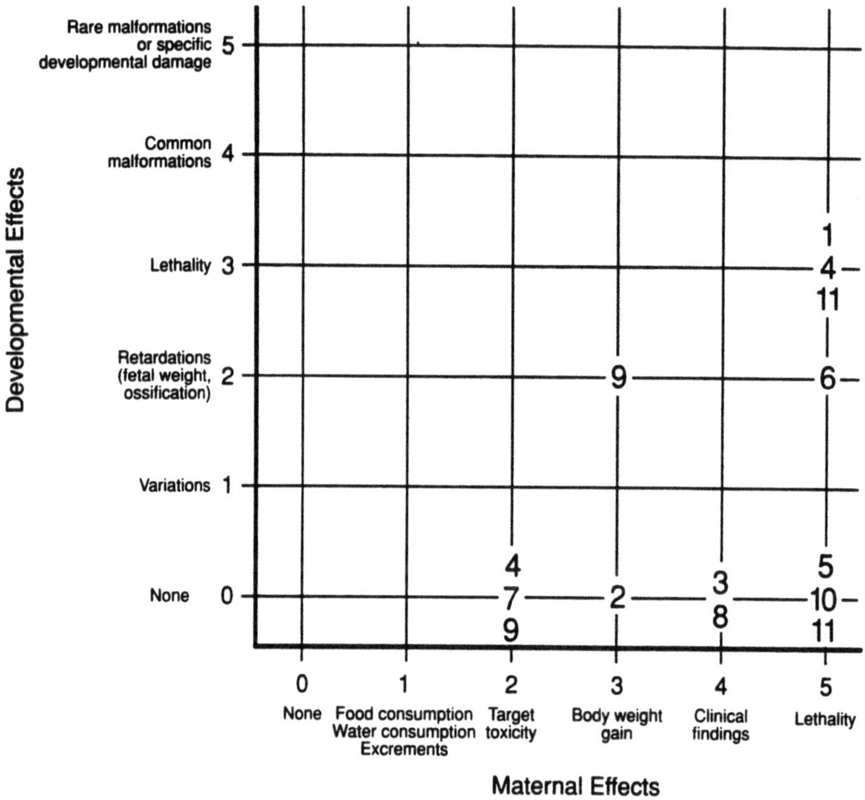

Fig. 1. OP esters. Correlation between maternal and developmental effects in rats obtained with 11 organophosphorus esters. Each number represents one organophosphorus ester. If a number is shown twice, two doses of the same compound are represented

At this point we should comment on the presentation method selected. There are no generally accepted rules regarding assessment of the severity of effects. For this, reference was made to publications which describe a ranking of effects, e.g.

those of Frankos (1985), and Wang and Schwetz (1987). We also attempted to take the experiences of our own laboratory into account. In the case of organophosphorus esters the target toxicity, i.e. the cholinesterase inhibition, should undoubtedly be regarded as being of lesser significance than clinical symptoms, as these presuppose high inhibition of acetylcholinesterase activity by at least 50%. As regards developmental toxicity, growth retardation appears to be less significant than malformations, and an increase in otherwise frequent spontaneous malformations might be less significant than rarely occurring or highly specific malformations.

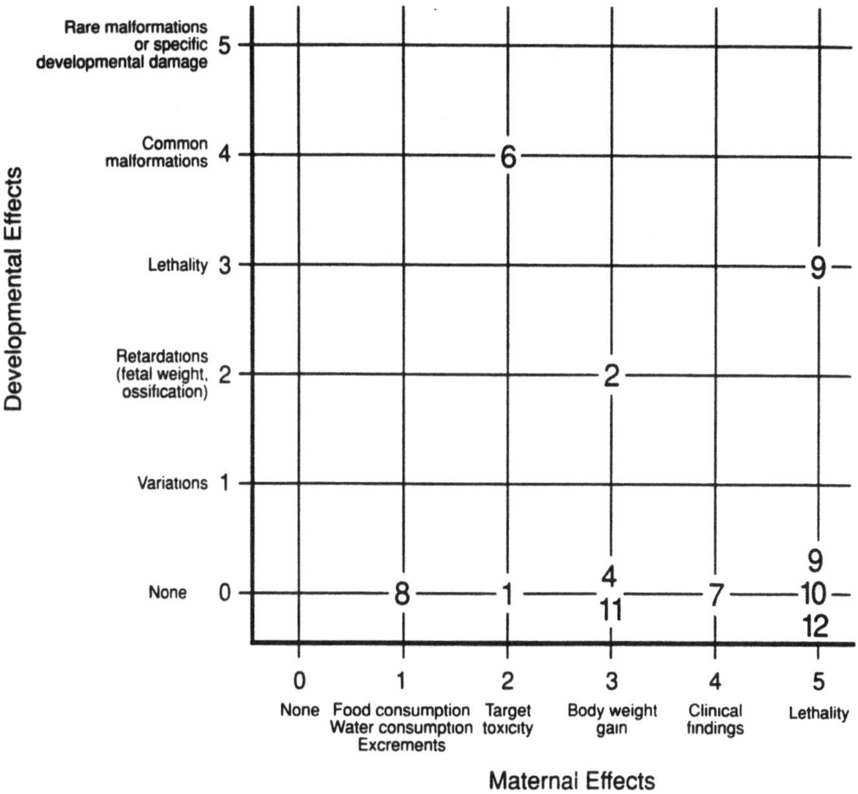

Fig. 2. OP esters. Correlation between maternal and developmental effects in rabbits obtained with 12 organophosphorus esters. Each number represents one organophosphorus ester. If a number is shown twice, two doses of the same compound are represented

The results with organophosphorus esters in rats and rabbits reveal that this substance class in general possesses no specific developmental toxicity potential. Highly maternal toxic doses - as a rule - have to be administered to induce developmental toxic effects; even then the effects are non-specific, and result in growth retardation or the death of the embryo.

Our second example, shown in Tables 3 and 4, is a *fungicidal triazole, HWG 1550 N.* Both the rat and the rabbit show a association between the maternal toxic and the developmental toxic responses. The rat is the more sensitive species in as far as it shows maternal toxic and developmental toxic effects at lower doses. The foetuses also show malformations, although at doses that induce severe symptoms and lethality in the dams. With this compound distinctly maternal toxic effects, e.g. weight loss or reduced weight gain in the dam, are associated with only slight or non-specific developmental toxicity in these two species.

Table 3. Tiazol HWG 1550 N. Embryotoxicity study in rats. Correlation between maternal and developmental toxicity in rats obtained with HWG 1550 N

Dose (mg/kg p.o.)	Maternal Toxicity	Developmental Toxicity
10	--	--
40	Body weight gain (↓)	Variations ↑ (supernumerary ribs)
160	Restlessness, ataxis, rough fur, tremor, salivation, self inflicted injuries, mortality Food consumption ↓ Water consumption ↑ Enlargement of spleen	Resorptions ↑ Fetal weight ↓ Malformations (palatoschisis)

-- : No effect observed

Table 4. Tiazol HWG 1550 N. Embryotoxicity study in rabbits. Correlation between maternal and developmental toxicity in rabbits obtained with HWG 1550 N

Dose (mg/kg p.o.)	Maternal Toxicity	Developmental Toxicity
10	--	--
40	--	--
160	Body weight loss Food consumption ↓	Resorptions ↑ Fetal weight ↓ Delayed ossification

If we consider the triazoles as a class and if we select the same presentation as in the case of the organophosphorus esters, the relationship between maternal and developmental toxicity in rats and rabbits is as shown in the Figures 3 and 4, which depicts *7 triazoles*. The following statements can be made: Triazoles do not show uniform behaviour. While there is by and large an association between maternal and developmental toxicity, individual substances appear to show sometimes pronounced developmental toxicity at doses where maternal toxicity is relatively low. This class of substances demonstrates teratogenic potential, the rat being more sensitive than the rabbit. As a rule, however, these effects occur at doses which impair body weight, produce symptoms, or are already lethal. In the majority of cases there is no appreciable dissociation between the maternal toxic and developmental toxic dose ranges.

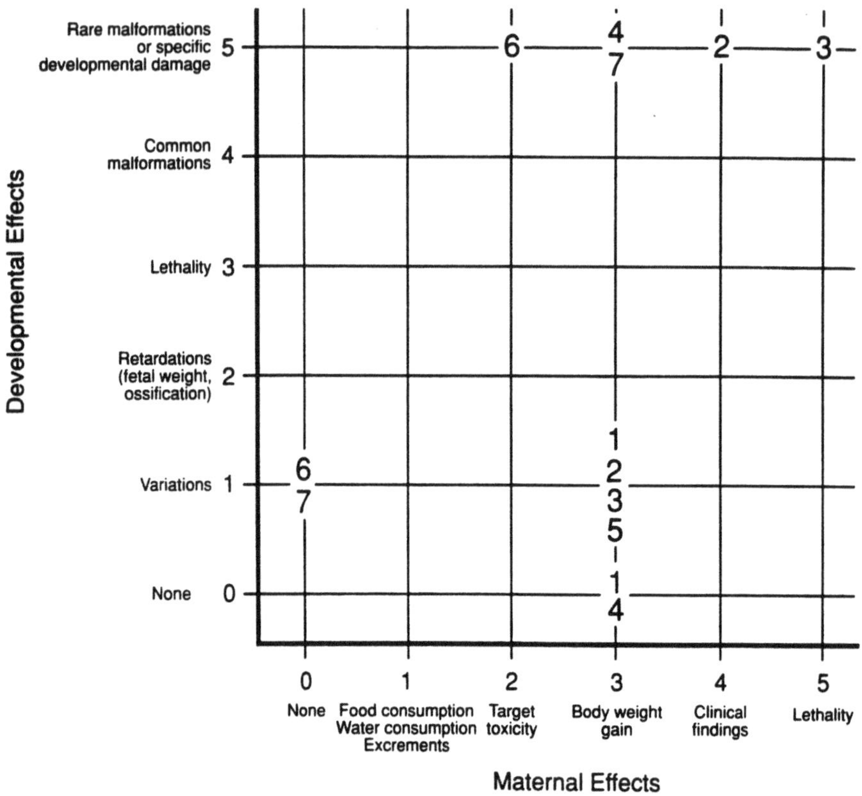

Fig. 3. Triazoles. Correlation between maternal and developmental effects in rats obtained with 7 triazoles. Each number represents one triazole. If a number is shown twice, two doses of the same compound are represented

Our last example shows the results for the *herbicidal active substance SLA 3992, a phenylpyrazole* (Tables 5 and 6). Firstly, the results in the rat show that minimal effects are observed on the liver of the dam, the main toxicological target organ, without developmental toxic effects. Influences on maternal parameters are observed over a wide dose range, but are slight overall. Severe malformations on the other hand, occur at doses which are tolerated by the dams without marked toxicity.

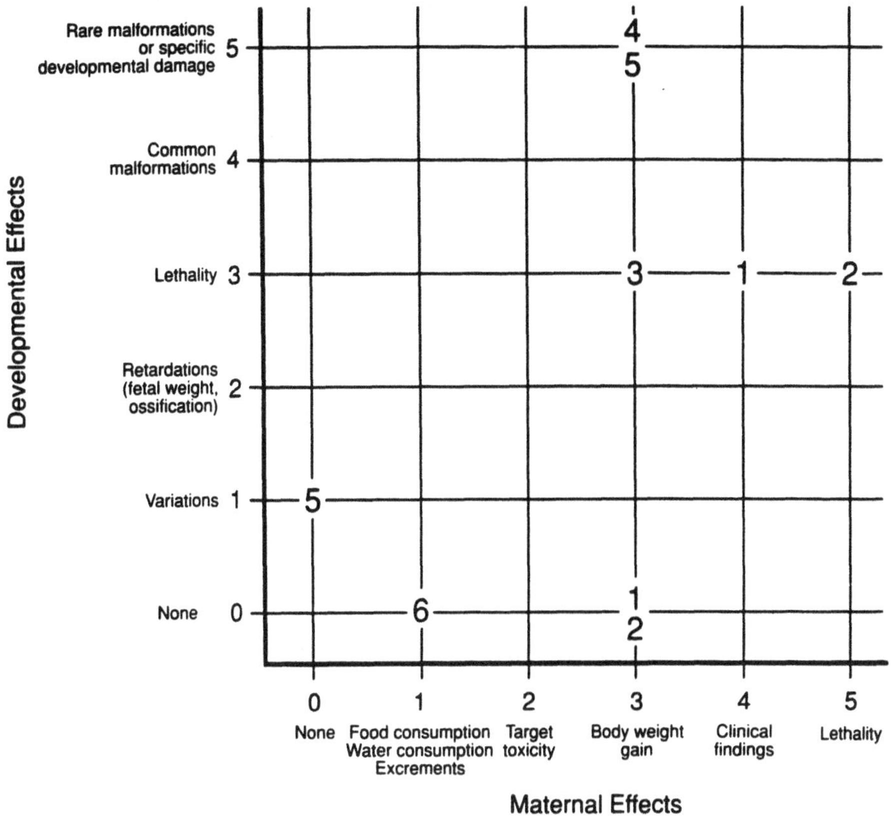

Fig. 4. Triazoles. Correlation between maternal and developmental effects in rabbits obtained with 6 triazoles. Each number represents one triazole. If a number is shown twice, two doses of the same compound are represented

In the rabbit the malformations that are known to occur in the strain used became more frequent at high doses. In parallel with this an impairment of maternal body weight gain was observed.

Specific developmental toxic potential was suggested above all by the results in rats. We therefore studied the damage mechanism of this active substance in more detail, and performed corresponding investigations in rat dams and foetuses.

Table 5. Phenylpyrazol SLA 3992. Embryotoxicity study in rats. Correlation between maternal and developmental toxicity in rats obtained with SLA 3992

Dose (mg/kg p.o.)	Maternal Toxicity	Developmental Toxicity
0.5	N demethylase ↑ Triglycerides (plasma) ↓	--
1	Minimal microvesicular Fatty change in hepatocytes	--
5	Lipid deposits in liver ↓ Heinz bodies ↑	Fetal weight ↓ Malformations (oedema, cryptorchism, dysplasia of scapula and long bones)
10	Food consumption ↓ Spleen weight ↑	Resorptions ↑ Necrosis of placenta
20	Diuresis ↑, Erythrocytes ↓, HB ↓, MCHC ↓, reti. ↑, MCV ↑, normobl. ↑, Spleen: erythropoiesis ↑, urea ↑	Pregnancy rate ↓ Skeletal retardations ↑ Placental weight ↓

Table 6. Phenylpyrazol SLA 3992. Embryotoxicity study in rabbits. Correlation between maternal and developmental toxicity in rabbits obtained with SLA 3992

Dose (mg/kg p.o.)	Maternal Toxicity	Developmental Toxicity
5	--	--
10	--	--
20	Body weight gain ↓ (arthrogryposis)	Malformations

As in the case of some other herbicides (Matringe et al. 1989), the mechanism of SLA 3992 is based on inhibition of protoporphyrinogen oxidase (Fig. 5). In damaged liver cells accumulated protoporphyrinogen undergoes non-enzymatic oxidation to protoporphyrin IX. The abnormal subcellular position of the protoporphyrin IX prevents the next stages of synthesis by the metal chelatase to the magnesium- or iron-containing porphyrins such as chlorophyll (in plants) and cytochromes (in mammals). This leads to an accumulation of protoporphyrin IX. Protoporphyrin IX is the key porphyrin in both plant and mammalian cells, and is the starting point for two fundamental biosynthesis pathways. Following the incorporation of iron, one of these gives rise to haem, which represents the prosthetic group of haemoglobin, catalases, peroxidases, and cytochromes. The other leads to the formation of chlorophyll following the inclusion of magnesium and a number of other biosynthetic stages.

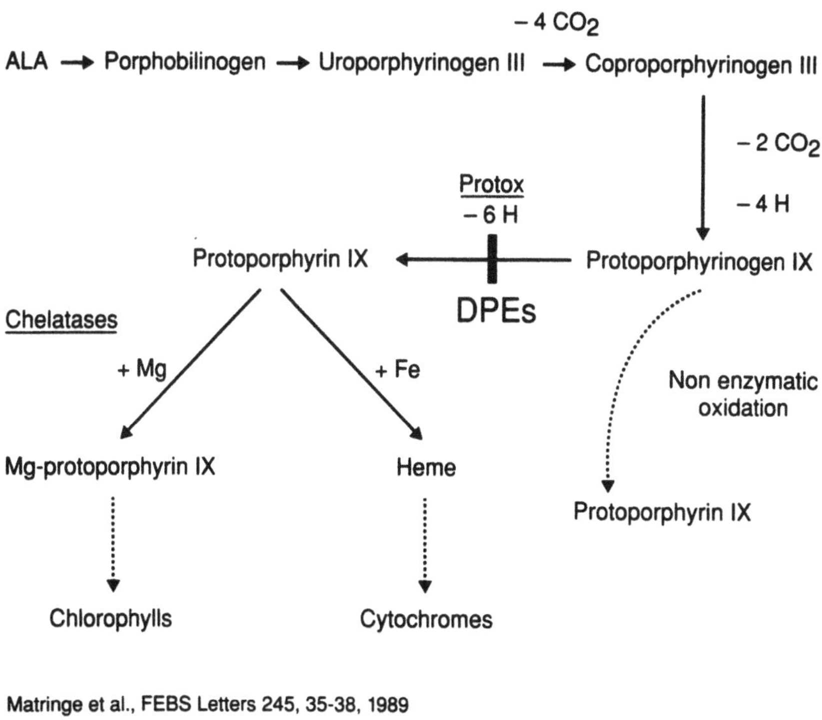

Fig. 5. Postulated mechanism of action of diphenyl ether-type herbicides (DPEs), according to Matringe et al. (1989). This mode of action is also to be assumed for the phenylpyrazol SLA 3992

We therefore tested for a link between this mechanism and embryotoxicity in rats by studying maternal and foetal liver (Fig. 6).

Inhibition of protoporphyrinogen oxidase gives rise to a much more intensive accumulation of protoporphyrin IX in foetal liver than in maternal liver. On the basis of this elevated sensitivity of the foetal organism one can assume a specific developmental toxic hazard potential, also manifested in the distinction between the slight maternal toxic and the pronounced embryotoxic effect in the rat.

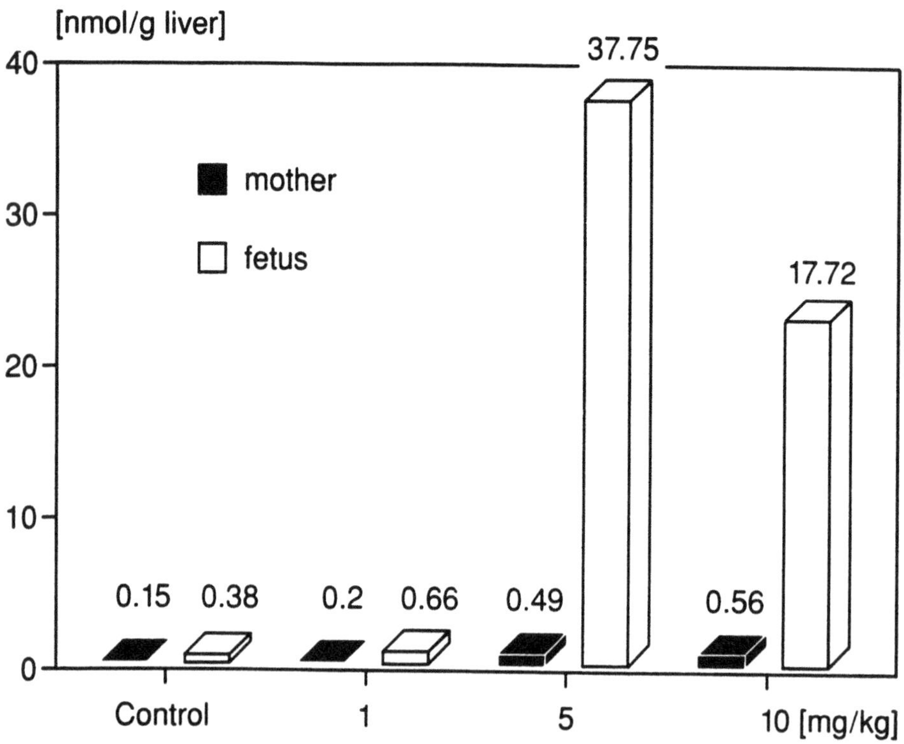

Fig. 6. Accumulation of protoporphyrin IX in maternal and fetal liver on day 20 of pregnancy of rats treated with SLA 3992 during days 6 to 19 of pregnancy

Risk Evaluation

Substances to which the embryo can be shown to be particularly sensitive as compared with the maternal organism can be regarded as causing a specific developmental toxic hazard. As a rule developmental toxicity is manifested either in the absence of maternal toxicity, or there is a clear discrepancy between the responses of the females and the offspring. These compounds may accordingly be labelled as embryotoxic.

On the other hand, substances should not be regarded or labelled as specifically embryotoxic if there is an association between embryotoxicity and maternal toxicity. Admittedly, even at maternally toxic doses specific developmental effects may occur since the mechanisms for maternal and developmental toxicity may differ. However, if elevated sensitivity of the embryo as compared with the maternal organism cannot be demonstrated, no specific developmental damage potential needs to be assumed for the purpose of risk evaluation. The reason for this is that, for example, in the case of plant protection agents, when health aspects for user and consumer are assessed, protection of the maternal organism with regard to health risk is, of course, also aimed at. The starting point is therefore essentially the no-effect level for the most sensitive parameter in all toxicity investigations, with incorporation of an adequate safety margin in the estimated tolerable exposure. In the absence of elevated embryo sensitivity, adherence to this safety margin for the maternal organism ensures corresponding protection of the embryo. In the case of the usually steep dose-response curve for developmental toxic effects, the safety margin of 100 normally selected for conventional toxic effects is usually adequate for protection of the embryo, too (Neubert et al. 1987).

The crucial factor for the registration of a developmental toxic substance should be that the anticipated or foreseeable level of exposure when the compound is used as prescribed in the practical situation represents no health risk, i.e., it is the use-related risk calculation, and not inherent substance properties or classification of a product, that should be decisive for registration.

References

DFG Deutsche Forschungsgemeinschaft (1990) Kriterien zur Beurteilung von Studien zur Reproduktionstoxizität mit Pflanzenschutzmitteln. Mitteilung XVIII der Kommission für Pflanzenschutz-, Pflanzenbehandlungs- und Vorratsschutzmittel, VCH Verlagsgesellschaft mbH, Weinheim

Frankos VH (1985) FDA Perspectives on the use of teratology data for human risk assessment. Fund Appl Toxicol 5: 615-625

Fraser FC (1963) Experimental teratogenesis in relation to congenital malformations in man. In: Congenital Malformations. Papers and Discussions presented at the Second International Conference New York, July 14-19, The International Medical Congress Ltd, pp 277-287

Hart WL, Reynolds RC, Krasavage WJ, Ely ThS, Bell HR, Raleigh RL (1988) Evaluation of developmental toxicity data: A discussion of some pertinent factors and a proposal. Risk Analysis 8/1: 59-69

Jusko WJ (1972) Pharmacodynamic principles in chemical teratology: dose-effect relationships. J Pharmacol Exp Ther 183: 469-480

Jusko WJ (1973) Pharmacokinetic principles in chemical teratology. Proc Study of Drug Toxicity XIV: 9-19

Karnofsky DA (1965) Mechanism of action of growth-inhibiting drugs. In: Wilson JG, Warkany J (eds) Teratology: Principles and Techniques, Univ. of Chicago Press, Chicago

Matringe M, Camadro J-M, Labbe P, Scalla R (1989) Protoporphyrinogen oxidase inhibition by three peroxidizing herbicides: Oxadiazon, LS 82-556 and M&B 39279. FEBS Lett 245: 35-38

Neubert D, Chahoud I, Platzek T, Meister R (1987) Principles and problems in assessing prenatal toxicity. Arch Toxicol 60: 238-245

Schwetz BA, Tyl RW (1987) Consensus Workshop on the Evaluation of Maternal and Developmental Toxicity, Work Group III Report: Low dose extrapolation and other considerations for risk assessment - models and applications. Teratogen Carcinogen Mutagen 7: 321-327

Setzer RW, Rogers JM (1991) Assessing developmental hazard: The reliability of the A/D ratio. Teratology 44: 653-665

Wang GM, Schwetz BA (1987) An evaluation system for ranking chemicals with teratogenic potential. Teratogen Carcinogen Mutagen 7: 133-139

Discussion of the Presentation

Neubert: Your talk has really taken my interest. Perhaps the whole trouble rests in our inability to define "maternal toxicity". It would also be impossible to define "toxicity" as such in adults in general. Reduction in weight gain may have a variety of different causes with quite different consequences on metabolic functions, uterine perfusion, blood pressure in the dam, etc. For this reason it is not surprising that some "maternal toxicity" is largely compatible with normal prenatal development, while other (possibly rather subtle) effects on the maternal organism will lead to abnormal prenatal development. As long as we are unable to define "maternal toxicity" we will be faced with the difficulty of interpreting results on the conceptus induced by maternally toxic doses. The only way out is to avoid using such doses. However, it is essential to clearly define, e.g., with preceding dose-finding studies, the toxic range for the adult animal. For this purpose also other data than those from studies on reproductive or developmental toxicity have to be used.

Machemer: I agree. Unfortunately, all major guidelines for embryotoxicity studies require the use of maternally toxic doses. As a toxicologist for the industry I cannot neglect the guidelines for studies thought for registration purposes. I also agree to the statement that beside dose-range finding studies with pregnant animals, all available data from other toxicity studies have to be used when selecting dose-levels or evaluating results of embryotoxicity studies. I think if we understand the term "toxic" in the sense that a dose causes some (adverse) effect in the maternal organism it would help much in practical terms. As I mentioned, in the case of plant protection agents, protection of the maternal organism is required in the context of general toxicity, too. The so-called no-effect-level is always used as a starting point for calculating tolerable exposure. In my presentation I put emphasis on the point that if no discrepancy between the dose which affects the embryo and those which affect the maternal organism is observed, no elevated embryo sensitivity has to be assumed. This seems to me to be the crucial point for safety consideration, for the selection of safety factors, for labelling, and for regulatory decisions.

Neubert: Thus, guidelines are not really "guidelines", but restricting rules.

Palmer: We cannot always avoid encountering maternal toxicity. It is not the studies that are at fault but the interpretation and labelling.

For organophosphates a multigeneration study is much more interesting than embryotoxicity and can give rise to a variety of direct and indirect responses.

Machemer: In my presentation I restricted myself to the period of organogenesis. I did not report on results of organophosphates in multigeneration studies. I know that some stages of the reproductive cycle are particularly sensitive to OP compounds. In general, according to our experience, expression of parental and reproductive effects are mediated by the anticholinesterase effect and thus, correlated with that basic mode of action.

Kavlock: In discussing the imprint of "maternal toxicity" on developmental outcomes, it is important to note that there is little experimental evidence in the literature that the magnitude of maternal responses called for in regulatory testing guidelines has not been shown to correlate with effects in the fetus. That is not to say we could not better characterize particular manifestations of maternal toxicity and how they might relate to developmental toxicity.

Krowke: I would like to ask an easy question, but I fear it is very difficult to answer: What is really meant when speaking about the "maternal toxic dose range"? There are many examples of chemicals and drugs where we do not find signs of "toxicity" in the adult animal, but orders of magnitude below the lowest dose range where you have visible toxic effects (e.g. body weight loss, neurological effects, etc.) you can find other abnormal effects which might affect normal development (e.g. changes in blood glucose, enzyme induction, etc). Would you like to comment on that?

Machemer: First of all, I would like to refer to my answer to Dr. Neubert. Then I would repeat what I said in my presentation, namely, in order to come to conclusions regarding developmental toxic hazard potential it is vital to identify and evaluate as accurately as possible, both the maternal effects and the effects on the embryo. The "toxic" dose to the mother animal should be found out prior to an embryotoxicity study or at least in satellite groups on a case-by-case basis. Yet, there is no single parameter for maternal "toxicity" which fits for all cases. The knowledge of the toxicological profile and target organs is a prerequisite for identifying the so-called maternal "toxic" dose range.

Stahlmann: The problem may be much more complex than struggling with the term "toxic". In the case of medicinal products "pharmacologic" actions may also cause problems. I am not necessarily referring to exaggerated actions, but, e.g., to the fact that medicinal products may (predominantly) be given to humans with pathological conditions: when, for example, treating hypertension with a drug which lowers blood pressure, this variable is not expected to fall below the normal range, but treatment of normotonic individuals (e.g. experimental animals) with comparatively high doses of the same drug may lead to severe hypotension. Is the intention of studies on possible reproductive toxicity to analyse such a situation?

Risk Assessment as Practiced in Various Countries

Assessment of Reproductive Toxicology of Drugs at the Food and Drug Administration

Judi Weissinger

Abstract

The Food and Drug Administration is interested in the most appropriate scientific approach to evaluation of reproductive and developmental toxicity. The Food and Drug Administration is also interested in an internationally harmonized approach to evaluating reproductive and developmental toxicity. The evolving harmonization guideline, that contains input from many sources, is as flexible a document as possible. It addresses multiple approaches, advantages and disadvantages of various protocols, and allows for selection of study protocol based on the scientific rationale and the uniqueness of the compound under development. There are areas in the studies for reproductive and developmental toxicity where the scientific community does not necessarily subscribe to a common philosophy. These areas, relating to duration of pretreatment of males in fertility studies, appropriate day of sacrifice after mating in segment I, appropriate dose, recognizing alterations in exposure associated with pregnancy, utility of behavioural studies, and considerations for dosing across segments will be discussed in this presentation.

Introduction

Studies to evaluate reproductive and developmental toxicity of drugs in the United States are currently conducted using the guidelines for reproduction studies for safety evaluation of drugs for human use that were published (D'Aguanno 1966). The currently submitted studies are based on that three segment approach. Segment I is designed to evaluate male and female fertility and general reproductive performance, segment II is designed to evaluate teratologic potential in two species, and segment III is designed to evaluate late fetal development, parturition, and neonatal development.

The guidelines have apparently been successful, as currently known human reproductive and developmental toxicants have been identified during the nonclinical studies. The guidelines were developed after the thalidomide crisis, and were based on a very conservative approach to identify any adverse effects on any phase of reproduction. Utilizing this approach, we probably can point to only one drug that was not identified as a teratogen over that time, diethylstilbestrol. It probably could have been identified by these studies, as subsequent experimental data have indicated, but it was originally used prior to routine use of these studies and not identified as a risk until the human epidemiological evidence suggested additional evaluation in animals would be useful.

Scientific approaches to evaluation of reproductive and developmental toxicity, after the thalidomide tragedy, were developed not only in the United States but elsewhere. Scientists have been trying to find the perfect assay, and that quest still goes on. Several versions of the "three segment guidelines" have been developed, with unique modifications. This manuscript will address the scientific basis for the approach to regulatory review of reproductive and developmental toxicity focusing first on the disharmony that exists in current guideline approach to evaluation of reproductive and developmental toxicity, and scientific issues that may still need to be addressed, that have been proposed for inclusion in a harmonized document.

Scientific Considerations

There are six issues that can be highlighted as major differences in the approach to evaluating reproductive and developmental toxicity by the three segment protocol (Weissinger 1991a,c,d). These include 1) the appropriate day of sacrifice after administration during segment I (13 vs. 20), 2) the appropriate level of maternal and paternal toxicity to accurately evaluate the effects of the drug on reproduction and development, 3) the opportunity to miss a finding by not dosing across segments, 4) the duration of time males should be treated premating, 5) the utility of kinetic and metabolic data in study design, and 6) the utility of behavioural studies in evaluating developmental effects.

It is important to consider the issues proposed for further investigation, and the potential experimental outcomes if these issues were not addressed in each instance.

1) In segment I, sacrifice on day 13, while adequately characterizing implantation and resorption sites, does not offer an opportunity to identify additional information related to contribution of treated males to potential for neural tube defects, late embryo loss, fetal deaths, morphologic malformations. These are currently observed, where present, with a day 20 sacrifice. A day 20 sacrifice also allows for the detection of possible malformations resulting from direct genetic damage to sperm, although, dominant lethal genotoxicity studies may be used as an indicator. Opponents of day 20 sacrifices are concerned that early resorptions may not be evident.
2) Considerable attention has been focused on the appropriate maternal dose for use in reproductive and developmental toxicity studies. Use of the maximally tolerated dose may lead to observations of toxicities not related to drugs but rather to altered maternal physiology or toxicity. Observation of minimal maternal toxicity may not always be consistent with an optimal dose to use throughout the study. Additionally, a study using doses that produces no toxicity is least desirable because it can offer little confidence that a drug does not have the capability to affect reproduction. Given the current recognition that dose may not be the appropriate index at all, exposure associated with a toxic effect should also be a consideration. Knowledge of the concentration of drug or its metabolites in the blood or target organs may be useful in considering potential relevance to clinical use. Where the therapeutic index is small, the minimal toxicity and maximally tolerated toxicity may be the same value, but for drugs with a wide therapeutic window, there may be quite a discrepancy in choosing the dose. Some excellent suggestions for evaluating maternal toxicity have been offered in the current draft of the guideline on "detection of toxicity in repro-

ducing animals" for pharmaceutical products (Baß et al. 1991). This document is currently circulating for world-wide input on the points to consider.

3) Recognition of the advantages and disadvantages of designing an extended study in which animals are treated from premating to weaning of offspring is essential to optimal study design. While it is recognized that certain drugs do not lend themselves to long-term treatment, such as some immunogenic products or drugs that, with time, affect their own metabolism, most drugs can be screened across the entire reproductive and development phases to evaluate effects of treatment related to accumulation that might be missed when dosing begins with each trimester. Triparanol was teratogenic when administered from day 4, but has been reported to show teratogenicity when administered from day 7 (Roux et al. 1973).

4) In the absence of published, peer reviewed studies demonstrating that there are no effects on sperm motility and fertility during the early period of spermatogenesis, and absent a mechanism for evaluating those effects short of administration for the entire sperm cycle, the current US approach is to collect these data. Several substitute experiments, employing other data gathered from the 90 day histopathology studies in males, and evaluating sperm motility and sperm maturation, have been utilized, still with no concrete data that effects would have been seen with a shorter dosing period. The potential for treatment effects to be limited to premeiotic and meiotic sperm and not affect post-meiotic sperm exists. It has been proposed that an adequate histologic evaluation of sperm after one to three months treatment, along with other indices of sperm viability, should be able to identify any changes that might affect fertility, potentially better than the 60 to 80 week pretreatment because a severe effect in sperm maturation is needed to detect an effect in fertility by current methodology. Validation of this latter hypothesis can save up to seven to eight weeks additional dosing in males. Alternatively, interest in evaluating animals treated throughout the complete sperm cycle under treatment may be accomplished by using the 90 day repeat dose male rats for an additional experiment.

5) One of the most highlighted areas for harmonization is the utilization of kinetic and metabolic data to select the appropriate dose for a desired exposure, and to select the appropriate species and route of administration. It

should be recognized that kinetic data are only useful when the exposure level is considered along with the pharmacologic or toxicologic response (Weissinger 1991b). The discussion of dosages in the proposed harmonized guidelines serves as an excellent discussion of the intent of the chosen dose. Interpretation of these considerations for any given drug should be considered in the context of the entire toxicity profile.

Experiments are generally conducted with exposures that represent an exaggeration of what humans might encounter under exposure, additionally considering changing physiological parameters such as might occur with repeated dosing, pregnancy, or lactation must be taken into account. In order to select an optimal dose and maintain the exposure associated with that dose over the experiment, drug disposition and the effects of prolonged high doses during pregnancy should be evaluated. If an optimal kinetic profile in normal animals was obtained prior to initiation of the study, a few pregnant animals could be evaluated concurrently with the study to confirm that no substantial changes have occurred in that profile.

Metabolic information might be useful to evaluate the adequacy of the animal species chosen or effects of high doses. If the human is exposed to two active metabolites, and the animals one, consider appropriate compounds to test, or appropriate species to select. Experimental considerations for metabolism also come into play when the animal biotransforms the drug by first pass metabolism and humans have a substantial amount of parent compound to which they are exposed. In this instance, questions regarding study design and value of data to be obtained are important to address prior to initiation.

6) There is a continuing debate regarding the utility of behavioural toxicity studies. Some have highly advocated these studies initially, and on discovering the lack of repeatable data, even with positive controls, have placed less emphasis on the need for data beyond sensory and locomotor evaluations. There may still be a place for learning and memory function studies if one considers that there are some drugs that are expected to affect neurological development. Perhaps certain classes of drugs such as neuropharmacologic and psychotropic agents will offer us some predictive data, and decisions to incorporate behavioural studies can be based on known effects

of the drug. These tests are listed in the proposed harmonization guidelines in segment II rat and segment III.

Harmonization

There are at least three approaches to harmonization under consideration. The first is that the EEC, Japan, and the US accept studies conducted by any of the three segment protocols suggested by those regulatory bodies. While this is a viable approach, where reproductive and toxicity studies have been conducted that do not address the such scientific issues offering assurance that continuous dosing was not indicated, behavioural toxicity would not be at issue, there cannot be genetic or motility effects on sperm with prolonged administration that were not evaluated, and that the appropriate dose was selected, they may be found inadequate to offer substantive confidence that a reproductive and developmental toxicity would have been observed if it is a property of the drug.

Another approach to harmonization of guidelines is just that. The twelvth draft of a multiple approach guideline is currently circulating and it points out the advantages and disadvantages of many approaches, and suggesting the use of scientific rationale for choosing the appropriate approach. This is an excellent idea, at least from the FDA perspective, because sponsors can select the most appropriate approach and ask for FDA concurrence prior to conduct of the study. This may not be as feasible an approach with other regulatory bodies that do not have a similarly structured system for preclinical and early drug development interactions. A potential drawback of the Eurodraft points to consider is that if the option chosen employs erroneous assumptions, then the studies may not be those directed by the guideline.

Lastly, the CMR sponsored a workshop, the proceedings of which are being developed (Lumley 1991). The participants in the workshop proposed combinations of existing regulatory protocols that may be used to address the considerations presented by regulatory bodies, industry researchers, and academics.

It may also be useful to develop a reproductive and development (RAD) screen, based on the desire to identify the 80% of drugs developed that are not reproductively toxic early in development. We have heard our colleagues from Eli Lilly and Company and Huntington Research Laboratories introduce screening concepts being used in their laboratories at previous symposia.

Conclusions

We are here to consider an optimal approach to evaluating reproductive and developmental toxicity. This approach should be based on scientific knowledge and flexibility. Alternative approaches to any guideline that is accepted should be considered in light of the information available. Where the six scientific issues that have been described above as "needing additional study to properly evaluate toxicity" have not been addressed during conduct of the study, they should be addressed using scientific rationale and augmenting data from other toxicity and kinetic studies. Providing the scientific issues have been appropriately included, a variety of study designs, such as proposed in the Eurodraft 12, should be acceptable. Eurodraft 12 (Bass et al. 1991) is still available for comment in the context of excellent science and harmonization.

Note: The scientific and regulatory content of this presentation were presented in several fora and are published in several proceedings, as noted in the references.

Acknowledgements: The author gratefully acknowledges the FDA/CDER Pharmacology/ Toxicology Reproductive Guidelines Committee for their participation in the identification of areas for further study.

References

Bass R, Ulbrich B, Hildebrandt A, Weissinger J, Doi O, Baeder C, Fumero S, Harada Y, Lehmann H, Manson J, Neubert D, Omori Y, Palmer A, Sullivan F, Takayama S, Tanimura T (1991) Draft guideline on detection of toxicity to reproduction for medicinal products. Adverse Drug React Toxicol Rev 9: 124-141

D'Aguanno W (1966) Guidelines for reproduction studies for safety evaluation of drugs for human use. FDA Guidelines, pp 41-46

Lumley CE (1991) Proposal for international guidelines for reproductive and developmental toxicity testing for pharmaceuticals. Adverse Drug Reactions Toxicologic Reviews: in press

Roux C, Aubry M, Dupuis R, Horvath C (1973) Comparative teratogenic effects of triparanol in the Wistar rat and the Sprague-Dawley rat. CR Soc Biol (Paris) 167: 1523-1526

Weissinger J (1991a) Philosophy behind the regulations: USA perspective. Proc J Jap Teratol Soc: in press
Weissinger J (1991b) Utility of kinetic, dynamic, and metabolic data in nonclinical pharmacology/toxicology studies. Proc Conf "Integration of Pharmacokinetics, Pharmacodynamics, and Toxicokinetics in Rational Drug Development". AAPS/ASCPT/FDA, Alexandria, Va., in press
Weissinger J (1991c) Commentary on proposal for mutual acceptance and proposed alternative approaches. Proc Internat Conf "Harmonization", Brussels, Belgium, Nov. 1991, in press
Weissinger J (1991d) The regulations and the underlying concepts and philosophy - the USA perspective. In: Lumley C and Walker S (eds) Current Issues in Reproductive and Developmental Toxicology, Proc Workshop held at Ciba Foundation, London, May 1991, Quay Publ, pp 57-64

Discussion of the Presentation

Krowke: What do you mean by harmonization of guidelines?

Weissinger: It is difficult to harmonize the guidelines when the meaning of the word "guideline" varies with each regulatory body. In the U.S.A., "guideline" was meant to be an example of a way to approach a study, with alternative proposals acceptable. While the same meaning exists in the U.K., the expert committee is appointed too late in development to consider alternate approaches prior to initiation of studies. Traditionally, in Japan, the word "guideline" has been synonymous with "requirement". Much of the progress in harmonization comes from acceptance of "scientifically valid approaches", "flexibility" in consideration of studies conducted in support of drug approval, and not from the development of identical "guidelines".

Nau: I understood from your presentation that the species should be used which most closely relates to the human in regard to pharmacokinetics and metabolism. However, this is an ideal(istic) case. As a rule, species are very different in regard to drug half-life, metabolism pattern and AUC values. Don't you agree that pharmacokinetics can provide a way to work with a less than ideal species as long as the levels of drugs and relevant metabolites are measured and used for interspecies comparison? Would your office accept such an approach?

Weissinger: Yes, this was the intent of the proposal - that knowledge of kinetics, dynamics and metabolism can be used to assess validity of species. Rather than to imply that inappropriate species are not utilized, the suggestion is to use your knowledge to alter route, schedule, or compound administered to a given species to most closely approximate the human exposure, where possible.

Neubert: I do not know of any example for your statement that treatment starting on day 3 of pregnancy in the rat may produce gross-structural abnormalities in contrast to treatment started on day 6 or 7, except for some alkylating (i.e. mutagenic) agents (studies of Nagao). Unfortunately, you did not provide a reference for this statement.

A C-section on day 20 with full skeletal evaluation in the segment I study, as you recommended it, seems to be repetitious. Is the additional information obtained by treating during preimplantation and implantation and during the fetal phase worth the considerable effort? If the additional treatment period is felt to be necessary an easier way would be to incorporate it into the segment II study.

New Approaches to Developmental Toxicity Risk Assessment at the U.S. Environmental Protection Agency

Robert J. Kavlock and Carole A. Kimmel

Introduction

The United States Environmental Protection Agency (US EPA) is charged under a variety of statutes, including the Federal Insecticide, Fungicide and Rodenticide Act, the Toxic Substances Control Act, the Clean Air Act, and the Clean Water Act, with evaluating the potential health risks of chemicals present in the human environment. In considering the implementation of these laws, it is important to note that the situations to be evaluated range from workers occupationally exposed during the manufacture of industrial chemicals or during the preparation and application of pesticides, to widespread, but low level, exposure of large segments of the general population through the food, water or air. Such situations are in direct contrast to the health risks associated with pharmaceutical agents, where the exposure is direct and controlled and the benefits accrue to the exposed individual.

The risk assessment process employed by the US EPA is closely modelled after that described by the National Research Council in the document "Risk Assessment in the Federal Government: Managing the Process" (NRC 1983). The risk assessment process as defined by the NRC consisted of four basic components: 1) hazard identification, 2) dose-response assessment, 3) exposure assessment, and 4) risk characterization (the last being an integration of the former three). Once the available scientific information has been considered in characterizing a particular risk, then the specific regulatory action to be followed is determined under the risk management process. Key considerations in risk management include the nature, magnitude and uncertainty of the presumed risks, the technical feasibility of mitigation, socioeconomic factors, as well as political impact.

While a general consensus has existed for more than 20 years on the basic design features for toxicological studies targeted at effects on the developing conceptus (e.g. FDA 1966; US EPA 1982, 1985), the interpretation of such studies for the purpose of human risk assessment has been less widely agreed upon. In an effort to provide advice on data interpretation and extrapolation to scientists within and outside of the Agency, the US EPA began a process in the early 1980's to describe the principles and procedures that it would use in conducting risk assessments. In publishing such guidance, the Agency took steps to ensure that health risk assessments would be conducted uniformly within its jurisdiction, to inform the public and external scientific community about what data would be examined and how it would be interpreted, and to identify the major gaps and uncertainties in the risk assessment. The resulting risk assessment guidelines were not intended to supersede any specific testing guidelines issued by the various Program Offices within EPA, but were specifically aimed at promoting the interpretation, rather than the execution, of developmental toxicology studies.

Guidelines for risk assessment in the area of developmental toxicity were first issued in draft form in 1984, and after a public comment period were published in 1986 as the Guidelines for the Health Assessment of Suspect Developmental Toxicants (US EPA 1986). That document broadly defined developmental toxicity as:

> "... adverse effects on the developing organism that may result from exposure prior to conception (either parent), during prenatal development, or postnatally to the time of sexual maturation. Adverse effects may be detected at any point in the life span of an organism. The major manifestation include (1) death of the developing organism, (2) structural abnormality, (3) altered growth, and (4) functional deficiency."

One important aspect of the 1986 Guidelines was the inclusion of a research needs section that was intended to promote research and analyses to ultimately improve the overall risk assessment process. Over the last several years, the EPA has sponsored several workshops targeted at particular aspects of interpreting data from developmental toxicity studies (e.g., Kimmel et al. 1988; Kimmel et al. 1989) and developed more standard procedures for extrapolating non-cancer health effects from animal models to humans (Barnes and Dourson 1988; Kimmel 1990). In addition, a large research effort has been underway within EPA to address a number of risk assessment issues. As a result of these activities and research advances in related topics, proposed revisions to the 1986 Guidelines were published in 1989 (US EPA 1989). These revisions went through a public comment period, as well as review by internal and external scientific panels, and were recently published in the Federal Register (US EPA 1991). The remainder of this paper will address a few of the major highlights of the revised guidelines, with emphasis on more quantitative approaches to developmental toxicity risk assessment, particularly the benchmark dose concept.

Assumptions of Developmental Toxicity Risk Assessments

Recognizing that a number of unknowns exist in the extrapolation of animal test data, the guidelines list five general default assumptions that form the basis for the use of animal test data in human risk assessment: 1) an agent that causes adverse effects in experimental animal studies poses a hazard to humans following sufficient exposure during development; 2) a biologically significant increase in any of the four manifestations of developmental toxicity is indicative of developmental hazard; 3) the types of developmental effects observed in experimental animal studies are not necessarily those that would result from sufficient exposure of humans; 4) in the absence of knowledge of the most appropriate species on which to base a human risk characterization, the most sensitive species is used; and 5) a threshold is generally assumed to exist for the induction of developmental effects.

Implications of Maternal Toxicity

Testing guidelines for developmental toxicity routinely require the use of a high dosage level that induces some minimal degree of overt maternal response (e.g. a 10% decrease in body weight gain to a maximum of 10% maternal mortality). While it is clear that agents which cause adverse developmental effects in the absence of maternal effects are of particular concern, developmental effects in the presence of minimal maternal toxicity also warrant the attention of the risk assessor. These effects should not be discounted as being secondary to maternal toxicity, as there is little evidence to conclude that such developmental effects result only from maternal toxicity. On the other hand, doses causing excessive maternal toxicity may make interpretation of the developmental effects problematic and require additional toxicological studies. It is also important to consider the maternal response in the overall evaluation of a chemical's toxicity profile, as the maternal toxicity data may identify pregnant or lactating females as a susceptible population in their own right. Lastly, the guidelines caution against the use of approaches for ranking the relative maternal and developmental responses. Not only do difficulties arise in defining each element of ratios used to relate maternal to developmental effects, but such schemes are highly dependent on the thoroughness of the examinations, on the number and spacing of dose levels, on the relative shapes of the dose-response curves, and are not consistent among species (Daston et al. 1991; Setzer and Rogers 1991). Therefore, such ranking systems are not useful in the risk assessment process.

Characterization of the Health-Related Data and Procedures to Derive the RfD_{DT}

In judging the critical effect (i.e., the effect seen at the lowest dose level) in either the maternal organism or the conceptus, the NOAEL (or No Observed Adverse Effect Level) is identified. The NOAEL is defined as the highest dose at which there is no statistically or biologically significant increase in the frequency of adverse effects when compared to controls. In establishing a NOAEL for a particular endpoint, consideration must be given to evidence of a dose-response relationship, and for this reason, the hazard identification and dose-response components of the risk assessment process for developmental endpoints are done concurrently. By merging these two elements, hazard is assessed in the context of the species, dose, route, timing and duration of a particular study. In addition, this approach de-emphasizes the labelling of chemicals as hazardous on a purely qualitative basis.

Having established the NOAEL for the critical effect for the entire available data set, the data are characterized as sufficient or insufficient to move into the quantitative part of the risk assessment process. This determination is based upon whether a minimum amount of evidence is available, with the volume necessary to consider a chemical relatively non-hazardous greater than that to suggest a potential hazard. Briefly, *sufficient evidence* would include 1) epidemiology studies that provide convincing evidence that a causal relationship is or is not supported; 2) a single, appropriate, well-conducted animal study demonstrating an adverse effect; or 3) data from well-conducted studies in at least two species that show no developmental effects at doses that are minimally toxic to the adult. The category of *insufficient evidence* would pertain to chemicals for which there is less than the minimum sufficient evidence. While a great deal of scientific interpretation is required to make the distinction, a risk assessment would proceed only if the data base were deemed sufficient.

If the data warrant, the next step in the risk assessment process is the determination of the reference dose (RfD_{DT}) or reference concentration (RfC_{DT}) for developmental toxicity (the DT subscript denotes that the RfD or RfC is specifically for developmental toxicity). An RfD or RfC is an estimate of the daily exposure to the human population that is likely to be without appreciable risk of adverse non-cancer health effects during a lifetime (Barnes and Dourson 1988). In the case of developmental toxicity, the exposure duration is usually over a much shorter period. The RfD_{DT} is determined by dividing the NOAEL for the critical effect by the product of uncertainty factors that account for inter- and intraspecies extrapolations (usually tenfold factors for each) and a modifying factor (a value greater than zero and less than or equal to ten but generally one) that reflects the professional judgement of the assessor in evaluating the overall data base for a particular chemical (e.g., pharmacokinetics, systemic toxicity and study quality). An additional ten-fold uncertainty factor is applied when only a LOAEL (Lowest Observed Adverse Effect Level) is available, but no adjustment for duration of exposure is made (i.e., less than 24 hours per day or less than a major fraction of the life-span) unless suggested by pharmacokinetic considerations. This lack of time-weighted adjustment is in contrast to the procedure used for most non-cancer health endpoints. The RfD_{DT} is then communicated to the risk manager together with a summary of the scientific judgements used to derive the estimate; a discussion of the shape and slope of the dose-response curves; the exposure conditions under which a chemical poses a specific hazard to humans; the likelihood of those exposures occurring in the human population; the magnitude of the estimated human exposure; and the size and characteristics of the exposed population.

The Benchmark Dose Concept

The use of the NOAEL from an animal bioassay as the entry point into the extrapolation process has been the focus of much recent debate within and outside the EPA, as efforts to introduce more quantitative approaches to risk assessment have gained momentum. Criticisms of the NOAEL approach include the fact that the value is limited to one of the experimental dose groups, and is therefore strongly influenced by the number and spacing of the dose groups; that it excludes information on the slope and variability of the dose-response relationship; that NOAELs are not equivalent between studies (the true response at the NOAEL has been estimated to range from 0 to 7% in developmental toxicity studies); that due to statistical factors related to the ability to detect effects, NOAELs are inversely related to the quality of the study; and finally, that the RfD process provides no information regarding the potential risks should exposures exceed that value (although this last comment does not pertain particularly to the NOAEL derivation itself).

The benchmark dose (BMD) concept was introduced by Crump (1984) as a simple procedure to address many of the above stated criticisms. The benchmark dose is derived from a mathematical model which operates within the experimental range. Specifically, the BMD is defined as the lower confidence limit on dose that results in a particular response level (e.g. 1%, 5%, or 10%). Because the model operates within or close to the experimental range, it has been argued that the result should be independent of the particular model employed, and because it utilizes all available experimental data, the impact of slope and variability are implicitly included, while the impact of dose spacing is minimized. Higher quality studies, with multiple dose levels and larger sample sizes, would receive the benefit of producing a better estimate of the true pattern of toxicological effect. In addition, because it is derived from a mathematical function that describes the dose-response curve, the BMD is an objective measure. This is in direct contrast to the subjective nature of the NOAEL, which may be defined differently by equally qualified experts. It is important to stress that the BMD as proposed here is not a low dose extrapolation tool, and that it is not a panacea for interpreting data. For example, judging whether or not an effect is adverse will not benefit by the application of a statistical model, nor will studies that produce ill-defined dose-response patterns yield information well suited for mathematical modelling.

Before the present day use of the NOAEL as the departure point in the extrapolation process for animal toxicology data can be supplanted by a more objective, mathematically derived value such as the BMD, several key questions must be addressed. For example, what is the minimum data set that can be used in a mathematical model? Which effects should be modelled, and how would one handle continuous versus quantal responses? What level of response should be selected (e.g., 10, 5 or 1% increases in risk for a particular effect), and is the lower confidence level on that value appropriate or too conservative (if not, what confidence level should be used)? What is/are the most appropriate model(s) to use? What constitutes a good fit of the model to the data? Can any of the uncertainty factors be reduced or eliminated by using the BMD approach? And finally, can it be done incorrectly?

In order to address these generic issues, as well as to explore particular aspects of the benchmark concept as it applies to developmental toxicity data, the US EPA is supporting research to develop and computerize several statistical dose-response models for developmental effects, and to apply them to data from 200 to 300 Segment II studies obtained from diverse laboratories that employed various species, strains, examination techniques, and chemicals from diverse classes. By applying statistical models to the various experimental designs and dose-response patterns contained in these data sets, we will obtain a rigourous evaluation of their utility and efficiency.

The specific models selected for comparison are the "NCTR" model (Kodell et al. 1991) and generalizations of the Rai and van Ryzin (RVR) model (Rai and van Ryzin 1985), and the log-logistic model (Kupper et al. 1986). The structure of the models is displayed in Figure 1, and the reader is referred to the original references for detailed descriptions of any particular model. Briefly, the three models have several elements in common: 1) they estimate a probability of effect for quantal variables as a function of dose and litter size; 2) they assume beta-binomial variation to account for intra-litter variation (i.e. the litter effect); 3) they derive maximum likelihood estimates for the model parameters; 4) they can be used to estimate a statistically based threshold value; 5) they allow a power (or Weibull) function to modify the dose parameter; 6) they can incorporate a non-zero background rate; and 7) they estimate the lower confidence limit on dose for particular levels of risk (in the current application the lower 95% confidence limit is estimated). Preliminary results of this project have been recently presented (Allen et al. 1992).

RVR MODEL:

$$P(d,s) = [1 - \exp\{-(\alpha + \beta(d-d_0)^w)\}] * \exp\{-s(\theta_1 + \theta_2(d-d_0))\}$$

NCTR MODEL:

$$P(d,s) = 1 - \exp\{-[(\alpha + \theta_1(s-s_m)) + (\beta + \theta_2(s-s_m))(d-d_0)^w]\}$$

LOG-LOGISTIC MODEL:

$$P(d,s) = \alpha + \theta_1(s-s_m) +$$

$$[1 - \alpha - \theta_1(s-s_m)] / [1 + \exp\{-(\beta + \theta_2(s-s_m) + \gamma\log(d-d_0))\}]$$

Fig. 1. Model structures under consideration for use in quantitative dose-response analyses for developmental endpoints. The "RVR" model is that of Rai and van Ryzin (1985), the "NCTR" that of Kodell et al. (1991), and the log-logistic that of Kupper et al. (1986). Models have been implemented in their most general form. Alpha, beta and sigma parameters refer to intercept, dose rate, and intra-litter correlations, respectively. The threshold (d_0) and litter size (s and s_m) parameters are optional. See text for more information

To visualize the model results, the dose-response functions obtained by applying the models to a developmental toxicity study of trichloroacetic acid (TCA; Smith et al. 1989) are presented in Figure 2. This particular data set was selected because of the wide range of dose levels employed and the graded nature of the incidence of malformations as a function of dose. As judged by Chi-square tests and log-likelihood values, all models fit the data equally well, although the RVR model provides higher maximum likelihood estimates of risks for a given dose (a probable result of the particular procedure for handling litter size in this model).

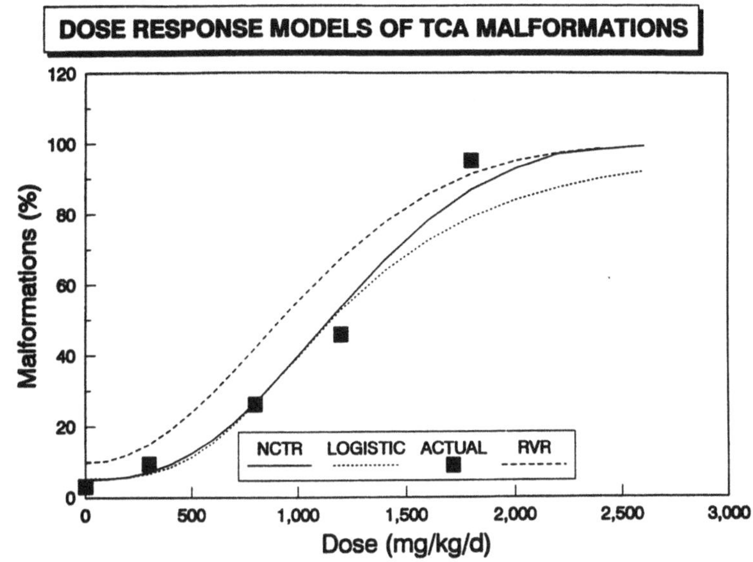

Fig. 2. Application of dose-response functions listed in Figure 1 to the data for trichloroacetic acid-induced malformations (Smith et al. 1989). Actual data points are depicted by square symbols. All three models adequately describe the data as evidenced by chi-square and log likelihood values

Of particular interest is the fact that a NOAEL for malformations could not be identified in the TCA study, as the incidence of malformation (primarily cardiovascular in nature) was significantly elevated at all dose levels. Previous work by Gaylor (1989) has suggested that the NOEL may be similar in magnitude to the 5% benchmark level (BMD(5)). As can be seen in Figure 3 for the various benchmark estimates from the NCTR model, the BMD(5) was estimated to be 201 mg/kg/d. This value contrasts with lowest dose tested in the TCA study (330 mg/kg/d) which, as noted, was an effect level. Applying the traditional 100-fold uncertainty factor to the BMD(5) would result in an estimated RfD_{DT} of 2 mg/kg/d. In contrast, had traditional RfD procedures using 330 mg/kg/d as a LOAEL been employed, either an additional ten-fold safety factor for the lack of a NOAEL would have been invoked (making the estimated RfD_{DT} 0.330

mg/kg/d), or the study would have had to been repeated at lower doses to experimentally define the NOAEL (at a cost of additional resources and animals). Both of these undesirable options are eliminated through the use of modelling procedures.

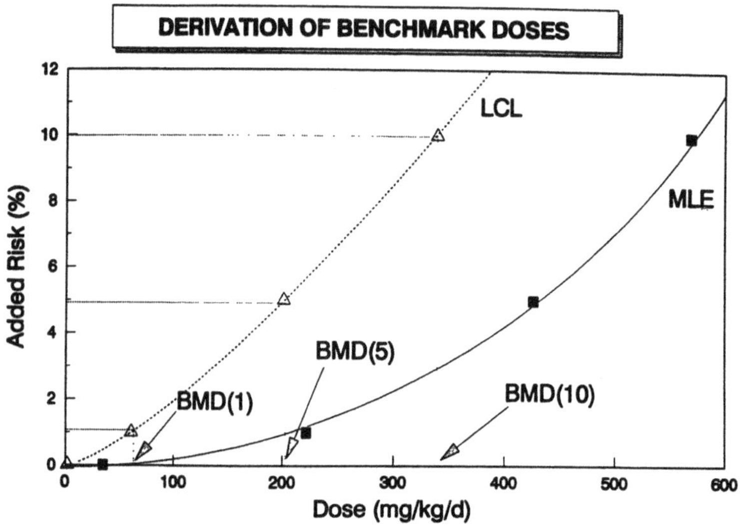

Fig. 3. Benchmark dose estimates from the NCTR model for TCA induced malformations as described by Smith et al. (1989). MLE is the Maximum Likelihood Estimate for the added risk over control (the square symbols representing the MLEs for 10, 5 and 1% response), the LCL is the Lower Confidence Limit on each of the MLE doses. BMD(10), BMD(5) and BMD(1) are the benchmark dose values for 10, 5 and 1 percent added risk, respectively. The dose values are 414, 201 and 61 mg/kg/d, respectively, for BMD(10), BMD(5) and BMD(1)

When completed during the next year, the project to evaluate quantitative dose-response models for developmental toxicity will have assessed the consistency of the outputs for three different models, compared the various risk estimates with traditionally obtained NOAELs, evaluated the impact of litter size adjustments and threshold parameters on the estimates, and assessed the impact of study design (number, spacing, group size) on the model estimates. This last element will be approached using data simulations based on values from actual data sets. At that point, the EPA intends to issue guidance on the use of the BMD approach for developmental endpoints as a supplement to the Guidelines. Readers seeking the latest information on the modelling effort may contact either author.

Summary

Risk assessment should be viewed as a dynamic process, with new information being incorporated into the principles and procedures as they gain acceptance in the scientific community. Only by encouraging this process will we gain increased confidence in our effort to protect public health. The rationale and history for the creation of formal guidelines for the risk assessment of developmental toxicants by the US EPA has been reviewed. Key changes that appear in the 1991 revisions have been indicated, with emphasis on the emergence and utility of quantitative dose-response models for animal bioassay data. The use of statistical dose-response models in deriving the BMD may be viewed as an intermediary step toward the development of more biologically oriented dose-response models that integrate information on pharmacokinetics, mechanisms of action, and pathogenetic processes, with information on the adverse biological outcomes at birth and postnatally. Such models, which require a much greater understanding of both normal and abnormal development, offer great promise of improving the predictive nature of intra- and interspecies extrapolations. However, due to the complexity of development and the effort needed to collect the data required for biologically motivated models, such models are a long way from implementation.

Acknowledgements: The evaluation of the benchmark dose concept is being done in collaboration with Bruce Allen of Clement International and Dr. Elaine Faustman of the University of Washington; their assistance in modelling the trichloroacetic acid data is particularly appreciated. Dr. Kate Smith of the US EPA kindly provided the individual litter data from the trichloroacetic acid study.

References

Allen BC, Van Landingham C, Howe R, Kavlock RJ, Kimmel CA, Faustman EF (1992) Dose-response modeling for developmental toxicity. The Toxicologist 12(1): 300

Barnes D, Dourson M (1988) Reference dose (RfD): Description and use in health risk assessment. Regulatory Toxicol Pharmacol 8: 471-486

Crump KS (1984) A new method for determining allowable daily intakes. Fund Appl Toxicol 4: 854-871

Daston GP, Rogers JM, Versteg DJ, Sabourin TD, Baines D, Marsh SS (1991) Interspecies comparisons of A/D ratios are not constant across species. Fund Appl Toxicol 17: (in press)

FDA (1966) Guidelines for reproduction and teratology studies of drugs. Bureau of Drugs, Rockville, MD

Gaylor DW (1989) Quantitative risk analysis for quantal reproductive outcomes. Environ Hlth Perspect 79: 243-246

Kimmel GL, Kimmel CA, Francis EZ, eds (1988) Evaluation of Maternal and Developmental Toxicity. Teratogen Carcinogen Mutagen 7: 203-338

Kimmel CA, Wellington DG, Farland W, Ross P, Manson JM, Chernoff N, Young JF, Selevan SG, Kaplan N, Chen C, Chitlik LD, Siegel-Scott C, Valaoras G, Wells S (1989) Overview of a Workshop on Quantitative Models for Developmental Toxicity Risk Assessment. Environ Hlth Perspect 79: 209-215

Kimmel CA (1990) Quantitative approaches to human risk assessment for non-cancer health effects. Neurotoxicol 11: 189-198

Kodell RL, Howe RB, Chen JJ, Gaylor DW (1991) Mathematical modelling of reproductive and developmental toxicity effects for quantitative risk assessment. Risk Analysis 11: in press

Kupper LL, Portier C, Hogan MD, Yamamoto E (1986) The impact of litter effects on dose-response modelling in teratology. Biometrics 42: 85-98

NRC (1983) Risk Assessment in the Federal Government: managing the process. Committee on the Institutional Means for Assessment of Risks to Public Health. Commission on Life Sciences, National Research Council. Washington D.C., National Academy Press, pp 17-83

Rai K, van Ryzin J (1985) A dose-response model for teratological experiments involving quantal response. Biometrics 41: 1-9

Setzer RW, Rogers JM (1992) Assessing developmental hazard: The reliability of the A/D ratio. Teratology, in press

Smith MK, Randall JL, Read EJ, Stober JA (1989) Teratogenic activity of trichloroacetic acid in the rat. Teratology 40: 445-451

US EPA (1982) Pesticide Assessment Guidelines, Subdivision F. Hazard evaluation: human and domestic animals. EPA-540/9-82-025

US EPA (1985) Toxic Substances Control Act Test Guidelines; Final rules. Fed Reg 50: 39426-39428 and 39433-39434

US EPA (1986) Guidelines for the Health Assessment of Suspect Developmental Toxicants. Fed Reg 51: 34028-34040

US EPA (1989) Proposed Amendments to the Guidelines for the Health Assessment of Suspect Developmental Toxicants; Request for Comments; Notice. Fed Reg 54: 9386-9403

US EPA (1991) Guidelines for developmental toxicity risk assessment; Notice. Fed Reg 56: 63798-63826

Discussion of the Presentation

Baß: Why two species? Is there a difference in approach, species selection, etc., comparing pharmaceuticals with non-pharmaceuticals?

Kavlock: If we had the perfect model for humans, we would need to test chemicals and drugs in only that one model. Unfortunately, our models are not perfect, and we cannot rely on one single test species including primates, to evaluate the risks to humans. Our historical evidence suggests that two species, be they rat and rabbit or cat and mouse, provide a reasonable prediction of effects known to occur in humans. Schardein's and Kulei's recent review covers the question of concordance quite extensively.

Krowke: In one of your slides you demand to choose the most sensitive strain. Beside the fact that this will not be known at the initiation of the studies and may vary for different chemicals, it seems more important to have ample information (i.e. historical controls) on a single strain to be used than to use different strains for different substances. I doubt whether it is wise to, e.g., use a mouse strain very sensitive to acquiring cleft palates or limb defects (if there were such a strain) for routine studies on reproductive toxicity.

Kavlock: Aspects of strain and species selection are covered by individual testing guidelines, and they do provide requirements for which strains or species would be appropriate. However, I would say that in the absence of data that a strain was inappropriate for use with a certain chemical, I would see no reason to exclude it. It is important to note that the strains you refer to are not universally more sensitive to all chemicals, and that strains with high background rates of malformations may make actual detection of an increase more and not less difficult.

Neubert: I doubt whether the data base with respect to reproductive and developmental toxicity is really sufficient to attempt any meaningful statement on the predictive value of animal experiments in this area. In fact, I am afraid that such a concordance is really no good, and comparative statements are worth nothing, up till now, in spite of other claims. You may end up with any conclusion you like. In my opinion, the value of animal studies in the field of reproductive toxicology is to elucidate suspicions on toxic potentials and possibly on potencies, in order to avoid unnecessary exposure of humans. Whether man would ever exhibit such an effect at relevant exposures is quite a different question, and cannot be answered unless studied in man (which often would be unethical). However, this problem is not confined to assessing reproductive toxicity, it also holds true for cancerogenicity, etc. Furthermore, cats and mice certainly should be considered as quite inadequate species for risk assessment on reproductive toxicity with relevance for man: the data base for the former is by far too small, and for the latter there is ample evidence that it often responds too sensitively, even to "unspecific" noxae. Even for the widely used species: rats and rabbits, the concordance with respect to pathological outcome and effective doses is far from being satisfactory. There is really no good scientific argument why two species are better than one (or why three or four would not be even better). Even studies with three species (cat, mouse, rat) would be quite inadequate to reveal the potential of the most notorious human "teratogen" (thalidomide), or that of warfarin. There is a strong argument for putting more effort into studies with a single species, e.g. rats (much more extensive kinetic studies, using more dose levels, treatment with higher doses for limited periods of time, evaluating more postnatal manifestations, etc.), which may provide more relevant information than today's routine and highly compromised studies with two species (i.e the additional information from the rabbit).

Neubert: You suggest that using ED_X-values is superior to a NOAEL approach and it is easy for me to agree to that. We suggested such a procedure several years ago, and have used it since. There is, of course, another intrinsic difficulty in assessing ED- or similar values at the low dose level: around the ED_{50} the deviation is rather small, but it becomes much larger the lower the ED-value is. Thus, e.g. ED_{10} may be measured accurately, but not calculated without a large error. The main difficulty I see is that the usual data sets from routine studies may not be appropriate for a more sophisticated assessment. Would you suggest that the testing strategy would have to be changed (e.g. more doses to be tested)?

Kavlock: Your question addresses one of the major concerns of our study, that is, how well do these quantitative dose-response models operate in "real world" situations, where effects may be observed only at the highest dose level (the "L-shaped" response mentioned by Dr. Palmer in his presentation). We expect to see a wide variety of dose-response patterns in the more than 200 segment II data-sets that we have accumulated, and hopefully these will represent a realistic spectrum of actual study results. Therefore, we will have a good handle on how well these models work with conventional study designs. Naturally, if there is no dose-response relationship, one will not be able to utilize quantitative risk assessment models, and those data-sets will be culled from our analysis. From the remaining data-sets we will determine how well the various models fit the data, and how variable are the benchmark estimates. You are probably familiar with the problems encountered in cancer risk assessment, where a bioassay developed for hazard identification purpose is now routinely used for low dose extrapolations using the multi-stage model. We need to avoid those same problems, which is one reason to stay within the observable range in choosing the actual benchmark doses. More dose levels are probably better for characterizing the dose-response pattern, and in a latter part of this project we will be using simulations to determine the extent to which alternate study designs might improve the estimates of the benchmark doses.

Scientific Basis for Risk Assessment (Reproductive Toxicity) for Pesticides as Practiced by the Bundesgesundheitsamt (BGA)

Wolfgang Lingk, Maged Younes, Rudolf Pfeil and Roland Solecki

Introduction

It is the task of public health control agencies to weigh the risks of a given chemical against its benefits and the need for it. The German pesticides law regulations call for decisions based on safety standards which imply acceptable degrees of risk, without providing any precise definition of permissible degrees of risk. The margin of judgement provided by such legislation confers upon public health agencies the difficult task of making discretionary decisions against a background of conflicting interests.

At what point does a chemical become adequately safe in terms of consumer protection? This question alone makes it apparent that safety decision-making contains two basic elements: scientific estimation and prognosis of public health. Evaluation of chemical substances thus becomes more than a mere question of scientific deduction processes; it also implies an element of general social interests based on the results of testing and on scientific data. Evaluation of such results implies setting standards of compatibility and acceptability - a thing is considered safe if it is judged to be safe on these grounds.

The Federal Health Office was involved in risk assessment of pesticides long before the scientific basis of reproductive toxicology was established. Until it became known that pesticides might exert developmental toxicity, the risk evaluation was based on their known toxic effects. Even following the thalidomide tragedy, pesticides were not suspected of being able to cause similar effects. Studies with the herbicide 2,4,5-T in chick embryos which were conducted after the Vietnam war showed for the first time that herbicides were also capable of inducing teratological effects. As a consequence, the development of a scientific basis for risk assessment of reproductive toxicology for pesticides was started. In the following, the single steps of risk assessment of reproductive toxicology for pesticides will be reviewed from the point of view of actual problems arising from the daily practice of pesticides registration.

Experiments with Laboratory Animals

In Germany the risk assessment of reproductive toxicity of pesticides is mainly based on data from standardized experiments with laboratory animals. Usually, studies conducted in two animal species are submitted.

The different endpoints of reproductive toxicology studies include the fertility of parental animals and maternal toxicity as well interference with prenatal, perinatal and postnatal development. In this respect, it should be noted that studies on peri- and postnatal toxicity are not given enough consideration in the OECD guidelines for toxicity testing of pesticides. From the point of view of risk estimation, practical problems arise from differences in test standards required by governmental authorities, as well as from differences in the definition of the evaluated parameters by the submitting companies. Therefore, harmonization of test standards for risk identification is ultimately needed. Evaluation of reproductive toxicity studies in our standard requirements include a multigeneration test in the rat which allows for an identification of reproductive toxicity, as well as a test for embryotoxicity and teratogenicity in the rat and/or another species. The tests must be performed according to the OECD guidelines.

The multi-(two-)generation reproduction toxicity test is designed to provide general information concerning the effects on male or female reproductive behaviour. The study may also include further endpoints and provide information about developmental toxic effects. Characteristic endpoints are shown in Fig. 1.

A reproduction study may serve as a guide for subsequent tests. The study is not designed to determine specific cause-effect relationships in all cases. The OECD guideline requires at least three dose groups and a control group. Ideally, the highest dose level should induce toxicity but not mortality in the parental animals, and the low dose should not induce any observable adverse effect on the parents or offspring.

The first step in the evaluation of a multigeneration study is to determine if and at which dosage the test substance is toxic to the parental animals and/or the offspring. The main aim of the multigeneration study is to determine the effects of toxic chemicals on reproductive behaviour as well as on the development of the offspring. Therefore, the interpretation of the results should differentiate between findings related to unspecific systemic toxicity of the test substance and such effects which clearly show an influence on the reproductive ability of the animals. Often, both categories overlap, e.g. an extreme body weight loss leads to an impairment of gonadal function.

The teratogenicity test is designed to provide information on the potential hazards to the unborn which may arise from exposure of the mother during pregnancy. According to the OECD guide-line, testing of pesticides for teratogenicity also requires the use of at least three dose groups and a control group. To select the appropriate dose levels, a pilot study may be advisable. The highest dosage level is required to induce some overt maternal toxicity, such as slight weight loss, but not more than 10% maternal deaths, while the lowest dose level should not induce observable effects. Among the parameters of maternal toxicity evaluated, the endpoints shown in Figure 2 are routinely estimated.

With respect to the requirement for a dose level causing "slight maternal toxicity", there is an ultimate need for an exact definition of the term "maternal toxicity". While the interpretation of numerical data, such as body weight, feed consumption, etc., is generally easy to perform, difficulties may arise in attributing some clinical or pathological changes to the test substance.

Endpoints used for the estimation of embryo-/fetotoxicity include the parameters listed in Figure 3. Functional abnormalities (physical and behavioural) can only be detected when the offspring are also observed after weaning. While the assignment of parameters such as fetal weight or fetal death to the effect of a certain substance is generally not difficult, the interpretation of data obtained from organ and skeletal examinations is more complex and sometimes requires the use of data from historical controls.

Essential Endpoints	*Further Endpoints*
Gonadal function	Clinical chemistry
Oestrus cycle	Hematology
Mating behaviour	Histopathology
Conception	Toxicokinetics
Parturition	Metabolism
Lactation	Teratogenesis
Weaning	Neonatal morbidity
Growth of the offspring	Mortality
Gross pathology	Behaviour

Fig. 1. Multi-generation reproduction toxicity test

Essential Endpoints	*Further Endpoints*
Mortality	Clinical chemistry
Gestation	Hematology
Body weight/	
body weight gain	Histopathology
Feed and water consumption	Organ weights
Clinical observations	Metabolism
Gross pathology	Toxicokinetics

Fig. 2. Teratogenicity test - Maternal toxicity

Essential Endpoints	*Further Endpoints*
Fetal death	Physical abnormalities
Fetal weight	Behavioural abnormalities
Organ malformations	Immunological defects
Skeletal alterations	Fetal histopathology
Developmental retardations	

Fig. 3. Teratogenicity test - Embryo-/fetotoxicity

Difficulties in the interpretation of such findings also arise from the fact that differences exist in the classification of malformations and variations between different laboratories. It is generally accepted, however, that an increased incidence of morphological changes that seldom or never occur spontaneously should be treated as an indication for a teratogenic potential of the test substance. In the final evaluation, it must be considered if the increased rate of morphological changes was evident in the absence of maternal toxicity or if it became apparent only at doses which were toxic to the mother animals.

Alternative Methods

In vitro methods and short term assays of reproductive toxicology can yield only partial information. This is caused by the fact that the complexity of the materno-fetal-unit cannot be simulated at present. Therefore, alternative methods should be regarded as screening tests or as additional secondary stage studies. In this respect, they allow for a better characterisation of the mechanism of action of an agent which was shown to be a teratogen *in vivo*. They cannot substitute for animal experiments, however.

Apart from experimentally based alternative methods the prediction of a possible teratogenic action on the basis of structure-activity relationships using special software is being established at present. As long as such predictions are not sufficiently validated, however, computer simulation in this sense will not be of practical importance for risk assessment.

Consideration of the Threshold Concept for Developmental Toxicity

In general, the reproductive toxicity of pesticides registered in Germany is well documented in animal studies. Therefore, the risk of adverse effects on the fetus can readily be extrapolated to man. This is of special importance since epidemiological studies, if available, can seldom establish a causal relationship between human exposure to the substance and subsequent developmental toxic effects in the progeny.

Risk estimation is based on the toxicological profile of the test substance on one the hand and on the probability for the occurrence of a possible damage on the other hand. The quantitative evaluation, i.e. the characterisation of the risk, must consider all the information relevant for an estimation of the dose-response effect and the level of exposition. The determination of the dose-response relationship is an important criterion for the estimation of the dose with no adverse effect which allows for the calculation of a tolerable human exposition. In practice, however, risk estimation is conducted using the data of standard studies with three dose groups. Therefore, the exact determination of the dose-response curve is seldom possible. In some cases, an effect is only seen at the highest dose.

While the existence of thresholds for the dichotomous responses of malformations and embryolethality is generally accepted, the safety factor approach is used assuming that the specified acceptable exposure level for a developmental toxicant is below the threshold dose for the most part if not all exposed members of the population.

Safety factors are generally derived from the No-Observed-Effect-Level (NOEL) in the most sensitive species. In this respect, it is assumed that the substance that shows a toxic potential in laboratory animals will cause similar effects in man. In the case of reproductive toxicology, a difference must be made between substance-specific effects and unspecific effects. Unspecific effects may be caused by a great number of agents if the dose level is high enough. Such effects include, for example, inhibition of body weight gain, retardation of development or lethality, and are often related to maternal toxicity or cytotoxicity. In such cases, it appears safe to treat reproductive toxicity data in the same way as data from other toxicity studies and to apply a comparable safety factor (e.g. 100).

In the case of specific effects such as post-implantation fetal losses, malformations or irreversible functional defects, safety factors should be adapted to the higher sensitivity of the embryo or fetus. This is especially true with those agents that affect the embryo or fetus in the absence of maternal toxicity. Empirical data revealed that, in general, a safety factor of 100 is sufficient. But there might be instances when the traditional safety factor of 100 may be inadequate. For example, if the slope of the experimental dose-response curve is shallow or if there is evidence that the animal threshold is well below the lowest experimental dose considered, then a safety factor larger than the traditional value may be necessary.

At present, mathematical models of risk estimation are not employed due to the fact that, in most cases, the data base does not allow for an exact calculation of the dose-response relationship.

At the EC level there are two categories for the classification of substances harmful to reproduction at present (Fig. 4).

Category	*Registration*	*Effects*
1	NO	Sufficient evidence for developmental toxicity
2	YES/NO (see examples)	Clear evidence of adverse reproductive effects in one or more species

Fig. 4. Categorisation of pesticides at the EC level

Category 1

Substances known to cause any developmental toxicity in man. There is sufficient evidence to establish a causal relationship between human exposure to the substance and subsequent developmental toxic effects in the progeny. In the practice of classification, these substances are classified in category 1 only on the basis of epidemiological data. In Germany, substances which are classified in category 1 are not registered for use as pesticides.

Category 2

Substances which should be regarded as if they cause developmental toxicity to man. There is sufficient evidence to provide a strong presumption that human exposure to the substance may result in developmental toxicity, generally on the basis of:

- clear results in appropriate animal studies have been observed in the absence of other signs of marked maternal toxicity,
- other relevant information.

In practice, substances are classified in category 2 on the basis of animal data. Data from *in vitro* studies or studies on avian eggs are regarded as "supportive evidence" and would not normally lead to classification in the absence of *in vivo* data.

For classification into category 2, there should be clear evidence of adverse effects in well conducted studies in one or more species. Since adverse effects which occur in pregnancy or postnatally may result as a secondary consequence of maternal toxicity, reduced feed or water intake, maternal stress, lack of maternal care, specific dietary deficiencies, poor animal husbandry, intercurrent infections, etc., it is important that the effects observed should occur in well conducted studies and at dose levels which are not associated with severe maternal toxicity.

The route of exposure is also important and should be relevant to potential human exposure. In particular, the injection of irritant materials intraperitoneally may result in local damage to the uterus and its contents, and the results of such studies should be interpreted with caution.

The science of neurobehavioural teratology/toxicology is not as well developed as other branches of toxicology, and isolated effects observed in such studies may not be sufficient to warrant classification. In common with most other types of toxic effects, apart from carcinogenesis, substances demonstrating development toxicity will be expected to have a threshold below which adverse effects would not be observed. Thus, even when clear effects have been demonstrated in animal studies the relevance for humans may be doubtful because of the total doses which were administered. Similarly, the route of administration may be inappropriate. For these or similar reasons classification may not be warranted in some cases.

In general, no classification should be awarded on the basis of studies which have inadequate numbers of animals or control groups, or where only high toxic doses have been used. The same applies to studies in which the only effects recorded are small changes in the incidences of spontaneous defects, small changes in the proportions of common variants such as those observed during skeletal examinations, or small differences in postnatal developmental assessments.

Examples

The most problematic group is represented by those pesticides which are classified in category 2. The first example (Fig. 5) is a dinitrophenol herbicide, dinoterb (2-tert-butyl-4,6-dinitrophenol). Teratogenicity studies in rats and rabbits revealed that severe malformations of the head occurred after oral and dermal application. In the more sensitive species, the rabbit, major malformations were evident after oral treatment with 5 mg/kg body wt per day or after dermal application of 6 mg/kg body wt per day. Minor malformations were also observed at the lowest dose group of rabbits (1.25 mg/kg body wt per day, dermally). Due to the high absorption rate of dinoterb following dermal application (ca. 40%), the total uptake by applicators was estimated to be about 0.1 mg/kg body wt per day. The safety factor between this estimated total uptake and the lowest teratogenic dose level is much lower (ca. 25-60) than that which is required with respect to the type of defects observed (100-500 compared with the NOEL). Therefore, dinoterb, like other dinitrophenols, is no longer registered in Germany.

	Rat	*Rabbit*
Oral application:	Severe malformations of the skeleton	Malformations of skeleton and organs
NOEL:	2.5 mg/kg body wt	< 1.25 mg/kg body wt
Dermal application:	Severe malformations of the skeleton	Malformations of head skeleton and organs
NOEL:	10 mg/kg body wt	3 mg /kg body wt
Classification:	Category 2	
Total exposition:	0.1 mg/kg body wt per day for the applicator	
Skin absorption:	40%	
Low margin of safety for dermal exposition		
-> Registration	NO (in Germany)	

Fig. 5. Reproductive toxicity of Dinoterb

The second example (Fig. 6) is also a herbicide: bromoxynil (3,5-dibromo-4-hydroxybenzonitril). Teratogenicity studies in rats and rabbits revealed major malformations of the head and ribs after oral application at moderate maternally toxic doses. Subsequent dermal studies revealed only an increased number of ribs in rats and no effects in rabbits. The No-Observed-Effect-Level (NOEL) in the more sensitive species, the rat, was found to be about 1.5 mg/kg body wt per day orally and 10 mg/kg body wt per day dermally. Assuming a total exposition of around 0.1 mg/kg body wt per day of the applicator and considering the low absorption in human skin of 1% per hour, there is a large margin of safety as compared to the NOEL of the dermal studies. Therefore, although bromoxynil and related compounds are presently classified in category 2, a registration was granted.

	Rat	*Rabbit*
Oral application:	Major malformations of the head and ribs	Major malformations of the head and ribs
NOEL:	1.5 mg/kg body wt	30 mg/kg body wt
Dermal application:	Increased number of ribs	NO EFFECT
NOEL:	10 mg/kg body wt	---
Classification:	Category 2	
Total exposition:	0.1 mg/kg body wt per day for the applicator	
Skin absorption:	1%	
Large margin of safety for dermal exposition		
-> Registration	YES	

Fig. 6. Reproductive toxicity of Bromoxynil

These few comments on current problems in assessing reproductive risks reflect the severity of the present conflict between the unconditional demand for safety in society in general on the one hand, and the ethically founded demand for a minimisation of animal experimentation on the other hand. This conflict cannot be resolved by the scientific community alone. Ongoing development of alternative methods - aimed at the replacement of experiments in animals - in conjunction with the defining of new regulatory approaches and standards can make a contribution towards resolving this conflict. It will be necessary, however, to develop alternative methods that permit a comprehensive assessment of the potential risks to consumer health, in order to avoid further increases in the number of routine toxicological tests required.

The tremendous rise in the number of toxicological tests and research results, particularly over the past three decades, appears to be more the result of political, and thus public, interests than of unadulterated scientific curiosity. Put differently, it looks as if society has developed a problem awareness which has now become codified in law but which often eludes critical examination. Health risks arising from chemical substances are often no longer subjected to a rational comparison with other risks associated with life in a modern society, and it is here that we find the justification for the demand that every chemical substance be subjected to intensive toxicological testing in order to reduce the ever present "remaining risk" to the absolute minimum. However, both our resources in toxicological expertise and our financial means - factors that ought not to be underestimated - are limited. Thus, we are confronted with the necessity of seeking a compromise to promote the maximum utilization of available resources in providing that level of safety in public health administration that currently appears achievable.

This is, of course, a difficult compromise to reach. This is not the forum to discuss the pros and cons of the need to develop new test guidelines for regulatory purposes in detail. It is, however, indisputable that by drawing up test guidelines we also determine at least some of the remaining risks for public health. Inasmuch as the selection of methods is involved, we are also defining the degree of safety in assessing chemical risks.

Discussion of the Presentation

Chahoud: Do you perform quantitative risk assessment using mathematical models? Which models do you use?

Lingk: As a rule, the Bundesgesundheitsamt does not perform quantitative risk assessment using mathematical models because the data base (dose-response relationship) for teratogenic effects of pesticides is often limited. Therefore, the safety evaluation is normally based on the conventional approach with a safety factor.

Chahoud: If you use a NOAEL-uncertainty-factor approach, are you using a fixed uncertainty factor? Which one? Why?

Lingk: For pesticides the Bundesgesundheitsamt uses the conventional approach with a safety factor of 100 applied to the NOAEL in animal studies as a general rule, and it has worked well. However, in the case of severe developmental effects at dose levels which are not maternally toxic and inadequate data on the mechanism of action, higher safety factors (500-1000) can be applied.

Krowke: Do you suggest a labelling on a qualitative basis of animal data, i.e. automatically when you see clear-cut gross-structural abnormalities at doses lower than maternally toxic ones? Do you take the extent of human exposure into account for labelling?

Lingk: Your question suggests that the extent of human exposure should be taken into account for the classification and labelling of chemical substances (pesticides). However, according to the EEC directive the classification and labelling of dangerous substances has to be based on the "substance-inherent characteristics". Classification and labelling is required when results in appropriate animal studies provide evidence that the substance causes development toxicity in the absence of overt signs of maternal toxicity. In our opinion, pesticides should not be classified and labelled when teratogenic effects in animal experiments occur only at dose levels grossly in excess of the likely human exposure or when the teratogenic effects in animal experiments are induced with exposure routes which are not relevant to those of man.

Neubert: The situation is not the same in the case of medicinal products and for the evaluation of environmental chemicals, because the data base is so different:

(1) For medicinal products mostly rather extended data on reproductive and developmental toxicity are available, and also, almost always reliable data on pharmacokinetics in the experimental species and for man. In such a case the possible risk should be assessed on the comparative basis of plasma concentrations (concentration-effect relationships).

(2) For environmental chemicals the data base on reproductive toxicity is often limited (e.g. only data from a multi-generation test are presented), and almost without exception no pharmacokinetic data for man are available. For this reason an evaluation on the basis of concentration-effect relationships is not feasible, and one has to be content with dose-response relationships as a basis for risk assessment. In such a situation information on the slope of the dose-response curve is indispensable, when a teratogenic effect has been observed. This automatically leads to the question whether three dose groups (as mostly used up till now), with one being maternally toxic (in our opinion nonsense!) and the lowest not inducing adverse effects, are satisfactory for the purpose anticipated. We feel that this is not the case. With information on the steepness of the dose-response available an extrapolation to an incidence of 1% or 0.1% is feasible, and using the probit of the effect vs. the log of the dose does not seem to be a bad choice.

(3) Asking for a "safety factor" of 1,000 has no scientific background, and it only signals that reliable data for a risk assessment are not available. Since, in most instances, the dose-response for a teratogenic effect is rather steep, extrapolation to an incidence or 0.1% and applying an additional "uncertainty factor" of 10 seems to provide a sufficient margin of safety for most substances.

(4) Some sort of "biomonitoring" would also improve the assessment of the possible exposure in the case of environmental chemicals. Therefore, attempts should be made for further improvement in this direction. Whatever toxicological data are to be used for risk assessment, there is no alternative for a reasonable measure of exposure or dose in man.

Peters: To what extent do you take occupational exposure into account in your risk assessment of pesticides in comparison to consumer intake?

Lingk: Occupational exposure during production, handling, and uses of environmental substances (pesticides) is normally much higher than consumer intake. Therefore, the risk assessment of prenatally-induced adverse health effects is usually based on the occupational exposure data, but the estimated dietary exposure is taken into account as well.

Scientific Basis for Risk Assessment (Reproductive Toxicity) in Ecotoxicology as Practiced by the Federal Environmental Agency (Umweltbundesamt)

Wolfgang Heger

Introduction

All pesticides and chemical substances which are new on the market are tested and assessed with regard to their environmental impact and to their toxicity for organisms. Ecotoxicology aims at not only acute or chronic toxic effects in one species, but also considers interspecies relationships. This means that the members of an animal community and the balance between populations is an important factor to be examined. The problems of ecotoxicological risk assessment consist in few laboratory test data derived from selected species and a poor basis of experimental investigations on the exposition of wildlife.

One example will demonstrate the situation:

The reproductive success of several great blue heron colonies in British Columbia, Canada, have been monitored since 1983 (Hart et al. 1991). One colony located near a pulp mill failed to fledge young for the first time in 1987. The bleaching process of the paper mill produced polychlorinated dioxins and furans as waste products. Measurements of dioxins in eggs showed a sharp increase of polychlorinated dibenzo-p-dioxin (PCDD) and polychlorinated dibenzofuran (PCDF) levels. Therefore, some people came to the conclusion that the dioxins killed the birds during development. In 1988 the hypothesis whether PCDD or PCDF caused the reproductive failure was tested: Heron eggs were collected from three different colonies with low, intermediate and high levels of dioxin contamination (Nicomekl, Vancouver and Crofton colony) and incubated under laboratory conditions.

The study revealed that:

- all incubated eggs were fertile,
- all low level contaminated eggs were able to hatch.

But, some deviations from normal development were observed:

- subcutaneous edema were only observed in contaminated chicks,
- fewer down follicles were present on the contaminated chicks,
- a small negative regression related to the TCDD level of the plasma calcium concentration, body weight, skeletal growth (tibia length), beak length, kidney, and stomach weight was calculated.

Hence, dioxins alone at the levels found in the environment did not cause the mortality of the heron embryos *in ovo*, the complete reproductive failure in 1987 must have had other reasons.

(The depression of growth and the presence of edema suggest that dioxins, at the levels found in the environment, have an adverse effect on the *development* of great blue heron embryos.)

This example demonstrates two major difficulties of field observations:

- to derive a clear causality between effects observed and the action of a chemical
- to correlate a monitored dose to an observed effect.

Ecotoxicity

The ecotoxicity risk assessment is performed in order to predict whether or not a substance (or a preparation) is dangerous for the environment and whether regulatory actions are needed. The Federal Environmental Agency (Umweltbundesamt) of Germany calls for acute, subacute and reproduction toxicity tests using mainly aquatic organisms as implemented by the Chemicals Act (ChemG) and the German Plant Protection Act (PFLSchG).

The environmental risk of a substance depends on

- the *toxic effects* of the substance itself or of reaction products with other substances or of its degradation products and
- the *exposition* of biological or other systems.

Thus, exposure assessment is of crucial importance. If it would be possible to completely exclude environmental exposure, it would not be necessary to assess the potential toxic effects. Due to the ecosystems' complexity and variety, the effects of a substance derived from studies in one species are not representative for the community of all organisms. Therefore, for precautionary reasons, high exposure itself must already be considered an indication of environmental hazardousness (UBA Texte 1990).

Ecotoxicity of Chemicals

According to the Chemical Act the notifier has to submit a notification dossier which includes results on ecotoxcity from experimental investigations. It seemed to be more feasible to request toxicity tests dependent on the production quantities of the chemical and independent on toxic effects, although it is not scientifically justified. Three levels have been introduced: Basic level, level 1 and level 2. Results of reproduction toxicity studies are requested in level 1 and level 2 notifications. The test proofs of Basic level comprehend no reproduction toxicity study.

Risk Assessment Schedule

1.) The test results of laboratory investigations on the toxicity and specific effects of substances on organisms are evaluated.

2.) The test results are transferred from one species to the specific class in the ecosystem.

3.) The prediction of the *potential environmental concentration* (local concentration) of a substance is made by calculation.

4.) The ecotoxicological risk assessment is stated. If there are indications for a high risk which is not definitely proven, additional toxicity studies can be requested.

The majority of the notified chemicals entering into the environment will arrive at the water in extremely low concentrations. Acute or subacute effects are therefore rarely observed, even with strongly bioaccumulating substances. Since in many cases the duration of the exposition determines the extent of toxic effects, the reproduction rate often indicates a chemical impact on the environment. Therefore, in recent years the reproduction toxicity studies in aquatic organisms have become more important (e.g. Gillet and Roubaud 1983; Görge and Nagel 1990; Ward et al. 1981). But, up till now, only the acute ecotoxic effects of many chemicals on various organisms are well documented (e.g. Schafer 1972; Bringmann and Kühn 1977), and there is still a crucial deficit of reproduction toxicity data for aquatic organisms.

For evaluation of the reproduction toxicity data in ecotoxicology, the Federal Environmental Agency uses the EC_{50}-value together with the slope of the dose-response curve. The EC_{50}-value together with the slope of the dose-response curve only gives sufficient information for risk assessment at lower concentrations. The risk calculations at low doses taken out of the data presented for permits for pesticides and for notification of chemicals, revealed a considerable risk if the dose-response curve was shallow. Even when using a safety factor of 1000 between EC_{50} and exposition concentration, the calculated risk could reach in those cases a value of 10% mortality. These results were verified by biometric analysis of data on new chemicals (Meister 1990). In the case of a shallow dose-response curve, the concentrations which cause no effect (NOEC) and a low level effect (LOEC) are calculated and used as assessment criterion to evaluate the risk for ecotoxic effects.

The basis of our risk assessment, in general, is the lowest toxicity data available. This procedure is in agreement with the risk assessment procedures discussed at a Workshop on Environmental Hazard and Risk Assessment of Chemicals held in 1990 at Ispra, Italy. When the slope of the dose-response curve is very steep (slope $\geq$ 6), the EC_{50}-value alone is used. Since only data of few or even only one species are available, the use of "safety factors" or "uncertainty factors" are frequently recommended. The quality of interspecies extrapolation of the toxicity results depends on the chemical and its mode of action in the metabolism. The agency discussed the application of an uncertainty factor of 100 for application on long-term toxicity data, but up till now this value is for many substances, especially for reproduction toxicity in aquatic organisms, not sufficiently justified. Therefore, we refuse to apply safety factors at the present stage of assessment development.

Ecotoxicity of Pesticides

In the case of exposition to pesticides the "granting permits" are dependent on the capability for local extermination of fungi, weeds or insects. On the other hand, the risk for the extermination of a whole species must be ruled out.

Since pesticides are spread into the area in considerable amounts, low level contamination of the environment and long-term exposition cannot be excluded.

Pesticides can produce death *in utero* followed by resorption or abortion. Well-known are the embryotoxic or teratogenic effects of dipterex or hexachlorocyclohexane in pregnant rats (Martson and Voronina, 1991; Dikshith 1991). Related products show similar effects on reproduction in fish (Nebeker et al. 1989). For the highly toxic pyrethrum insecticide, deltamethrine, the teratogenic action could not be verified. Görge and Nagel (1990) found an elevated rate of external malformations in zebra fish at 0.5 and 0.8 μg/l deltamethrine. Since no dose-response relationship was established, this pesticide is not classified as teratogenic for fish.

The evaluation of many pesticides and chemicals revealed that aquatic organisms are generally more susceptible for reproduction toxicity effects than terrestric organisms, as demonstrated by Kulzer et al. (1983). The indication of reproduction toxicity in mammals or other terrestric organisms is an indication for possible reproduction toxicity in aquatic organisms. It must be kept in mind, however, that the dosage of aquatic tests are given in mg test substance per litre water volume and are not based on mg per kilogram body weight. One reason for the high sensitivity of aquatic organisms might be that the whole organism is continuously exposed. Orally given substances are usually applied to mammals in sequential doses. Therefore, metabolism and excretion, beside other effects, may influence the plasma levels to a large extent. The comparison of water concentrations in aquatic tests with oral doses in mammals are not easy to perform. The extrapolation of reproduction toxicity data from one aquatic animal class to another frequently reveals wrong values The extrapolation of the results of fish early life-stage toxicity to crustaceans reproduction is, for many substances, especially for pesticides, not sufficiently justified because of specific toxic effects. Therfore, the ecotoxic risk is assessed on the basis of the lowest available NOEC, which takes into account sublethal effects such as impairment of reproduction, growth, survival and behaviour. If the predicted environmental concentrations differ less than a factor of 10 to the NOEC-concentrations, we appraise that adverse effects may occur. This factor is derived from the limited number of animals in the experimental study, the variability of the laboratory tests and the biological variance, but it does not consider possible differences in sensitivity to pesiticides of species not tested, although being representatives in the ecosystem. These data have been proven to stand the weighing up of the German Administrative Court.

If predicted water contamination approaches the concentration at which subchronic fish toxicity cannot be excluded, and if, in addition, there are indications for a thyroid activity or elevated bioaccumulation, we call for an early life-stage test in a fish species, or for reproduction test in amphibians (*Xenopus laevis*). However, we are aware that full life cycle fish reproduction studies are more sensitive than early life cycle tests (Chorus 1987). These tests are only called for when clear-cut scientific indications exist for harmful activity to the environment (such as exposition, shallow dose-response curve, etc.).

Concerning the reproduction in birds, Hoffman and Albers (1984) published a review on the toxicity of 42 substances. Organophosphorus pesticides (diazinon, pirimiphos-methyl, and methyl-carbamates) produce various defects in avian embryos such as skeletal malformations, beak defects or decreased hatchability. The data indicate that birds seem to be more susceptible to toxic effects of pesticides on reproduction than mammals. A good correlation of the chicken egg test to the

EC_{50} in the early life-stage test in zebra fish (results based on mnol/egg and nmol/l, respect.) was demonstrated by Van Leeuwen and coworkers (1990). Since a reproduction toxicity is a sensitive indicator for ecological risk by chronic toxicity a guideline for a 60-day reproduction test in Japanese Quail is in preparation.

Conclusion

- Reproduction toxicity data of animal species give important information about the ecological risk of chemicals. This information is not given in the basic-set information for notification of chemicals nor in the permit of pesticides. For chemicals, this information can be asked for in the level 2 notification according to the Chemical Act.
- Reproduction and embryonic development is a sensitive parameter for chemical action.
- The significance of the test results rely on a clear causality and a dose-response relationship.
- The Federal Environmental Agency uses EC_{50} in connection with the slope of the dose-response curve for risk assessment.
- The extrapolation of experimental data from one animal species to other species needs very careful consideration. Because of the very limited knowledge on mechanisms of reproduction toxicity in aquatic organisms, we refuse at present to apply safety factors based on probable species differences.

References

Bringmann G, Kühn R (1977) Befunde der Schadwirkung wassergefährdender Stoffe gegen Daphnia magna. Z Wasser- und Abwasserforsch 10: 161-166

Chorus I (1987) Literaturrecherche und Auswertung zur Notwendigkeit chronischer Tests - insbesondere des Reproduktionstests - am Fisch für die Stufe II nach dem Chemikaliengesetz. UBA F&E-Vorhaben Gesch-Z. I 4.1-97 316/7

Dikshith TSS (Ed) (1991) Toxicology of pesticides in animals. CRC Press, Boca Raton

Gillet C, Roubaud P (1983) Prehatching survival of carp eggs after treatment during fertilization and early development with the antimitotic fungicide carbendazim. Water Res 17: 1343-1348

Görge G, Nagel R (1990) Toxicity of lindane, antrazine, and deltamethrin to early life stages of zebrafish (*Brachydanio rerio*) Ecotox Environm Safety 20: 246-268

Hart LE, Cheng KM, Whitehead PE, Shah RM, Lewis RJ, Ruschkowski SR, Blair RW, Bennett DC, Bandiera SM, Norstrom RJ, Bellward GD (1991) Dioxin contamination and growth and development in great blue heron embryos. J Toxicol Environ Health 32: 331-344

Hoffman DJ, Albers PH (1984) Evaluation of potential embryotoxicity and teratogenicity of 42 herbicides, insecticides and petroleum contaminants to mallard eggs. Arch Environ Contam Toxicol 13: 15pp

Kulzer E, Fidler M, Bastian HV (1983) Embryotoxische und teratogene Auswirkung von Herbiziden bei Amphibien und Fischen. Forschungsbericht UBA SIGN CH51-0180

Martson LV, Voronina VM (1991) Experimental study of the effect of a series of phosphoro-organic pesticides (Dipterex and Imidan) on embryogenesis. Environ Health Perspect 13: 121 pp

Meister R (1990) Biometrische Analyse von Konzentrations-Wirkungsbeziehungen bei neuen Stoffen. UBA F&E Gutachten

Nebeker AV, Griffis WL, Wise CM, Hopkins E, Barbitta JA (1989) Survival, reproduction and bioconcentration in invertebrates and fish exposed to hexachlorobenzene. Environ Toxicol Chem 8: 601-612

Schafer EW (1972) The acute oral toxicity of 369 pesticidal, pharmaceutical and other chemicals to wild birds. Toxicol Appl Pharmacol 21: 315-330

UBA Texte (1990) Chemicals Act - Principles for the assessment of new chemicals under the Chemicals Act. Federal Environmental Agency (Umweltbundesamt) Heft 28e

Van Leeuwen CJ, Grootelaar EMM, Niebeek G (1990) Fish embryos as teratogenicity screens: a comparison of embryotoxicity between fish and birds. Ecotox Environ Safety 20: 42-52

Ward GS, Parrish PR, Rigby RA (1981) Early life stage toxicity tests with a salt-water fish: Effects of eight chemicals on survival, growth, and development of sheepshead minnows (*Cyprinodon variegatus*) J Toxicol Environ Health 8: 225-240

Discussion of the Presentation

Baß: You have outlined an important aspect which is often ignored, because most risk assessments are anthropocentric. How many substances have been tested for reproductive toxicity in fish?

Heger: If full life cycle and the early life-stage in fish species are accepted as reproduction toxicity tests, according to my knowledge, about 100 chemicals have been tested in fish reproduction. Beside the published data on fish reproduction studies in the literature, additional data are available at the Federal Environmental Agency which were presented mainly for registration of pesticides. The ecotoxicity database on pesticides at the Agency contains test results of 51 substances assessed in the fish reproduction toxicity. The toxicity of an additional 38 chemicals on early life-stages in fish are noted in the AQUIRE database. None of the reproduction toxicity studies in fish has been required for the notification of a new chemical.

General Discussion: Choice of Species

Chahoud: After Dr. Kavlock's paper we started discussing the problem of a second species to be used in studies to reveal possible reproductive or developmental toxicity. However, there are many more aspects to this problem. There are situations in which ample information from one species may be sufficient, and others in which information from two is insufficient, and data from another is warranted.

An important aspect is to critically review the contribution of data from studies with the rabbit. It would be interesting to learn, at what percent or in which special situations information from this species was crucial for a risk assessment. In my opinion it is a considerable drawback that almost no further toxicological information is available for the rabbit (e.g. chronic toxicity, cancerogenicity, kinetic data, etc.). Furthermore, it is difficult to reconcile why 70 fetuses in this species should be sufficient for a risk assessment, but for rats > 200 are found necessary. I have the feeling that the excuse for choosing the rabbit is still that some effect was seen with thalidomide; not a very good argument.

If we do not limit our discussion to the evaluation of prenatal toxicity the situation will still be different: for assessing possible effects on fertility the rat may be the most inappropriate choice, because of the huge excess of sperm. Possible effects on male fertility can be more easily and more significantly assessed in rabbits (e.g. by using spermatograms). To my knowledge, no serious attempts have been made to assess possible effects on male fertility in non-human primates.

Lehmann: Many experiments demonstrate differing results in the two usual species used in segment II studies, e.g. a compound being teratogenic in our species but with no embryotoxic effects in the other. Very often pharmacokinetic reasons cannot be detected to explain the differences observed. As our knowledge concerning the underlying mechanisms is limited in most cases we cannot decide which species is the right model for man. Therefore, a rationale to use only one species in segment II studies does not exist.

When discussing whether the rat is a suitable species in reproduction toxicity studies it should not be forgotten that a huge bulk of data is available for this species recording pharmacodynamics, pharmacokinetics, and general toxicity. All this information may be helpful for the interpretation of results from reproduction toxicity studies. Therefore, the rat should generally be regarded as the species of choice if not contradicted by specific reasons.

Chahoud: The finding of a divergence in the outcome of studies on reproductive and developmental toxicity in two species is in itself not a good argument for using more than one species, especially if the routine testing is performed in a very schematic and simple way (3 doses, 10 days of treatment). Unless reasons for such a divergence are provided, this kind of risk assessment is more a gamble than a scientific enterprise. What is the argument for two and not three species? Money? And why choose the rabbit and not a non-human primate?

Sterz: With respect to alternative species for reproductive toxicology: in general toxicology we will see that the mini-pig will replace the usual non-rodent species to a certain extent. Reproduction toxicology should join in and validate this species for segment II studies.

Nau: Another reason for needing several animal species is the interspecies differences in placental structures and their changes during development. The yolk sac placenta during early organogenesis in rodents may be different from the structures present during that time in the human. How these differences affect placental transport of drugs is quite unknown.

Peters: For some drugs information is provided with respect to the passage of pharmaceutical products and/or their metabolites through the placenta. This information might come from human case studies, taking cord blood at or just after delivery. At that time the placenta is a "geriatric" organ, just at the end of its existence. The morphology and physiology of the placenta might change during pregnancy. Moreover, during early genesis, at least until the neural tube formation and primordial gutformation, a placenta does not exist. The human embryo is completely dependent at that time on the yolk sac. It is quite unlikely that data on old-placenta passage indicate enough for. e.g., the stage of histiotrophic nutrition.

Stahlmann: Perhaps someone can correct me, but I do not know of a single case in which placental passage was the crucial issue in risk assessment with respect to various species. Prof. Beck (now in Australia) had introduced the ferret as an experimental species, with the aim of providing a carnivore with a very different placenta than that of rodents (*see: Methods in Prenatal Toxicology, 1977, Thieme Verlag*), but this argument was apparently not convincing, since this species is no longer used extensively.

Merker: I would very much agree with what Dr. Peters has said. It seems not so much a matter of the type of placenta, but the problem of the very different rates of transfer to the conceptus during different stages of development. There are many examples for this changing property of the placenta during gestation in the literature, and no attempt is made to routinely obtain more information in this respect. Again this argues not for more species, but for more subtle investigations in a single species, in order to obtain a basis for a better interpretation of the data.

Neubert: A crucial question is: Have we overlooked a lot of toxic substances with respect to reproductive and developmental toxicity in man. Although we all realize that this question is difficult to be answer since specific observations in humans at risk would be required and these are largely lacking, there certainly has not been a second thalidomide disaster. The fear of the beginning 60's that we might be surrounded by many "thalidomides" is no longer justified with respect to gross-structural abnormalities. In fact, there seem to be very few "thalidomides", and we seem to be able to effectively control the release of dangerous substances in this respect with the animal studies.

I am not convinced that we can state the same for congenital dysfunctions, e.g. effects on the immune or the hormonal system, because it is much more difficult to reveal them in clinical studies, and we do not in any way take into account or control such possibilities in the studies designed up till now.

For such studies, e.g. on the immune system, rodents may not be the experimental animals of the first choice, or insufficient as the only species to be used, because we know that they do not respond to several classes of substances in a typical and sensitive way (*see: R. Neubert et al., this book*).

Scientific Basis for Risk Assessment (Reproductive Toxicity) for Medicinal Products as Practiced in Germany and the European Community (EC)

Rolf Baß and Beate Ulbrich

Introduction

It is often said that regulatory agencies keep their secrets about the interpretation of risk, and risk assessment, and we shall try to find out if this is really true.

There is not only a scientific basis which needs to be considered for risk assessment, but also and even beforehand, a legal basis. We find the following in the German Drug Act (article 5, paragraph 2):

"As unsafe shall be considered medicinal products which, in the light of scientific knowledge currently prevailing, under correct and stipulated use, are justifiably suspected of having harmful effects that exceed the bounds considered reasonable in the light of medical knowledge available."

Although giving the impression of a "rubber band stretchability" on first sight, the boundaries of risk do have a quite stable background, which are defined by the terms "...scientific..." and "...medical knowledge...". This implies that risk assessment and decisions to be taken from risk assessment can only be as (un-) precise and (un-)precisely described as allowed by science and those scientists guaranteeing or "making" the scientific state of the art. Any other reason leading to "flexibility" in this area should be avoided. The result of risk assessment should, therefore, be reliable as the scientific state of the art, recognizable, reproducible and open for understanding.

When Does Risk Assessment Take Place?

The first overall review of data and results towards achieving risk assessment by regulatory authorities occurs when the application for marketing authorization is evaluated. This process has become harmonized for the member states of the EC, whereas the situation is different during the clinical trial stages. The "second" review will occur as necessitated after marketing in the form of pharmacovigilance. At both stages the behaviour of the pharmaceutical manufacturer and of the regulatory authority have to be considered. Risk assessment in all cases will lead to decisions concerning and changing (or not) the marketing status of the medicinal product. As depicted in Figure 1, portions of this process appear quite simple, while others remain areas of uncertainty, due to overlap or multiplicity.

Do We Make Scientific Concepts for Risk Assessments Available?

Generally and often compound safety is circumscribed as "GRAS" (generally recognized as safe). This does not seem to be an appropriate statement to be made for medicinal products and their use in relation to reproduction.

Quantitative risk assessment can be attempted to describe safety:
- under the actual conditions of intended or inadvertent exposure
- considering the severity of potential toxicity (e.g. can a "critical toxic effect" be defined?).

Qualitative risk assessment can be attempted to describe safety:
- in the form of a "yes" or "no" answer
- to be dependent on safety factors.

The flow of thought in risk assessment in reproductive toxicity has been laid down in "working instructions" of the EC (Guidelines for the evaluation of reproductive toxicity CEC/V/E/3/ LUX/17/86) as spanning from collecting information about the use of a product (risk situation), over actual exposure resulting from use, assembling data available on reproductive toxicity, and evaluation of such toxicity to including this evaluation into an overall risk evaluation with the perspective of describing and defining risks and risk populations for defined uses of the medicinal product. It has to be pointed out that the ideal situation of having available all the necessary data and information occurs only very seldomly. This cannot be overcome by mathematically or otherwise unscientifically stressing and stretching the few data available.

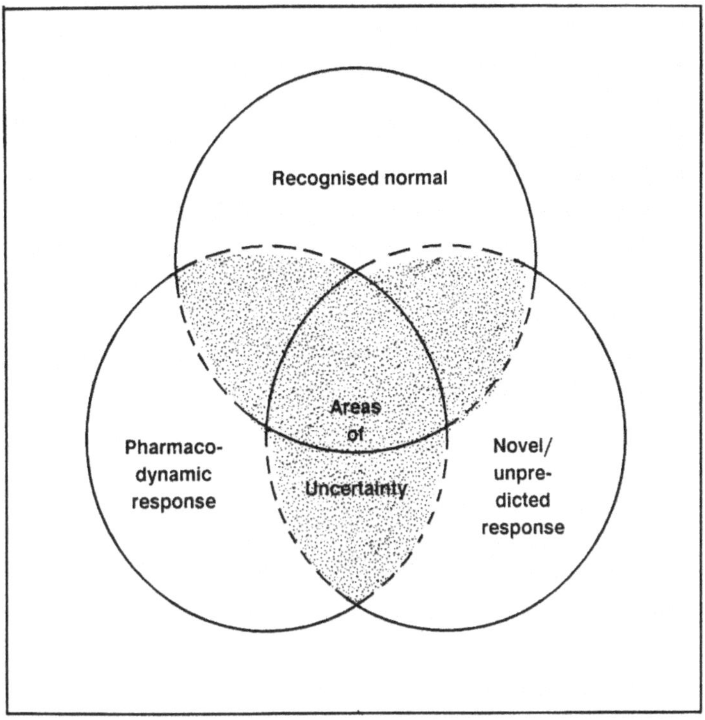

Fig. 1. Basics of risk assessment

Questions we will try to answer include:

- Do the studies give us a NOEL?
- How does this relate to human therapeutic conditions?
- Are we dealing with effects which are likely to be relevant for humans?
- How do pharmaco- (toxico-) kinetics compare for experimental systems employed and for humans?
- How can this be translated into labeling?

The concept/procedure employed for risk assessment of medicinal products before marketing is laid down in Figure 2. From this procedure we would like to highlight three items:

- If the "normal" requirements are insufficient to allow the necessary degree of risk assessment, -"special" requirements are generated and will have to be fulfilled,
- risk assessment will be incorporated into (or hidden by?) a risk benefit assessment and decision,
- labeling will be required for products to be marketed.

Which Data Are Normally Available Through "Normal" Requirements?

Data from reproductive toxicity studies are available for medicinal products containing "new" active substances:

- usually three doses (plus control)
- usually rat (plus rabbit)
- usually segments I, II, and III (regardless of protocol used)
- "usually" pharmacokinetics (single doses more often than repeated doses, later stages of development more often than earlier stages, comparison between pregnant and non-pregnant animals more often for rats than for rabbits, often data on lactation in the rat, very often very little information on rabbits)
- usually expert report.

For those products containing "known" active substances usually less information is available, which only very seldomly is counterbalanced by epidemiological studies or case reports.

Good or bad studies will be used for risk assessment, bad studies, however, will lead to more restrictive labeling.

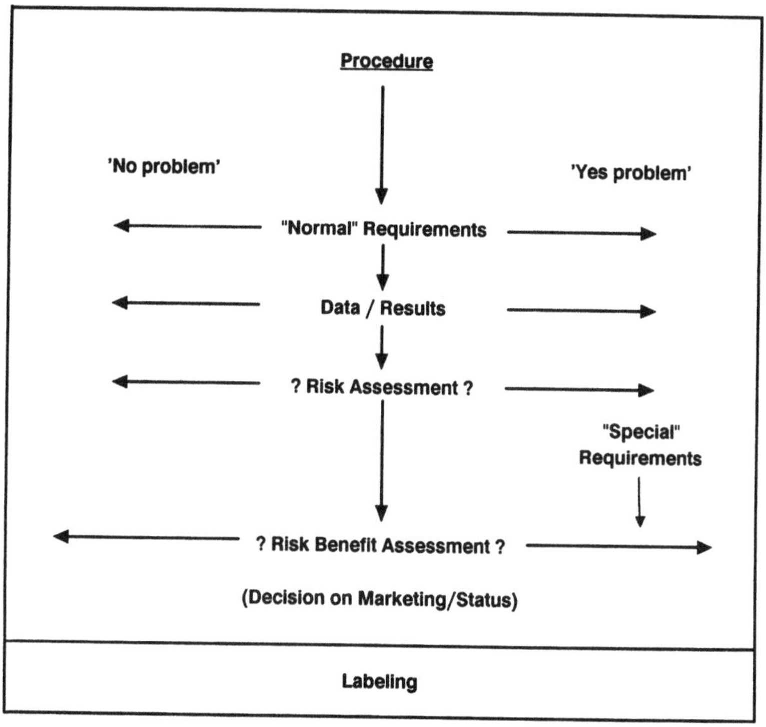

Fig. 2. Risk assessment - application for marketing

Which Results Are Normally Available From The Data?

The evaluation of the data submitted to the BGA (medicinal products) show that "toxic doses" have not been administered quite often (for possible reasons we like to refer to the discussions led here on the first day of the symposium) (cf. also Fig. 3).

Whereas until now we have discussed the situation when first marketing decisions are to be taken, we also have to take into consideration the importance of which kind of information becoming available from pharmacovigilance after a marketing authorization has been granted. For many other types of toxicity (e.g. organ toxicity) clinical experience gained during the post-marketing stage will let us deviate from the risk assessment taken from animal data before marketing. This is only very seldomly the case for reproductive toxicity. It could be the case that through handling risk assessment very restrictively, adverse reactions occur very seldomly indeed. It could also be the case that data on the use of medicinal products in humans and in relation to reproduction are not readily available, e.g. two years after marketing (experience report) or after each five year period following marketing (renewal of license). It is now well recognized that the lack of epidemiological data is the major cause for the current situation which leads to continuing counselling on the basis of "non-human" data. Counterproductively this may lead to recommendations of use of medicinal products which are more restrictive than otherwise necessary. Since a "justifiable suspicion" arises quite easily from animal experimental data (but can hardly and seldomly be ruled out from risk assessment based solely on animal experiments), lower risks (reaching into the area of speculation) must lead to more restrictive action by regulatory authorities. This implies that by preventing to make available data from drug use, e.g. in pregnant women, the scientific element in risk assessment surely decreases.

What Is The Situation For Labeling?

In the EC the information assembled through the Summary of Product Characteristics (SPC) has to be made available in the formats of "patient" and "physician" information. Whereas in Australia, in Sweden and in the USA a concept for labeling classes has been developed and enacted, the EC is lagging behind. We are still discussing the usefulness of labeling classes, the hierarchy of information to be given, the degree of detail of such information, and whether or not recommendations should be given which can be used for decisions to be taken in a single case. We refer to the draft Note for Guidance made available for discussion (1990). In this draft proposal both risk and safety are to be described in a general manner, but precisely enough to allow interpretation of this "general risk assessment" for the "risk benefit decision" to be taken for the single patient.

Toxic dose (number of studies)	no	yes
Segment 1		
Rat	174	242
Mouse	11	7
Segment 2		
Rat	214	286
Mouse	64	38
Rabbit	180	264
Monkey	5	13
Segment 3		
Rat	128	169

Fig. 3. Results from reprotoxicity studies

Conclusions

The applicability and application of risk assessment tools in the area of reproductive toxicity is defined and is in line with the procedures to be taken for other areas of toxicity. Since even after long periods of marketing of medicinal products data on human experience remain scarce, risk assessment continuously depends on data obtained in animal experiments. This is counterproductive to the level of science which all of us would like to introduce into risk assessment. Labeling of medicinal products for use reflects this uneasiness.

Discussion of the Presentation

Neubert: Would a risk assessment from your office be different if pregnant women were exposed (a) unintentionally, or (b) medicinal products were given for treatment of maternal diseases (e.g. antimimetics, antiepileptics, antibiotics etc.), or (c) pregnancy as such was treated (e.g. tocolytica)? Or are only the risk-benefit considerations thought of here?

Baß: There is a difference between "unintended" and "intended" use, be it for maternal or pregnancy reasons. Where "unintended" use is expected, contraceptive measures will be recommended.

Neubert: What decision is made if, in a study to assess developmental toxicity, no maternally toxic dose is reached? Do you reject such a study?

Baß: Case-by-case either new studies, including pharmacokinetic studies, will be requested, or the data submitted are considered sufficient. In both cases, this is not a formal rejection of the study, but rather making the best use of it.

Neubert: Why did you state that the information on human data does not increase steadily after marketing?

Baß: After marketing ADR-reports can be expected. Data on the normal outcome of pregnancy are not usually reported to the pharmaceutical manufacturer or the regulatory authorities. At the BGA, we, therefore, request "experience reports" two and five years after marketing for those medicinal products which are intended for use in pregnancy.

Neubert: The problem of labelling and classification for possible developmental toxicity is in my opinion largely unsolved for a number of reasons. These include: (a) various ways of classification in different countries or even by different agencies or the manufacturer itself. (b) Little help for the physician except for information on a strong embryotoxic agent. (c) The vast majority of all medicinal products would have to be characterized as: "insufficient information in man and experimental animals". Several years ago, Ralf Stahlmann and I attempted to categorize all available mono-preparations in Germany (Red List) for developmental toxic effects. We soon realized that labelling using only a few categories is largely worthless. Therefore, we decided to (a) include information on group characteristics, and (b) to make footnotes elaborating on the possible specific hazards and problems. Would you like to comment on these problems?

Baß: We agree with your proposal and with what you are aiming at. Realization, however, must be achieved throughout the EC.

Neubert: Do you feel that additional animal studies are required during the stage of pharmacovigilance? Under which conditions? What kind of studies? To obtain what information?

Baß: If human use of a medicinal product leads to untoward effects which are "unexpected" from existing data on animal experiments, additional animal studies have very seldomly been requested for clarification.

Krowke: Are data on concentrations of chemicals in milk or data on the amount of transfer during lactation required or desired?

Weissinger: Lactation studies, with respect to drug or metabolite levels in milk and maternal plasma, may be conducted in humans. Information on levels to which neonates are exposed can be obtained by this mechanism. If rat milk levels and dam plasma levels are known and related to an effect on neonates, limited predictions to similarity of effects in human neonates may be considered with respect to dose.

Manson: What is the basis for studying transfer to the pup via milk in the rat?

Sterz: I think that scientists studying transfer of a drug into the milk of the rat do not have risk assessment for man in mind, but they would like to know whether the drug reaches the neonate and pup; so we simply add an interesting information to other data, important for evaluating a study's results and interpreting our findings.

Risk Assessment in Reproductive Toxicology as Practiced in South America

Francisco J.R. Paumgartten, Eduardo E. Castilla, Roque Monteleone Neto; Helena L.L. Coelho and Sarah H. Costa

Introduction

Comprehensive assessments of safety and efficacy of medicinal products and risk assessment of pesticides, food additives, industrial chemicals, and others, have rarely been, if ever, undertaken in South America. Moreover, when risk estimates are used for risk management in South America, those estimates are drawn, in most instances, from studies conducted in industrialized countries which do not necessarily take relevant characteristics of the particular exposure scenarios into account.

This situation is obviously far from satisfactory and expertise in health risk assessment is very much needed for guiding decision-making on food, drug and environmental chemical control on the South American continent. Bearing this fact in mind, considerable efforts have been made in recent years to develop expertise in this area at the National Institute of Health Quality Control, Oswaldo Cruz Foundation in Rio de Janeiro.

Data from animal experimentation and data from clinical and epidemiological studies form the scientific basis for qualitative risk assessment in reproductive toxicology, as well in any other area of toxicology. Up till now, since animal studies are still in the process of being established, hazard identification in reproductive toxicology in South America has relied mainly on clinical and epidemiological data. Therefore, our focus will be on this latter approach.

The Latin American Collaborative Study of Congenital Malformations

Background and Aims

The genetic and environmental causation of malformations and the epidemiology of congenital anomalies in Latin America have been studied by the Latin American Collaborative Study of Congenital Malformations (ECLAMC : Estudo Colaborativo Latino Americano de Malformacoes Congenitas). The director of ECLAMC is Prof. Dr. Eduardo E. Castilla, a human geneticist working at the Department of Genetics of Oswaldo Cruz Foundation, the institution where the data bank of the collaborative study is located.

ECLAMC started in 1967 as a research project in the Ministry of Public Health of Argentina. It rapidly grew up to its present status of a non-governmental, non-institutional, research project for the study of congenital anomalies, based on the voluntary collaboration of medical investigators from the area of perinatology. It now includes almost one hundred maternity hospitals from 12 Latin American countries: all ten South American countries plus Costa Rica and the Dominican Republic.

ECLAMC is essentially a research program in the mechanisms of dysmorphology, in the study of genetic and environmental causation of malformations, and in the epidemiology of congenital anomalies in Latin American populations. The clinical-epidemiological approach is used in an attempt to preserve the clinical quality of data within the populational scope of epidemiology.

Activities

The prevention of the congenital anomalies is sought through: Research, monitoring for new teratogens, technical assistance to governments, and education within the health system first, and to the general public later.

Research

The running of the ECLAMC as a research program provided several advantages that favoured its long-term survival. One of them was the possibility of obtaining funds from research funding institutions, usually not so much tied up to present priorities but rather to future situations. Another one was the lack of institutional links, allowing for the international handling of data on the individual basis, thus overcoming bureaucracy, national boundaries and political instability.

Monitoring

Since 1967 the ECLAMC has registered 60,000 malformed infants diagnosed in two and a half million births, systematically examined in consecutive series. Two different operating modes were run in parallel since 1982. The most recent mode, aimed at monitoring, recording fully described congenital anomalies, plus just a few variables needed for the interpretation of the recorded frequencies. Since these variables, usually found in the delivery room log-books, are recorded for all births a cohort type of analysis can be applied to them. These variables are: Hospital of Birth, Birthdate, Sex, Birth weight, Twinning, Maternal age, Gravidity, Survival, and Autopsy. The older operational mode is a case-control method which entirely includes the newer one and, in addition, registers a more sophisticated information about genetic background and prenatal environmental occurrences distributed in 50 risk factors.

The methodology used since 1982 by the ECLAMC for monitoring congenital anomalies can be outlined as follows:

1. Periodicity: Quarterly and annually. Quarterly analysis is performed 90 days after the occurrence of the latest birth while annual monitoring is performed on June 30th of the following year, including definitive and more complete data.

2. Diagnoses: 56 selected diagnoses are monitored quarterly, and all diagnoses (about 300) annually. Furthermore, since the ECLAMC is mainly run by clinical dysmorphologists, it relies more on the clinical approach of making correct diagnoses of real affected infants than on the handling of codes in a more or less sophisticated statistical context. Its approach is a clinical-epidemiological one.

3. Multiply Malformed: Special attention is given to the multiple malformed infants due to the fact that most known teratogens produce poly-malformed patterns (Castilla et al. 1985). When properly selected, true multiply malformed cases are rare. Just about 100 per quarter in the ECLAMC, a program with twice the annual birth population of Sweden. This limited number of cases to be monitored allows for a clinical approach, where each case is seen and discussed as an "interesting case". In doing this, there are good chances to identify the birth, within a given quarter, of two or more infants affected by a similar pattern of associated congenital anomalies, which is the essential aim of the art of monitoring.

4. Statistical Testing: The statistical significance of the observed versus expected value differences are analyzed by the method of Poisson confidence limits. This method was chosen because ECLAMC, being hospital-based (not population-based), has a variable denominator. Critical values of 0.05 are used just for suspicion and 0.01 for defining alarms. Expected birth prevalence values are derived from observations previously made during a baseline period. Special care is taken to match the populations as closely as possible.

5. Routine Treatment of Alarms: When there is an alarm for a given congenital anomaly type in a given quarter, all cases with that diagnosis are reviewed by a senior researcher with experience in the clinical and epidemiological aspects of congenital anomalies. Similarities among cases are analysed with respect to place of birth, sex, survival, birth weight, twinning, gravidity, maternal age, and descriptive details of the congenital anomalies. Isolated and associated cases are separated, and the patterns of association of the latter sub-group is reviewed. Further sub-divisions of the involved congenital anomaly type are attempted; i.e. extension and shape of the cleft of the palate, completeness of the extra digit in post-axial polydactyly, level and type of spina bifida. The occurrence of that congenital anomaly type during the next quarters is closely watched. The participating hospitals are notified of a given alarm only when it goes off for two consecutive quarters, which constitutes a very rare event.

Table 1 shows, as an example, the quarterly monitoring of birth defects by ECLAMC during 1987 and 1988. Fourteen out of the fifty-five diagnoses analysed are represented, and the previous four-year period was taken as baseline. Among the malformation types represented, an alarm was observed for hydrocephaly in the third quarter of 1988, and another for cleft lip in the first quarter of 1987. Both were probably due to random variations since in neither case an epidemic could be proven after follow-up.

6. *International Collaboration:* ECLAMC is a member of the International Clearinghouse for Birth Defects Monitoring Systems (International Clearinghouse for Birth Defects Monitoring Systems 1991). Quarterly and yearly exchange of data with other similar programs throughout the world enhances the sensitivity in detecting new teratogens. Alarms are reported to the Clearinghouse for comparison with other parts of the world. If the alarm is strongly suspected to be a real epidemic, the Clearinghouse is then called to intervene, following the steps of the so-called Situation-1 (Castilla et al. 1986).

Searching for Clusters

While the above mentioned monitoring of teratogens and mutagens is aimed at the detection of unusual changes in time, epidemics, another approach is used to keep track of unusual changes in space, endemics, or even time-space changes undetected by classical monitoring because of one or more of the following circumstances. 1. The unusual frequency may be limited to a geographic area uncovered by the monitoring system. 2. The unusual frequency may involve a congenital anomaly type undiagnosed during the perinatal period, as is the case for instance, for rubella embryopathy. 3. The unusual frequency may be longstanding, their high values also being included in the expected frequencies, and, therefore, unable to give an alarm for similar observed rates. This would be the case of most genetic isolates. 4. The unusual frequency may affect a very rare congenital anomaly type and the sample size of the monitoring system may be too small to be able to detect the change.

Table 1. Quarterly monitoring of birth defects by ECLAMC - malformation type (frequency/ 10 000 births)

	1982-1986 (baseline)	1987-1	1987-2	1987-3	1987-4	1988-1	1988-2	1988-3	1988-4
Births examined	869 750	41 857	44 156	46 508	42 214	45 533	45 397	45 608	40 106
Omphalocele	2.2	2.6	1.6	0.7	2.4	2.2	2.6	2.6	1.2
Anencephaly	6.7	7.4	6.1	5.8	9.2	6.2	6.0	6.4	5.1
Spina bifida	6.7	7.2	8.4	8.2	6.6	5.7	6.4	6.6	7.3
Hydrocephaly	4.0	4.5	4.1	4.7	4.3	4.6	6.2	9.0*	5.4
Cephalocele	1.7	1.4	0.9	1.3	1.2	1.1	2.9	2.4	2.7
Na/Microtia	3.7	4.3	3.4	3.7	2.8	4.6	4.6	5.1	4.9
Cleft palate	3.2	3.1	3.4	1.9	1.9	1.5	33	2.4	2.7
Cleft lip	9.2	15.3*	11.3	10.5	10.7	12.3	11	11.6	8.3
Esophageal atresia	2.6	3.8	2.5	1.9	3.1	3.1	35	2.8	3.2
Imperforated anus	3.7	2.9	3.4	3.0	2.1	0.9*	4.0	4.7	6.4
Hypospadias	9.4	6.7	8.2	8.4	5.0	7.0	8.4	5.8	11.4
Reduction limb defect	5.4	4.1	4.8	5.4	6.4	4.2	37	4.3	5.1
Hip subdislocation	25.3	18.4	23.1	20.2	20.4	21.5	26.7	18.6	24.4
Down's syndrome	15.7	12.2	18.8	15.9	17.1	13.6	15.9	18.0	14.7

* = Alarm

Adapted from Castilla and Lopez-Camelo, 1990

The main activity of ECLAMC within this context is called "Rumour", which is a routine for the systematic evaluation of suspected endemics, epidemics, or exposed sub-populations for congenital anomalies and their risk factors. This routine is quite similar to the one operating in the State of California (U.S.A.) by the California Birth Defects Monitoring Program, CBDMP (Grether et al. 1988).

The guidelines used by ECLAMC in the RUMOUR routine, include:

1) *Problem definition:* Consisting in a simple, but objective, statement of the problem. Many inquiries stop here because the Rumour has insufficient grounds to allow for its clear statement.

2) *Index case definition:* A list of cases is requested, stating: Identification (Name or Number), Date of birth, Place of birth, Sex, Survival, and Description of the congenital anomaly, as detailed as possible.

3) *Data analysis:* Expected frequency figures are derived from the ECLAMC files most of the time. Observed/expected numbers of case differences are usually tested by confidence limits for the Poisson distribution because the rule is to deal with small numbers. Since at this stage we are looking for strong suspicions rather than proofs, 90% limits are enough to continue the evaluation further. At this stage a conclusion is made about the Rumour having some solid ground or not.

4) *Further steps:* The large majority of rumours do not get beyond this first step of Rumour delineation. From here on we will no longer be dealing with a Rumour but with a strong working hypothesis, and the methodology will be the usual ones used for the study of clusters.

Future

Most South American countries are approaching the magic threshold of an infant mortality rate of 20/1,000 and some of them, as in Uruguay for instance, are already stepping on it (Castilla et al. 1991). From that point on, the congenital anomalies will occupy the first or second post among the causes of infant death. They will then become of public health interest and the ECLAMC will have to change its objectives or call off its activities.

Environmental Pollution and Congenital Anomalies: The Cubatao Study Example

The Cubatao study is a good example of the epidemiological approach used in Brazil to clear up a suspicion of an environmentally-induced increase in birth defect rates.

Located at the narrow seaboard area lying between the Atlantic ocean and the "Serra do Mar" ("Sea Mountain Range"), on the South Eastern coast of Brazil, Cubatao is a town with 80,000 inhabitants and a high concentration of chemical industries. Such a high concentration of industries which give off a large amount of pollutants, in a place where climatic conditions prevent dispersal, made the air in Cubatao the most polluted in the country, and perhaps in the world (Queiroz-Neto et al. 1982).

A cluster of anencephaly cases was reported in Cubatao causing general alarm in the early 80's. Such a report on an unusual occurrence of a birth defect, and concern about its environmental causation, motivated a clinical and epidemiological study aimed at investigating the actual prevalence of congenital anomalies in Cubatao. This study was carried out under the supervision of Prof. Monteleone Neto, from the Department of Genetics of Paulista School of Medicine in Sao Paulo (Monteleone Neto et al. 1985).

The ECLAMC methodology was used to start a systematic congenital anomaly monitoring in the three maternity hospitals of Cubatao. From June 1982 to December 1985 10,378 births (10,218 live and 160 stillbirths, weighing more than 500 g) were examined and the anomalies detected within the first 48 hours of life.

Among all the live births examined, 150 infants with one or more congenital anomalies were detected, and nine among the stillbirths presented malformations. The frequency of stillbirths (1.54%), as well as of malformed infants (1.53%), was not any higher than those observed in the country as a whole or in South America. As can be seen in Table 2 the frequencies of anencephaly cases and of other malformation types in Cubatao were not any higher than those recorded by the ECLAMC for South America (Castilla and Lopez-Camelo 1990), or in other parts of the world such as the USA, Italy and North Ireland using different monitoring systems (ICBDMS 1987).

In the Cubatao case-control study neural tube defects were found to be positively associated with consanguinity and also with acute maternal illnesses - mainly influenza - during the first trimester of pregnancy, but were not found to be associated with exposure to chemical and physical agents (Monteleone Neto 1986).

Table 2. Cubatao study: frequencies of 10 congenital anomalies as compared to the rates in South America, USA, Italy and North Ireland (frequency/10 000 births)

Study	Cubatao*	ECLAMC**	USA*** (Atlanta)	Italy***	North Ireland***
Period	June 1982- Dec. 1985	1982- 1986	1981	1981	1981
Births monitored	10 378	869 750	---	---	---
Anencephaly	6.75	6.2	3.0	3.7	17.4
Spina bifida	8.67	6.2	5.6	7.5	20.3
Hydrocephaly	2.89	3.9	9.7	3.1	4.7
Cleft palate	4.82	3.2	2.6	4.9	9.4
Cleft lip	9.64	9.6	10.4	8.4	10.9
Esophageal atresia	2.94	2.4	4.5	3.1	4.7
Imperforated anus	2.94	3.7	3.7	4.9	8.0
Hypospadias	7.71	8.3	25.0	---	8.4
Limb reduction	4.82	5.4	4.1	---	7.6
Down's syndrome	10.60	15.7	9.5	13.9	17.2

* Monteleone-Neto 1986
** Castilla and Lopez-Camelo 1990
*** ICBDMS 1983

Therefore, suspicion of environmental (air) pollution-induced increase in the frequency of birth defects did not find support in the results of the Monteleone Neto study in Cubatao. Nevertheless, it should be stressed that in the Cubatao study, as well as in the ECLAMC monitoring system, only congenital anomalies detectable at birth were covered. Other birth defects, including mental retardation and other adverse pregnancy outcomes such as miscarriages, were not studied yet in this area of Cubatao.

Misuse of Misoprostol for Abortion Induction in Brazil

In Brazil, a predominantly Catholic country, abortion is illegal except in cases of rape or when pregnancy implies risk of life for the mother. Since contraceptive methods, sexual education, and family planning are not available for most of the population, there is a high incidence of unwanted pregnancies. Moreover, since there is no strict control on pharmaceutical sales, prescription-only drugs are very often freely available over the counter in Brazil, as well as in most of the South American countries. Within this context, the massive and deliberate misuse of drugs with abortifacient properties is likely to occur.

Misoprostol (Cytotec, Searle) is a synthetic prostaglandin E 1 methyl ester analogue ((+/-) methyl 11 alpha, 16-dihydroxy-16-methyl-9-oxoprost-13E-en-1-oate) with a dose-related gastric antisecretory activity, and possibly also with a cytoprotective effect on gastric mucosa (Lewis 1985). Misoprostol is particularly indicated for preventing gastric ulcers caused by nonsteroidal anti-inflammatory drugs (NSAIDs) in patients with a high risk for complications (PDR 1990). The therapeutic dose recommended for adults is one tablet (200 μg) four times daily.

Abortifacient properties of prostaglandin analogues are well known (MacKenzie et al. 1978). Misoprostol induces uterine contractions, uterine bleeding and may cause expulsion of the products of conception. Therefore, this drug is contraindicated in women who are pregnant and in women of childbearing age if pregnancy cannot be excluded.

The misuse of misoprostol for abortion induction was first detected in Brazil by a drug consumption study carried out in Fortaleza, capital of Ceara state in the North East of the country (Coelho et al. 1991). A retrospective survey (from January 1986 to December 1990) on the records kept at the Maternity Hospital Assis Chateubriand, in Fortaleza, Ceara state, indicated that misuse of Cytotec started in 1988 with 20 cases, that is 12% of all cases of admissions because of complicated induced-abortions. In 1990 the number of abortions induced by Cytotec went up to 525, that is 73% of all cases of recorded induced abortions. It was also found that since misoprostol was introduced into the market in 1987, the number of cases of abortions induced by illegal helpers (traditional midwives) has decreased, whereas the total number of complicated induced-abortion cases attended to at the hospital increased seven times from 1986 to 1990. Doses used for abortion induction go up to 46 tablets (9,200 μg), but in most of cases approximately four tablets (800 μg) are taken as a single dose, by oral, intravaginal, or both routes. Misoprostol is usually taken during the first trimester of pregnancy but sometimes abortion induction with the drug is attempted even later.

A prospective study in progress in Rio de Janeiro has also provided data on the massive misuse of misoprostol for abortion induction. The research project, carried out under the supervision of Dr. Sarah H. Costa from FIOCRUZ, aims at investigating the determinants and health consequences of induced abortions. Two groups of women were studied: women who were admitted to hospital with diagnosis of abortion complications and women who were admitted to hospital for delivery. Seven public hospitals in the Municipal area of Rio de Janeiro were chosen for studying, and an effort was made to select one service in each administrative zone. Preliminary results of 470 cases of induced abortions, collected between April and September of 1991, indicated that misoprostol was used as an abortifacient method by 63.4% of the cases (Table 3). The total quantity of tablets (200 μg) used ranged from 1 to 64, but the mode was 4 tablets. Misoprostol was administered orally (64.5%), vaginally (4.4%), or by both routes (30.8%), and 64% of the women reported to have used the drug as a single dose. Furthermore, the majority of women who were admitted with abortion complications had taken misoprostol during the first trimester of pregnancy.

Table 3. Frequency distribution of abortifacient methods used by women admitted to hospital with diagnosis of abortion complications. Percentage of 470 cases collected in Rio de Janeiro between April and September 1991 (Costa 1991)

Cytotec only	51.1%
Cytotec plus other methods	12.3%
Oral medications not identified by name	12.1%
Herbal teas	8.9%
Intramuscular injections	5.5%
Abortion clinics (curettage and suction)	2.3%
Catheter insertion	1.9%
Violent exercises and pressure on abdomen	1.7%
Potassium permanganate tablets intravaginally	1.0%
Introduction of foreign objects	0.8%
Other methods	2.4%

An interesting finding from the study was the large proportion of delivery patients who reported to have attempted to abort with Cytotec (7.1%) (Table 4). An additional 3.8% of the women reported to have used oral medications which they were not able to identify by name. Since pharmacies are reported to be selling four tablets of Cytotec without packaging it is quite possible that many of these women had in fact used the drug without being aware of it, possibly increasing the exposure rate among delivery patients to about 10%. Ninety percent of the patients had attempted to abort with misoprostol during the first trimester of pregnancy.

Since abortion induction with misoprostol fails and the pregnancy continues to term, there has been increasing concern about possible adverse effects produced by prenatal exposure to such high doses of this PGE1 analogue.

Five cases of malformed children born to mothers who had taken misoprostol (400 to 600 μg by oral and/or intravaginal route) during the first trimester of pregnancy were reported in Fortaleza. According to the authors the malformation is a rare anomaly of the skull, consisting of a "localized frontal/and temporal defect with asymmetrical, well-circumscribed defect of the cranium and overlying scalp, exposing dura mater through which the cerebrum can be seen" (Fonseca et al. 1991). Only one of the reported cases, however, has been illustrated by the authors.

Table 4. Use of misoprostol by women admitted to hospital for delivering. Data collected in seven public hospitals of Rio de Janeiro from April to September 1991 (Costa 1991)

	N	%
All delivery patients interviewed	420	100
Patients who thought of terminating pregnancy	164	39
Patients who actually attempted to abort	82	19.6
Patients who took Cytotec to induce abortion	30	7.1
Patients who took oral medications which they were not able to identify by name	16	3.8
Patients who made use of other abortifacient methods	36	8.6

Adverse health effects which may result from prenatal exposure to high doses of misoprostol are cause for deep concern, but the teratogenic potential of this PGE1 analogue is still far from being understood. The congenital anomaly described by Fonseca et al. (1991) is consonant more with a disruptive fetal lesion either vascular or mechanical induced late in pregnancy, than with an embryopathy induced in the first trimester. Since, according to the authors, at least three out of the five mothers of the malformed infants denied any attempt to terminate pregnancy after the first trimester, a causal relationship between the use of misoprostol and the reported malformation seems to be unlikely. Furthermore, although misoprostol is massively misused all over the country, as well as in other developing countries, and the malformation described is rather conspicuous, no other cases outside Fortaleza have been reported up till now, even when malformation registers had been alerted world-wide on this suspicion through the network of the ECLAMC and the Clearinghouse as early as January 1991 (E. Castilla, personal communication).

Data from animal studies suggest that misoprostol has no effect on fertility and embryo development at an anti-ulcer dosage. In a preclinical three-segment study, no evidence of embryofetotoxic effects was found in rats at oral doses up to 10,000 μg misoprostol/kg body wt, and in rabbits at oral doses up to 1,000 μg misoprostol/kg body wt (Kotsonis et al. 1985).

The human pregnant uterus, however, seems to be much more sensitive to this PGE1 analogue than the rat or the rabbit pregnant uterus. In a double-blind placebo-controlled study carried out in Germany in cases of legally permitted first trimester abortions, misoprostol induced vaginal bleeding in 45% and complete abortions in 11% of the patients treated with two doses of 400 μg. In the same study half of this dose (2 x 200 μg) induced vaginal bleeding in 34% of the cases, and complete abortions in 9% of the patients (Rabe et al. 1987). Data on the misuse of misoprostol in Fortaleza and Rio de Janeiro revealed that the majority of the women who sought emergency care for vaginal bleeding had taken from 200 to 800 μg of misoprostol (Coelho et al. 1991; Costa 1991). Therefore, doses which are within, or at least are not very far from, the anti-ulcer therapeutic range seem to be effective in inducing vaginal bleeding leading to hospital admission and subsequent curettage in a significant proportion of the cases. Unfortunately, there are no data available on how effective misoprostol is in inducing complete abortion under present conditions of misuse in Brazil.

Large scale misuse of misoprostol by pregnant women in Brazil, who deliberately take high and maternally toxic doses of the drug, clearly illustrates the fact that exposure conditions in South America are very different from those prevailing in developed countries and may require additional studies for risk assessment.

Acknowledgements: The ECLAMC study was supported by the Consejo Nacional de Investigaciones Cientificas y Tecnicas (CONICET) of Argenttina, the Conselho Nacional de Desenvolvimento Cientifico e Tecnologico (CNPq) of Brazil, the FAPERJ of Rio de Janeiro, and the Pan American Health Organization (PAHO). The research project on the determinants and health consequences of induced abortion in Rio de Janeiro is being supported by PAHO. FJRP is recipient of a research fellowship from CNPq.

References

Castilla EE, Czeizel A, Kallen B, Mastroiacovo P, Oakley GP Jr, Takeshita K, De Wals P, Kuliev A (1986) Methodology for birth defects monitoring. Birth Defects Original Article Series 22 (5): 1-43

Castilla EE, Lopez-Camelo JS (1990) The surveillance of birth defects in South America. I. The search for time clusters: Epidemics. Adv Mutagen Res, Vol. II: 191-210

Castilla EE, Lopez-Camelo JS, Dutra GP, Paz JE (1991) Birth defects monitoring in underdeveloped countries. An example from Uruguay. Internat J Risk Safety Med 19: 1-19

Castilla EE, Orioli IM, Lopez-Camelo JS (1985) On monitoring the multiply malformed infant. I. Case-finding, case-recording, and data handling in a Latin American program. Am J Med Genet 22: 717-726

Coelho HLL, Vale RMG, Gonzaga SLP, Lopes MH, Araujo VM (1991) Uso do Misoprostol para Interrupcao Voluntaria da Gravidez em Fortaleza-CE: Estudo Retrospectivo. Proc VI Ann Meet Brazilian Fed Soc Exp Biol, pp 414

Costa SH (1991) The determinants and consequences of induced abortion in Rio de Janeiro, Brazil. Unpublished project report for Pan American Health Organization (PAHO)

Fonseca W, Alencar AJC, Mota FSB, Coelho HLL (1991) Misoprostol and congenital malformations. Lancet 338: 56

Grether JK, Harris JA, Hexter AC, Jackson RJ (1988) Investigating Clusters of Birth Defects: Guidelines for a Systematic Approach. California Birth Defect Monitoring Program. Berkeley, California

ICBDMS, International Clearinghouse for Birth Defects Monitoring Systems (1983) Annual Report, 1981. Swedish National Board of Health and Welfare

ICBDMS, International Clearinghouse for Birth Defects Monitoring Systems (1987) Annual Report - 1987. Bergen

ICBDMS, International Clearinghouse for Birth Defects Monitoring Systems (1991) Congenital Malformations World-wide. Elsevier, Amsterdam

Kotsonis FN, Dood DC, Regnier B, Kohn FE (1985) Preclinical toxicology profile of misoprostol. Digest Dis Sci 30 (11) suppl: 142S-146S

Lewis JH (1985) Summary of the 29th Meeting of the Gastrointestinal Drugs Advisory Committee, FDA, June 10, 1985. Am J Gastroenterol 80: 743-745

MacKenzie IZ, Davies AJ, Embrey MP, Guillebaud J (1978) Very early abortion by prostaglandins. Lancet 1: 1223

Monteleone Neto R (1986) As Anomalias Congenitas e as Perdas Gestacionais Intermediarias e Tardias no Municipio de Cubatao. Thesis submitted to the University of Sao Paulo. Ribeirao Preto SP Brasil

Monteleone Neto R, Brunoni D, Laurenti R, Mello Jorge MH, Gotlieb SLD, Lebrao ML (1985) Birth defects and environmental pollution: The Cubatao example. In: Marois M (ed) Prevention of Physical and Mental Congenital Defects. Part B: Epidemiology, Early Detection and Therapy, and Environmental Factors, Alan R. Liss Inc, pp 65-68

PDR, Physician's Desk Reference (1990) Product Information: Cytotec (misoprostol). pp 2056-2057

Queiroz-Neto JP, Monteleone Neto R, Marques RA (1982) A Situacao em Cubatao: Documento Sintese do Grupo de Trabalho sobre Cubatao da SBPC. Ciencia e Cultura 35: 1164-1175

Rabe T, Basse H, Thuro H, Kiesel L, Runnebaum B (1987) Wirkung des PGE1-Methylanalogons Misoprostol auf den schwangeren Uterus im ersten Trimester. Geburtsh Frauenheilk 47: 324-331

Discussion of the Presentation

Neubert: You made several very interesting contributions. I would like to ask two question: Firstly, in the case of misoprostol the possibility of misuse was discussed by Rolf Baß and myself several times after the drug was marketed in Germany for the treatment of ulcers, but apparently the chance of misusing this substance is small in our country. You have been cautious with respect to the skull defects reported in the case reports and causal relationships. They certainly appear rather atypical, and I would not exclude an artifact. Experimental studies to reveal whether gross-structural abnormalities may be triggered in surviving embryos may be interesting or even important, but it may be more desirable to prevent the misuse of the substance in your country.

Paumgartten: I have extensively discussed the skull defect reported by Fonseca et al. with Dr. Eduardo Castilla from the ECLAMC and we are very sceptical about any causal relationship with misuse of misoprostol.

First of all, the skull defect is rather more a disruptive fetal lesion which occurs late in pregnancy, e.g. a lesion due to an amniotic adhesion, than an embryopathy produced in the first trimester. Since, in this case, exposure to misoprostol was reported to have occurred during the first three months of pregnancy a causal relationship is very unlikely. Secondly, the authors have mentioned five similar cases in their letter to Lancet, but up till now they have presented a picture of only one of those cases. This is a crucial point: the defect is very rare and the probability of two or three fetuses with disruptive lesions exactly at the same place is extremely low. We are looking forward to seeing pictures of the other four cases. Until then we take into account the existence of just one case. Thirdly, except for this single case in Fortaleza no other similar case has been detected by the ECLAMC monitoring system up till now. Since misuse of misoprostol occurs all over the country, and in most of the South American countries covered by this monitoring system, other cases of such conspicuous and rare anomaly should have been observed outside Fortaleza.

On the other hand, misoprostol induces contractions of the pregnant uterus and vaginal bleeding, and is massively used by pregnant women in our county. Sarah Costa has shown that exposure rate among delivery patients in Rio de Janeiro may be as high as 10 percent. Within this context adverse health effects which may result from prenatal exposure to this PGE1 analog are cause for deep concern. However, the use of misoprostol for inducing abortion, independent of its teratogenic potential, is a misuse and as such must be controlled.

Stahlmann: The ECLAMC system is interesting and the large number of births monitored is very impressive. I wonder whether or not with the retrospective studies performed on the malformed children considerable questioning bias may exist. We strongly feel that anamnestic information should be available prospectively. In Berlin we have, with the aid of the BGA, initiated a system in which we obtain an extended anamnesis before birth and monitor adverse pregnancy outcomes rather carefully. However, the number of mother/child-pairs recruited in this way is rather small (1,000 to 3,000 per year) and unfortunately, it may not be possible to finance such a system in Germany longer than 1992.

Paumgartten: Your concerns are justified, but the alternative we have is either a case-control study or nothing, since the cohort approach is very expensive and not feasible for studying a large number of births in developing countries. It is important to mention that the bias depends very much on the variable that is being studied, for instance it is greater for drugs and acute illnesses than for chronic diseases. Moreover, the bias can be estimated and taken into account when a conclusion is to be drawn. One possibility in this way is to make use, in parallel, of a "sick control", that is an infant with another malformation type.

Goujard: When a great number of variables is compared with several groups of defects, the risk of chance correlations is high without a definite hypothesis. What do you do in the ECLAM program when a significant result is observed in your case-control studies?

Paumgartten: We use a multivariate analysis whenever this is feasible. Needless to say that the possibility of chance correlations has been taken into account in the ECLAMC so that we have been very cautious in drawing conclusions.

Reproductive Hazard Evaluation and Risk Assessment under California's Proposition 65

Gerald F. Chernoff and Steven A. Book

Introduction

California's Proposition 65, 'The Safe Drinking Water and Toxic Enforcement Act of 1986' (California Health and Safety Code, Section 25249.5 et seq.), ushers in a new approach to the management of developmental and/or reproductive toxicants (DARTs) and carcinogens (Kizer et al. 1988). This paper, which only addresses DARTs, will begin with a discussion on the background and implementation of Proposition 65, then provide a brief account of the methodologies employed in evaluating chemicals for reproductive and developmental toxicity, and end with a discussion on the impact implementation of the proposition has had within and outside the State of California.

Background and Implementation of Proposition 65

Proposition 65 became law in California by what is known as the initiative process. Reacting to what was perceived as a failure on the part of state agencies to provide adequate health protection against hazardous chemicals, and a frustration with the legislative process, the citizens petitioned for the placement of the proposition on the general election ballot. In November of 1986, the voters approved Proposition 65, and the law went into effect on January 1, 1987.

Although relatively brief, Proposition 65 contains several major and key provisions (Kizer et al. 1988; Totten 1989):

1. *The proposition sets forth a procedure for recognizing and listing chemicals as Developmental and/or Reproductive Toxicants (DARTs).* Specifically, the Governor is charged with publishing at least once each year a list of those

chemicals known to the state to cause developmental or reproductive toxicity. A chemical will be known to the state to be a DART if: (a) in the opinion of the state's qualified experts it has been clearly shown through scientifically valid testing according to generally accepted principles to cause developmental or reproductive toxicity; (b) a body considered to be authoritative by such experts has formally identified it as causing developmental or reproductive toxicity; (c) an agency of the state or federal government has formally required it to be labeled or identified as causing developmental or reproductive toxicity (California Code of Regulations, Title 22, Sections 12306 and 12902).

2. *The proposition prohibits the discharge of DARTs into sources of drinking water.* This provision includes the discharge of any significant amount of a listed DART into water or land from which the chemical may pass, or probably will pass into a source of drinking water.

3. *The proposition requires a clear and reasonable warning be given prior to exposing an individual to a listed DART unless there is a 1000-fold margin of safety.* This is the so-called "right to know" provision, in which no person in the course of doing business shall knowingly and intentionally expose any individual to a listed chemical without first giving clear and reasonable warning. Exceptions to this requirement are an exposure for which federal law governs warning in a manner that preempts state authority, or when the person responsible can show that the exposure will have no observable effect, assuming exposure at 1000 times the level in question.

4. *The proposition sets the allowable exposure level for DARTs at 1/1000 of the most appropriate NOEL.* The rigidity of this requirement has been the subject of debate (Russell 1989). Unlike traditional assessments which derive an uncertainty factor based on the species tested and quality of data, Proposition 65 mandates without exception, a 1000-fold safety factor (Kizer et al. 1988). While recognizing this limitation, it should also be noted that the mandate serves a useful purpose by setting an unambiguous number which allows one to ask the straight forward question, "Is the DART above or below the allowable exposure level?".

5. *The proposition places the burden of proof for warning on the responsible party*. This is a key provision in the proposition, and one that sets it apart from usual regulations. For statutes such as the Toxic Substances Control Act, enforcement cannot be carried out until a regulatory limit has been set by the US Environmental Protection Agency. This provides an incentive for delaying the setting of levels to avoid compliance (Roberts 1989). With Proposition 65, once a chemical is listed there is an incentive to determine levels that are in compliance with the statute. It is not required that the State set a level - business can develop their own. However, without a set number, the level of concern drops to any detectable amount.

6. *The proposition provides a mechanism for alternative enforcement by the state attorney general, local district attorneys, and private citizens*. This is the so-called 'bounty hunter' provision of Proposition 65. Utilizing a mechanism for alternative enforcement, a citizen, under certain circumstances, can bring action against a violator and collect 25 percent of the $2500 per day fine.

7. *The proposition sets forth a strict time frame for implementation*. Once a chemical is listed, compliance with the warning provision must occur within 12 months, and within 20 months for the discharge prohibition. Since this is a fixed timetable with no provisions for modification, it encourages action while discouraging delay.

The above provisions demonstrate that Proposition 65 is unique in several ways. Most notable, it shifts the burden of proof from the government to the responsible party. Less obvious but equally important in terms of assessment, the proposition covers only the first step of the risk assessment process: hazard identification (NRC 1983). The second step, reference dose determination, is not required to be performed by a regulatory agency. Responsibility for the final steps, exposure assessment and risk characterization, is transferred from the regulator to the party causing the exposure. Consequently, the decision of whether or not a product, workplace, or environmental release needs a warning is placed on those responsible for the exposure.

Chemicals may be placed on the Governor's list by one of the three mechanisms shown in Figure 1. Findings from an authoritative body may be used by the Office of Environmental Health Hazard Assessment, a department within the California Environmental Protection Agency, for listing without further deliberation. For example, pharmaceuticals classified as category X or D teratogens by the U.S. Food and Drug Administration may be put on the Governor's list through this mechanism. A chemical may also be listed if a State or federal agency requires it be identified as a DART. Finally, a chemical may be listed upon recommendation of the Proposition 65 Scientific Advisory Panel (SAP). The SAP is composed of recognized experts in the fields of reproduction, development, and carcinogenicity, who are on contract with the Office. This panel meets two times a year to deliberate data submitted by the technical support staff within the Office of Environmental Health Hazard Assessment, and any other interested parties which can include industry representatives, environmental advocacy groups, and the general public.

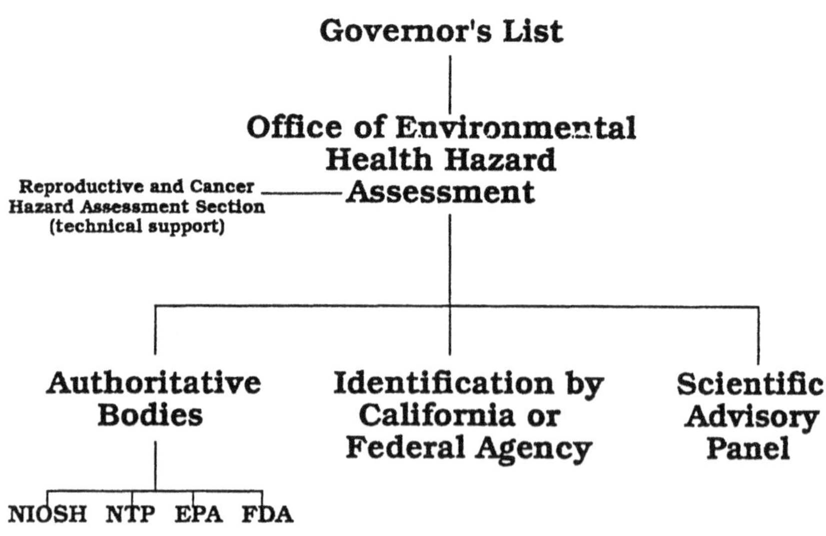

Fig. 1. Mechanisms for listing

With this background, it is now time to consider the methodologies employed by the technical support staff in evaluating and recommending chemicals for listing to the Proposition 65 SAP.

Evaluating Chemicals for Developmental/Reproductive Toxicity

Faced with the task of deciding in what order the many potential DARTs should be evaluated, a method of prioritization was developed (Donald et al. 1991). Briefly, this involved constructing a pool of candidate agents from which a high priority list of 15 chemicals was derived, based on their perceived hazard and potential for significant exposure.

Evaluations of agents from the high priority list are based on the draft guidelines for hazard identification and dose-response assessment of DARTs being developed by the Office of Environmental Health Hazard Assessment (1991). As shown in Figure 2, the hazard identification step begins with the evaluation of individual papers using criteria similar to that prescribed by the U.S. Environmental Protection Agency (1988a,b; 1989). Based on this evaluation, an individual study may be considered to provide data indicative of an adverse effect, data indicative of no effect, or data which is inconclusive. After evaluation of the individual studies, an evaluation is made for the combined animal data, and the combined human data. The combined evidence may fall into one of four categories: *Sufficient* in which the combined evidence establishes that an agent should be recommended to the SAP for listing as a DART; *Limited* in which the combinations of evidence are indicative of a DART, but not conclusive; *Deficient* in which combinations of evidence do not clearly indicate the presence or absence of developmental/reproductive toxicity; and *Null* in which the combinations of evidence fail to show developmental or reproductive toxicity. After making this determination for both the animal and human evidence, the two are combined for an overall evaluation. In those cases where there is adequate evidence to identify a developmental or reproductive hazard, the agent is placed before the SAP for deliberation. Using criteria developed by the panel's Reproductive Toxicity Subpanel, the panel may either approve or decline the recommendation that the agent be placed on the Governor's list.

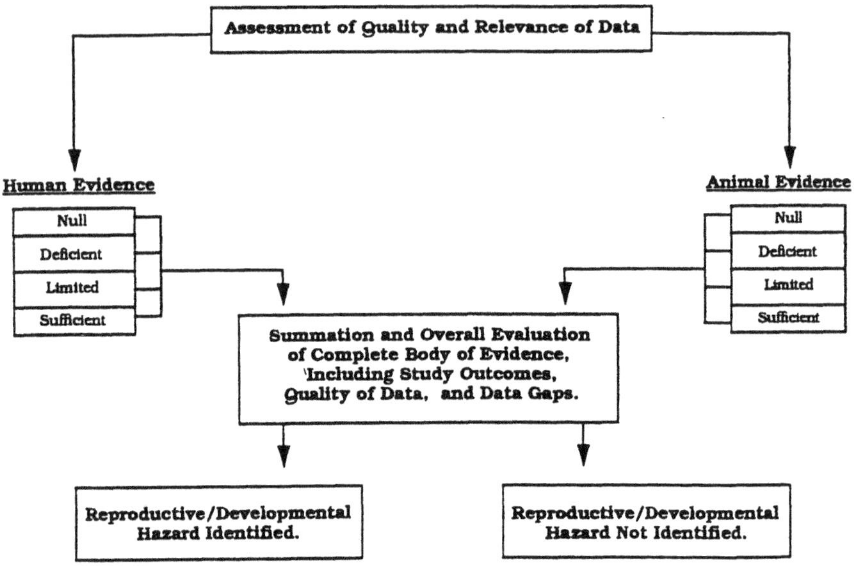

Fig. 2. The hazard identification process

The Impact of Implementing Proposition 65

Using the various mechanisms for listing, 114 chemicals have been identified as DARTs. As shown in Table 1, the vast majority are pharmaceuticals. Occupational and environmental chemicals such as lead, methyl mercury, and PCBs represent approximately 10% of those agents listed, while agricultural chemicals such as dinoseb and bromoxynil, and recreational agents such as alcohol, toluene, and cocaine represent another 10%.

The listings have resulted in three major outcomes. The first is the posting of warnings in places of business for workplace exposures, on or near products for consumer product exposures, and in newspaper advertisements for environmental exposures. The second outcome has been litigation. There have been approximately 20 enforcement suits by public agencies over the failure to provide a proper warning. Only three suits have been filed by private persons (the so-called bounty hunter provision). The third outcome has been termed "quiet compliance" (Totten 1989). Rather than face the marketplace consequences of placing a label on a product warning of possible birth defects or reproductive harm, some companies have chosen to reformulate a product, thereby eliminating the listed chemical. This form of compliance, which was apparently the intent of the proposition's authors, has been turned into a marketing asset by some companies who now advertise their products as new and improved, meeting California's Proposition 65 safe exposure level requirements (Smith 1990).

The impact of Proposition 65 has not stopped at the California borders. In addition to obvious effects such as the distribution of labeled products to other states and nations, some claim that the listings have prompted federal legislation, such as the enactment of a federal labeling requirement on alcoholic beverages (Roberts 1989; Russell 1989). Of greatest impact, however, is the incentive Proposition 65 has created for industry to develop alternatives to listed chemicals, thereby creating safer products. The impact of this extends not only to protection of the nations public health, but to a less toxic environment.

Table 1. DARTs listed as of April 1991

Pharmaceuticals	91
Occupational/Environmental Chemicals	13
Agricultural Chemicals	5
Non-medically used drugs	5
Total listed	114

Conclusions

It has been appropriately said that Proposition 65 represents a new era of *preventative strategies* in the management of toxic chemicals. By shifting the burden of proof for demonstrating a safe level of exposure to the parties responsible for the exposure, a listed chemical is in effect assumed to be a hazard until proven otherwise. This serves as an incentive for the responsible parties to collect data to demonstrate that either: a) the level of exposure does not pose a significant risk; b) a warning is required; or c) the product should be reformulated.

Through the warning provision, Proposition 65 has the potential to increase public awareness and the need for better education about toxic exposures. It should also serve as an incentive for industry to aggressively work at acquiring better data upon which we in the regulatory community can base our decisions. In turn, we in the scientific community, through meetings and monographs such as this, must attempt to resolve our differences and refine our methodologies to better serve the health interests of both individuals, and society as a whole. It is through such a combined effort that we can encourage the development of safe consumer products and workplaces, and preserve and maintain the quality of the water, land, and air which we all must share.

References

Donald JM, Monserrat LE, Hooper K, Book SA, Chernoff GF (1991) Prioritizing candidate reproductive/developmental toxicants for evaluation. Repro Toxicol: in press

Kizer KW, Warriner TE, Book SA (1988) Sound science in the implementation of public policy. A case report on California's Proposition 65. J Am Med Assoc 260: 951-955

NRC (1983) Committee on the institutional means for the assessment of risks to public health. Risk assessment in the federal government: managing the process. Commission on life sciences, National Research Council, National Academy Press, Washington DC, pp 17-83

Office of Environmental Health Hazard Assessment (1991) Draft guidelines for hazard identification and dose-response assessment of agents causing developmental and/or reproductive toxicology. April 3, 1991

Roberts L (1989) A corrosive fight over California's toxic law. Science 243: 306-309

Russell C (1989) Proposition 65 California's controversial gift. APF Reporter 12: 33-38

Smith RB (1990) California spurs reformulated products. Wall Street Journal (Nov. 1, 1990), pp B1, B4

Totten G (1989) Controversial Proposition 65 provokes industry outrage, 'quiet compliance' as it marks second anniversary as law. Chem Reg Report (May 5, 1989), pp 169-174

U.S. Environmental Protection Agency (1988a) Proposed guidelines for assessing female reproductive risk. Fed Reg 53: 24834-24847

U.S. Environmental Protection Agency (1988b) Proposed guidelines for assessing male reproductive risk. Fed Reg 53: 24850-24869

U.S. Environmental Protection Agency (1989) Proposed amendments to the guidelines for the health assessment of suspect developmental toxicants. Fed Reg 54: 9386-9403

Discussion of the Presentation

Anderson: Why did the state of California choose a 1000-fold safety factor?

Chernoff: I am not sure why that choice was made. I would like to point out that the issue for me is not the number, but the lack of flexibility. It does not matter if the NOEL is based on a rodent, rabbit, or primate study. In each case, the 1000 uncertainty factor would apply. This is not scientifically justified.

Neubert: In order to assess a possible risk, the human exposure has to be defined. Since exposure of subpopulations to different products will be quite different, labelling of each product would have to be considered separately. Is this correct?

Chernoff: Yes, this is correct. If a product contains a listed chemical at a concentration above the no significant amount level, a warning would be required. If another product contained the same listed chemical at a concentration below the no significant level, no warning would be required. The no significant amount level is defined as $^1/_{1000}$ of the most appropriate NOEL

Stahlmann: I was not aware that toluene was considered a clear-cut human hazard with respect to developmental toxicity. The meagre information for such a possibility comes from women abusing the substance by sniffing high concentrations (among other chemicals). It is difficult to exclude an alcohol symdrome. In fact, the German MAK commission after extensive discussions had expressed the view that insufficient evidence for such actions exist both in humans and from experimental studies, up till now, and the same conclusions were reached in an expert group of the European Community more recently. Could you please comment on this?

Chernoff: There are a series of case reports in humans associating toluene abuse (for narcotic effects) during pregnancy with a rather specific pattern of malformations in the offspring. In addition, both regulatory studies, and studies in the open literature have demonstrated toluene developmental toxicity in experimental animals. Based on these findings, the SAP recommended listing in November, 1990 and it was listed in January, 1991.

General Discussion: Risk Assessment

Neubert: Dr. Peters, it is certainly impressive how much progress has been made within the last years in the field of monitoring systems. This looks very encouraging. What is your feeling with respect to the problem of assessing causal relationships? I would assume that in general it is difficult to obtain prospectively anamnestic information on the women who later obtain abnormal children. I am very critical with respect to obtaining exposure data retrospectively. MADRE looks very interesting in this respect, but the number of mother/child-pairs will be much smaller. What is your opinion on this matter?

Peters: A drawback we have in risk assessment with respect to reproductive (but also for other areas of experimental and safety and efficient evaluation) within the European Economic Community is the fact that we are dealing with different Directorates General (authoritive bodies). For example, many committees exist, in O6 V a committee dealing with occupational hazards, in D6 III committees for animal feed, for drugs, etc. Health, as culture and education, are second-rate issues or even non-issues in the EEC. Economies prevail. I find this really remarkable since in my opinion health, culture and education belong to the most important items for a human being.

Neubert: There are many aspects one could comment on, I would like to mention only two: Although everyone likes the expression "reproductive cycle", there really is no reproductive *cycle*, but a sequence. My children or parts of them do not return to me or my wife in a cycle.

Peters: You are absolutely right. It is not a closed cycle. I must confess that I like the ying-yang sound so much that I deformed the reality in that picture.

Neubert: And, as you also realize, comparisons of the susceptibility of man versus experimental animals is very difficult and the statements are not very valid. In the majority of cases, data for man are so poor that only few conclusions can be drawn. Furthermore, just for these examples the animal models are rather poor: the rabbit for thalidomide, the rat for alcohol, no species for warfarin or folic acid antagonists, etc. The difference in susceptibility between man and monkeys in the case of thalidomide is certainly small, if it exists at all. With the marmoset we obtained clear-cut effects at doses of 1 mg/kg body wt, which may even be lower than the doses needed in women.

Peters: Again I agree. This comparison only has a limited significance.

Chernoff: As you know Dr. Ken Jones and I pioneered many of the concepts associated with teratogen registries in San Diego several years ago. It pleases me to see that you have taken many of these ideas and built a fabulous program throughout Europe. It saddens me to report that in this respect my government has not been as progressive in funding such efforts, and consequently, the California Teratogen Registry has not been able to expand beyond it's original, rather limited size, because of insufficient funding.

I also have two questions: Do you offer information on all exposures, or just pharmaceuticals? Is each child born examined by a physician?

Peters: To your first question: All GNTI'S programs give information on drugs and other exogenic risk factors, such as infections, drugs of abuse, radiation, occupational exposure and household products.

In answer to your second question: In our TIS - the Netherlands - each family has a general practitioner, who by habit pays a "congratulation"-visit to the home of the family with the newborn. In other TIS this will be different, but special follow-ups with careful newborn examination is possible.

Somogyi: Does regulation of pesticides in the Netherlands differ from the way described for Germany?

Peters: No, I don't think so. At least not so dramatically that we have great difficulties in cooperating or in recognizing each other's assessments within the European Economic Community.

Neubert: Prof. Yasuda, the picture of the Europeans and the Americans beating on the poor Japanese with a hammer is no longer true. Many consider the Japanese guidelines clearly superior to the European and the American ones, and many aspects of the Japanese guidelines are now being generally accepted.

The approach you mentioned of making use of the very large collection of human embryo specimens for risk assessment is very interesting and unique and certainly only possible in Japan. How is the assessment of the exposure performed in these evaluations? Are there records, and how good are they? What was the cause of hyperthermia? Was it infections? Can infections or a treatment be confounders?

Yasuda: Special record forms were filled out by obstetricians and provided with embryos. Not all of these records are good, of course, but they are still invaluable sources of information. The cause of hyperthermia was mostly infections.

Claude: I have two questions concerning risk assessment: Are very high doses in segment II studies relevant? How can one evaluate drugs given only one time in humans?

Yasuda: Any drug may be used inadvertently in a wrong way. Hence, maximum precaution is recommended for preclinical tests.

As to the first question: Dose range finding tests point out which dose levels are to be used in the segment studies, taking the human dose levels into account. To understand the results of segment studies a clear report from the dose range finding studies and a motivation for the levels of doses is necessary. It will be clear that high doses in segment II studies might give other results simply because the metabolization and excretion of the drug might take a longer time; therefore a more prolonged effective drug concentration in comparison to lower drug doses might be present.

Neubert: There are the interesting studies performed by Dr. Nagao in Japan: clear-cut malformations were observed when (short-lived) substances were given during the preimplantation period. Until now, I do not know of a single example of a substance acting only at the preimplantation stage and producing no gross-structural abnormalities during the organogenesis phase. Most of the agents acting in such an early stage seem to be mutagens, and such an effect has also been claimed to exist for X-rays. Would you consider this important enough to extend the treatment period?

Yasuda: No. However, I learned from Dr. Kavlock that there are possible pre-implantation-specific teratogens. If this is true, near term observation in segment I studies is warranted.

Kavlock: In reference to your request about pre-implantation specific "teratogens", I believe the work of Generoso on ethylene oxide exposure around the time of fertilization is appropriate. Also I would like to point out that weak estrogens such as methoxychlor (Cummings, et al.) can accelerate zygote transport in the fallopian tube, resulting in embryo delivery to the uterus prior to it's being receptive to them. Also, LH Release Inhibitors can delay ovulation, and cause "over-ripe" eggs to be released on the following day. These are several examples of critical aspects of the peri-fertilization period, although malformations are only part of the effects seen.

Secondly, in my opinion, every drug needs to be tested according to the indication for use, already assisting human and animal data, in the right animal with full justification of the (different) methodology used. In case of using a drug only once, I question whether the indication of the use of such a drug during pregnancy is that necessary. Furthermore, I think that animal experiments must give results as "reproductive profile", so that we can make a "well educated" guess what the effects might be in the human, irregardless of how seldomly or frequently the drug will be used.

Pritchard: In 1989 there were comments relating to extending segment I studies in Japan to look for postnatal effects due to "paternal teratogenicity". Would you please comment on the concept of "paternal teratogenicity".

Yasuda: There have been reports on behavioural abnormalities of offspring when a drug is administered to male animals before mating. Hence, it is recommended to observe postnatal development in segment I studies if postnatal developmental disorders are suspected to result from the drug treatment in segment I studies. I feel, however, application of postnatal observation should be considered case by case.

Chahoud: In my opinion evidence for paternally-induced teratogenicity is very weak. There is no example of an effect which has been reproduced in several laboratories. We have ourselves performed a very large study on possible paternally-induced effects of 2,3,7,8-TCDD in the offspring of untreated dams (*Chahoud et al. Arch Toxicol 63: 432-439, 1989*), because it was claimed that such an effects might occur in man. No adverse effects were found using doses with very high general toxicity.

Dose-Response Relationships and Quantitative Risk Assessment

Statistical Problems (and Some Solutions) Associated with Testing for Effects in Developmental Toxicology

R. Woodrow Setzer

Introduction

We have come a long way in testing for effects in developmental toxicology since the debates over the "unit of observation" in teratology (Weil 1970; Kalter 1974; Staples and Haseman 1974; Becker 1974; Haseman and Hogan 1975). Much of the recent biostatistical work on testing has been on devising powerful statistical tests that reflect the proper choice for the unit of observation. This paper will review some of the statistical problems posed by developmental toxicology data as well as principal solutions posed for some of those problems. Included in the review are several simple new methods that have yet to be applied in this field. The analysis of developmental toxicology data would be improved through increased utilization of biological content that these new approaches offer.

There is generally a duality between testing and estimation, as estimation methods generally can yield a test, and testing methods can generally be modified to yield both point estimates and confidence intervals. Testing answers questions such as, "Does group T differ from group C?", or more precisely, "Is there sufficient evidence to believe that group T differs from group C?". Estimation, in contrast, answers questions like "How large are the responses in groups T and C?", or "How large is the difference between group T and C?", and "How reliable are these estimates?". Since the two are subject to somewhat different criteria, and other papers in this symposium discuss aspects of estimation relevant to developmental toxicology, I will restrict the remainder of this discussion to testing. Furthermore, since most endpoints of interest to developmental toxicologists are quantal, I will restrict my discussion further, to quantal outcomes.

It may be useful at this point to discuss some terminology specific to statistical testing. First is "null hypothesis", which, in toxicology, is usually the hypothesis that there is no treatment effect. We often "accept" the "null hypothesis" unless there is sufficient evidence to the contrary, but, it is important to be wary of a too simplistic belief in the "truth" of an unfalsified "null hypothesis": the treatment effect may just be too small for detection with the experiment being analyzed. Thus, the null hypothesis is often a convenient fiction in which we do not really believe! The way to accommodate the often-stated importance of identifying a "biologically significant" (as opposed to statistically significant) effect, is to prespecify the magnitude of effect that would constitute biological significance, and pose as the null hypothesis that the treatment effect is smaller than that magnitude. The "alternative hypothesis", then, is what we think is true if the null hypothesis is false.

Two terms commonly used to describe the properties of a statistical test are Type I and Type II error. Type I error is rejecting the null hypothesis when it is true, and Type II error is NOT rejecting the null hypothesis when it is false. The probability of a Type I error (usually symbolized a) and of Type II error (ß) can be calculated (or at least estimated) for any given test and specific hypotheses, assuming a probability model for the data (for example, independent, binomially distributed observations). The complement of the probability of a Type II error (that is, 1 - ß) is "power". The ability of a test to identify specific treatment effects can be described graphically by plotting power versus treatment effect. In clinical testing, "specificity" has much the same meaning as a in statistical testing, and sensitivity means much the same as ß.

Some Problems with Testing for Developmental Toxicity

An early problem identified as special to testing for developmental toxicity is that littermates' responses are often correlated, or, for quantal responses, the variance among litter proportions of affected fetuses is often greater than expected for a binomial distribution: the "litter effect". The litter effect has received considerable attention in the biostatistical literature (see, for example, the review by Haseman and Kupper 1979), and several solutions are now available for dealing with it, some of which will be discussed in a later section.

Another issue peculiar to developmental toxicology data is the evaluation of multiple (often competing) endpoints. Multivariate analysis of variance (MANOVA) is appropriate for multiple continuous endpoints, but there has not been an analogue for quantal endpoints and mixtures of quantal and continuous endpoints, which is common in developmental toxicology. Recent work (Lefkopoulou et al. 1989) has applied the results on generalized estimating equations of Zegar et al. (1988) to the problem of estimating and testing effects in situations with multiple qualitative responses. Other new work (Catalano and Ryan, submitted) develops methods for handling mixtures of continuous and dichotomous traits, such as fetal weight and abnormalities.

Problems of interpretation increase when responses at one endpoint interfere with the observation of responses at other endpoints. For example, fetal death masks abnormality, a limb reduction defect might preclude observing digit defects, or exencephaly would preclude the observation of hydrocephalus. In these cases, we cannot estimate and test the unconditional probability of the masked response, only the conditional probabilities of the masked response, given the non-occurrence of the masking response.

Finally, most laboratories have developed a large body of information on background incidences of abnormalities. Formal methods for including such control information would not only enhance the precision of some of our tests by serving as a balance for the occasional extreme concurrent control group, it would also allow a formal testing procedure for rare abnormalities, for which conventional methods do not work as well. However, background incidences measured in the same laboratory at different times may differ, either randomly or with a definite secular trend. Although there has been some work on incorporating historical control data into cancer risk models in toxicology (e.g., Smythe et al. 1986; Makuch et al. 1989), as yet no published work treats this problem specifically for developmental toxicology endpoints.

Solutions Accommodating the Litter Effect

A primary statistical concern regarding the choice of the observational unit is that the probability of Type I error of a test be well known. Statisticians are most comfortable taking the litter as the observational unit because it is the unit of randomization, (Haseman and Hogan 1975). Nevertheless, if there were no litter effects, the fetus could serve well as the observational unit. The choice of unit can make a big difference in apparent sample size, especially to investigators who work with animals that have large litters. Part of the conflict over observational unit has certainly been driven by this difference, although, as Haseman and Hogan (1975) point out, the reduction in power attendant upon using litter-based methods even in the absence of litter effect is generally small, and not at all proportional to the difference in the apparent sample size. However, the consequences of ignoring litter effects can be severe. The probability of a Type I error can be substantially increased in methods that treat the fetus as the unit of observation. Among the examples in Haseman and Hogan (1975), the probability of a Type I error of a per-fetus Chi-square test ranged from 0.120 to 0.285 for a test with a nominal a of 0.05. Consequently, prudence would seem to favour choosing litter-based methods.

The simplest way to assure that the litter effect does not inflate the probability of a Type I error is to work solely with litter means, using t-tests, analysis of variance (ANOVA), or methods based on rank scores (Haseman and Kupper 1979). Ordinary t-tests and ANOVAs are based on the assumption that the data are normally distributed, and, although both are relatively insensitive to deviations from normality, it is advisable to transform litter proportions using either the standard arcsine transformation for proportions or the Freeman-Tukey binomial transformation before analyzing the data. Methods based on scores, in which each observation is replaced by its rank, and then a transformation (or score) applied to that rank, have commonly been used in lieu of t-tests or ANOVA when concerns about the normality of data have been paramount. Most commonly, Wilcoxon scores, in which each data point is simply replaced by its rank in the data, have been used, resulting in the Wilcoxon or Mann-Whitney U tests (Sokal and Rohlf 1981). Although there are formulas and tables for exact P-values for rank score tests, for even moderate sample sizes it is usually adequate to carry out ordinary t-tests or ANOVAs on the rank-score-transformed data.

It is possible for tests with the same Type I error to differ in their power to detect treatment effects. Most efforts to improve the power of methods using litter means have been based upon weighting the observations to reflect differences in their variances, which will be discussed more fully below. However, some other simple methods have been devised to improve the power of t-tests, ANOVAs, and rank statistics. Brownie et al. (1990) recently reported an approach that may increase the power of simple t-tests and ANOVAs. If the variance under the alternative hypothesis is expected to increase with treatment (as is usually the case in developmental toxicology data), it is valid (in the sense of having the correct probability of Type I error) to use as the denominator of the test statistic a quantity based on the error mean square from the control group alone, rather than a pooled mean square error from all treatment groups. The resulting test is very often more powerful than the conventional test. The increase in power comes because the loss of degrees of freedom in the denominator is more than balanced by the reduction of the denominator itself. Naturally, this approach would need to be selected *a priori* based on prior experience, and not applied solely because a test indicated that variances were heterogeneous. Furthermore, it is inappropriate to base confidence intervals for the treatment effect on this test; they should be based upon a variance estimate that assumes that the control and treatment group variances differ. This could occasionally lead to the apparent paradox that confidence intervals for the difference between treatment and control groups would include zero, even though the statistical test indicated a significant treatment effect.

The power of tests based upon scores can also be improved upon. Some authors have derived scores for the situation in which only a fraction of treated subjects respond (Good 1979; Johnson et al. 1987; Conover and Salsburg 1988). The derived scores weight extreme observations more heavily than do Wilcoxon scores (Fig. 1). Since a consequence of the litter effect is that some litters appear more responsive to treatment than others, these scores may be good choices for developmental toxicology data.

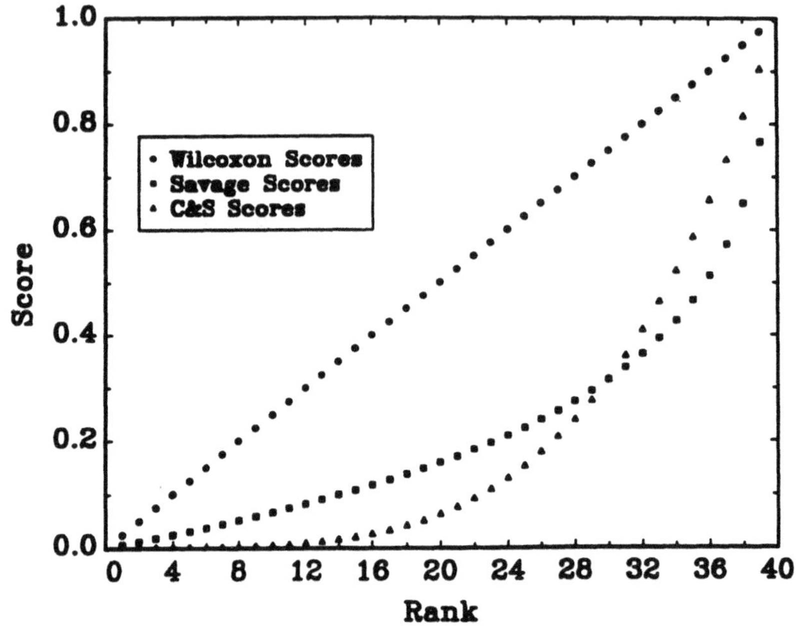

Fig. 1. Wilcoxon scores and two scores due to Conover and Salsburg (1988): inverse Savage (labelled Savage) and their $s_2(i)$ (labelled C&S). To facilitate comparison, all scores have been scaled so that the maximum value is 1. This shows how the spacing between scores increases with rank

Least-squares methods such as t-tests and ANOVAs assume that all observations have equal variances. They are less efficient when this is not true, as for developmental toxicology data when litter sizes vary, and in the presence of litter effects. Weighting observations to account for varying litter sizes and the litter effect should improve the power of tests based upon litter means. Several methods (differing in how the weights are calculated) first calculate weights for each observation from an estimate of its expected variance, a function both of the litter effect and litter size, and then carry out a weighted ANOVA (Cochran 1943; Kleinman 1973; Marubini et al. 1988). Tests are based upon the usual F-statistic. Gladen's (1979) jackknife procedure for estimating and testing the significance of treatment effects is similar in spirit to these methods.

The weight assigned to an observation can be thought of as a function of the mean of the treatment group, whose estimate, in turn, is a function of the weights. Thus, methods that alternate between calculating weights and means could improve power over those that iterate only once. In maximum quasi-likelihood methods (QLMs) (Williams 1982), litter means are treated as independent, and weights are based upon variances. The more general method of generalized estimating equations (GEEs) (Lefkopoulou et al. 1989) takes individual fetuses or even measurements of specific endpoints on those fetuses as the basic data unit, and uses weights which are based upon correlations. Testing is accomplished in both approaches by fitting successively more restricted models, and testing the change in the deviance, a goodness of fit measure (McCullagh and Nelder 1989).

Beta-binomial models (Williams 1975) accommodate the litter effect by assuming that the probability of a response in a litter is itself randomly sampled from a beta distribution (a probability distribution defined on the interval (0,1), with two parameters to determine the mean and shape). The observed number of affected fetuses in a litter is assumed to be binomially distributed, conditional on the litter size and the litter-specific probability of response. The mean responses and litter effects of the treatment groups are functions of the two parameters of the beta distribution. Testing is carried out much as in QLM and GEE models, based on changes in the likelihood function.

Simulation studies comparing the powers of statistical tests appropriate for developmental toxicology data (e.g., Haseman and Hogan 1975; Gladen 1979; Pack 1986; Marubini et al. 1988; Setzer and Rogers 1989) indicate that iterative weighting methods are generally more powerful than one-step weighting methods, which are, in turn, more powerful than non-weighted methods, including tests based upon Wilcoxon scores. However, preliminary simulation results (Setzer in prep.) indicate that combining rank scores (e.g., those due to Conover and Salsburg (1988)), with the testing approach reported by Brownie et al. (1990), can improve power over that of the iterative methods. The power differences seem to be small (Fig. 2), and other considerations, such as ease of computation or the desirability of estimates of effect sizes, should be weighed when deciding which tests to use.

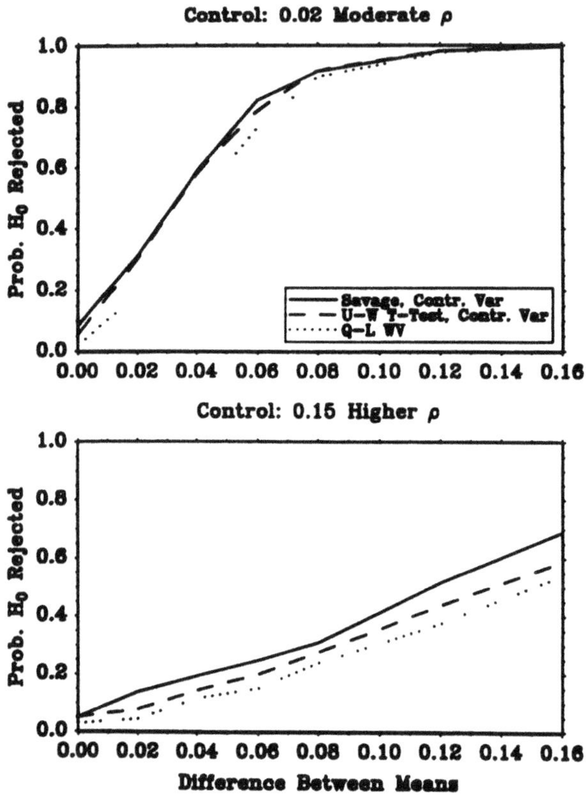

Fig. 2. Summary of simulations (Setzer, in prep) comparing power of statistical tests for developmental toxicology data. Tests compared were: t-tests on Freeman-Tukey transformed data, conventionally and using Brownie et al.'s (1990) modification; tests based on rank-score transformations using Wilcoxon, inverse Savage, and Conover and Salsburg's $s_2(i)$, using conventional t-test and Brownie et al. (1990) modified t-test; and four iterative methods: beta-binomial maximum likelihood (Pack's 1986 test H5); MQL test with variance proportional to binomial variance; MQL with variance of Williams (1982); MQL using Williams' variance formulation, but modified to estimate a separate litter effect for each dose group. Data were generated as beta-binomial random variates, 20 litters in each of two dose groups. Litter effect for a group (r) was a function of the mean of that group, and simulations were run using two expressions of r as a function of mean, one twice the magnitude of the other. The power of the most powerful test in each group is plotted versus the difference between the expected means of the treatment groups for two values of the control malformation incidence. Tests plotted are: inverse Savage scores using Brownie et al.'s (1990) modification (Savage, Contr. Var); Brownie et al.'s (1990) t-test on Freeman-Tukey binomial transformed variates (U-W T-test, Contr. Var), and MQL using Williams variance formulation and separate variance estimates for each group (Q-L WV)

Exploiting the Litter Effect

The methods discussed thus far have attempted to improve power by weighting observations to reflect the litter effect, by applying the non-Wilcoxon rank scores methods discussed above, or by basing tests of the null hypothesis on an estimate of the variance valid under the null hypothesis. Although, all these methods reflect some of the biology of the developmental toxicology situation, the biology is not explicit in any of the methods developed so far. Generally, the litter effect parameters are treated as nuisance parameters, and do not contribute substantially to tests. However, the litter effect tends to increase for low litter-mean responses and decrease at higher responses (e.g., Moore 1987; Williams 1988). Thus, the litter effect and the mean responses may be related causally, and therefore their relationship may contain some information about treatment effects. A simple model, inspired by the biology of the situation, that described both the response of the mean and of the litter effect to treatment might enjoy increased power over the statistical models presented so far.

Suggested causes of the litter effect include maternal variation of the delivery of the active metabolite to the conceptus (Kimmel and Young 1983); variation in timing due to mating time variability (Holson et al. 1976; Endo and Watanabe 1988; Fujinaga et al. 1990); and interfetal interactions or local intrauterine variation of conditions (Mankes et al. 1986; Setzer and Rogers 1988). Although all of these factors probably contribute to the litter effect and may need to be included in a comprehensive biological model of developmental responses to treatment, a relatively large component of inter-dam variability may be pharmacokinetically based. Kimmel and Young (1983) found that in rats treated with a single dose of salicylate, variation in the 45 minute maternal blood salicylate concentration accounted for about 52% of the among litter variance of the fraction of affected fetuses (note that, because of binomial sampling variation of the fraction responding, the maximum amount of variance that could be explained by variation of a maternal measurement is about 71%). Thus, in this model system, a very large part of the litter effect is due to inter-dam variability, and the dose-response (expressed as a function of delivered dose) participates in both the mean response (expressed as a function of administered dose) and the litter effect. A general model with just such a structure was suggested by Meister et al. (1991), and this may be similar to the latent variable models of Catalano and Ryan (submitted).

Conclusions

Although statistical models for developmental toxicology data have increased in complexity and sophistication over the last 20 years, gains in power (sensitivity) to detect treatment effects have been modest. Purely statistical improvements in test methods may well contribute further gains in power, but it seems worthwhile to incorporate more biological information into our test methods. In this way, it may be possible to construct tests that use more information about the response to treatment than our present tests do.

Disclaimer: This document has been reviewed in accordance with U.S. Environmental Protection Agency policy and approved for publication. Mention of trade names or commercial products does not constitute endorsement or recommendation for use.

References

Becker BA (1974) The statistics of teratology. Teratology 9: 261-262

Brownie C, Boos DD, Hughes-Oliver J (1990) Modifying the t and ANOVA F test when treatment is expected to increase variability relative to controls. Biometrics 46: 259-266

Catalano P, Ryan L (submitted) Bivariate latent variable models for clustered discrete and continuous outcomes. submitted to J Am Stat A

Cochran WG (1943) Analysis of variance for percentages based on unequal numbers. J Am Stat A 38: 287-301

Conover WJ, Salsburg DS (1988) Locally most powerful tests for detecting treatment effects when only a subset of patients can be expected to "respond" to treatment. Biom 44: 189-196

Endo A, Watanabe T (1988) Interlitter variability in fetal body weight in mouse offspring from continuous, overnight, and short-period matings. Teratology 37: 63-67

Fujinaga M, Jackson EC, Baden JM (1990) Interlitter variability and developmental stage of day 11 rat embryos produced by overnight and morning short-period breeding regimens. Teratology 42: 535-540

Gladen B (1979) The use of the jackknife to estimate proportions from toxicological data in the presence of litter effects. J Am Stat A 74: 278-283

Good PI (1979) Detection of a treatment effect when not all experimental subjects will respond to treatment. Biom 35: 483-489

Haseman JK, Hogan MD (1975) Selection of the experimental unit in teratology studies. Teratology 12:165-172

Haseman JK, Kupper LL (1979) Analysis of dichotomous response data from certain toxicological experiments. Biom 35: 281-293

Holson JF, Scott WJ, Gaylor DW, Wilson JG (1976) Reduced interlitter variability in rats resulting from a restricted mating period, and reassessment of the "litter effect". Teratology 14: 135-142

Johnson RA, Verrill S, Moore DH (1987) Two-sample rank tests for detecting changes that occur in a small proportion of the treated population. Biom 43: 641-655

Kalter H (1974) The choice of the number of sampling units in teratology. Teratology 9: 257-258

Kimmel CA, Young JF (1983) Correlating pharmacokinetics and teratogenic endpoints. Fund Appl Toxicol 3: 250-255

Kleinman JC (1973) Proportions with extraneous variance: single and independent samples. J Am Stat A 68: 46-54

Lefkopoulou M, Moore D, Ryan L (1989) The analysis of multiple correlated binary outcomes: Application to rodent teratology experiments. J Am Stat A 84: 810-815

McCullagh P, Nelder JA (1989) Generalized Linear Models. Second Edition. Chapman and Hall. London

Makuch RW, Stephens MA Escobar M (1989) Generalised binomial models to examine the historical control assumption in active control equivalence studies. The Statistician 38: 61-70

Mankes RF, Renak V, Fieseher J, LeFevre R (1986) Birthweight depression in male rats contiguous to male siblings *in utero* exposed to high doses of 1,3-butanediol during organogenesis. JACT 5: 189-196

Marubini E, Correa Leite ML, Milani S (1988) Analysis of dichotomous response variables in teratology. Biom J 30: 965-974

Meister R, Chahoud I, Jurgens M, Iverson F, Bochert G (1991) Biometrical analysis of strain differences and litter effects. To be published in the Springer Series (seen in manuscript)

Moore DF (1987) Modelling the extraneous variance in the presence of extra-binomial variation. Appl Statist 36: 8-14

Pack SE (1986) Hypothesis testing for proportions with overdispersion. Biom 42: 967-972

Setzer RW and Rogers JM (1988) The similarity of toxic response of neighboring fetuses in the same uterine horn in mice. Teratology 37: 491

Setzer RW and Rogers JM (1989) The relative powers of statistical tests for developmental toxicology data: maximum likelihood beta-binomial, maximum quasi-likelihood, and ANOVA based on the Freeman-Tukey binomial transform. Teratology 39: 481

Setzer RW (in prep) The power of statistical tests for developmental toxicology data.

Smythe RT, Krewski D, Murdoch D (1986) The use of historical control information in modelling dose-response relationships in carcinogenesis. Statist Probab Lett 4: 87-93

Sokal RR, Rohlf FJ (1981) Biometry. Second Edition. W.H. Freeman. San Francisco.

Staples RE, Haseman JK (1974) Selection of appropriate experimental units in teratology. Teratology 9: 259-260

Weil CS (1970) Selection of the valid number of sampling units and a consideration of their combination in toxicological studies involving reproduction, teratogenesis, or carcinogenesis. Fd Cosmet Toxicol 8: 177-182

Williams DA (1975) The analysis of binary responses from toxicological experiments involving reproduction and teratogenicity. Biom 31: 949-952

Williams DA (1982) Extra-binomial variation in logistic linear models. Appl Statist 31: 144-148

Williams DA (1988) Estimation bias using the beta-binomial distribution in teratology. Biom 44: 305-309

Zegar SL, Liang KY, Albert PS (1988) Models for longitudinal data: A generalized estimating equation approach. Biom 44: 1049-1060

Discussion of the Presentation

Sterz: If we agree that the rabbit is not always a better model species than the rat how can we accept that in segment II studies 12 to 15 rabbits are accepted as a sufficient number of test units/group whereas for the rat the numbers lie between 20 and 30?

Setzer: I have two things to say to that. The first is that there are indeed conventional sample sizes for each species commonly used in a bioassay. These seem to have resulted from a compromise between general sensitivity requirements on the one hand and economic and esthetic considerations on the other (rabbits are "cuter" than rats to many people), not from an explicit consideration of the sensitivity requirements of a specific bioassay. From a statistical point of view, once we specify the parameters that go into a sample size prescription (for example, minimum size of effect we want to be able to detect, control incidence, litter effect at different response rates, litter sizes), the same sample size prescription results regardless of the species used. If there is concern that the rabbit bioassay has more variability, say because of a greater litter effect than rats, or because of a greater propensity for resorption, then we should want more rabbits in a study than in a rat study, for the same precision. The second thing I want to say is that these considerations are really separate from the quality of the species as a model. If the rabbit is a bad model for what is going on in humans (since it is the human response we want to predict), then increasing the rabbits used per dosage group only lets us know that bad model's predictions more precisely.

Meister: We found that litter effects vary with *species*, substance, dose and endpoint. Therefore, sample size determination has to be performed on a case-by-case basis, using information on litter effects from control and pilot studies and also using data bases if possible.

Are tests available (Streitberg, Rommel, Biometrics?), e.g., which are uniformly more powerful than the Wilcoxon test which could use a second score function accounting for different intralitter-correlation? How could effects on litter effects be incorporated into testing for "hazard identification"?

Setzer: Two approaches come to mind. First, the ß-binomial and quasi-likelihood/GEE approaches allow the test directly using a two-degree-of-freedom test. However, Pack (Biometrics 42: 967-972, 1986) has shown that this approach has no power advantage over looking at means alone for ß-binomial data. The other approach would be to find a single parameter that summarized both the mean response and the intra-litter correlation. This *might* give us a more powerful test.

Scott: At yesterday's discussion Dr. Chernoff asked if we could test better by including more dosage groups containing a smaller number of animals. Biologically this makes good sense to me. Can you tell us if statistically this type of experimental design makes sense?

Setzer: There is probably no simple answer to this question, because it depends upon the specific goal of the testing and the (unknown!) true underlying dose-response. If you consider a series of experimental designs using the same number of animals but differing in the number of doses, designs with more doses give better resolution of the dose-response, those with fewer doses give better precision of individual estimates of effect. In the simple hazard identification testing situation, probably a few, well-placed doses will be best. For estimation of the shape of the dose-response curve (using, say, non-parametric regression or data smoothing methods). However, the precision of such an estimate may be disappointing. An alternative approach might be to specify a specific parametric model with a parameter to control the shape of the curve. For such a model, the optimum number of doses could be derived. However, for a dose-response model, the position of the doses is also important. Consider the following unfortunate situation: an investigator wants to fit a logistic model to her data, and has three dose groups with good N plus a control. However, at the end of the experiment, she finds that the dose placement was such that all of the pups were abnormal in the highest dose group, and none in the others. In this case, it is impossible to estimate the dose-response parameters (although, as Reinhard Meister has pointed out, lower confidence bounds on the slope can be calculated). Therefore, probably the optimum solution for dose-response estimation will be an iterative solution, with multiple experiments designed to successively improve our idea of the optimum placement of doses. This is very like the situation now in many studies, in which first you do a dose range-finding study, before the larger, say segment II study is planned.

Aspects of Concentration-Response Analysis

Reinhard Meister

Introduction

The use of quantitative methods for risk assessment is not a routinely applied approach in teratology or other fields of reproductive toxicology, although in our opinion these methods should be used by researchers and accepted by regulatory agencies. One reason for this attitude is, that the usefulness of animal experiments for prenatal risk assessment is always under discussion. However, these experiments are, up till now, the only way to find and to exclude some of the risks for the unborn before a drug is marketed. If animal experiments are performed, they should be analysed using the best methods available.

To state our position very clearly we would like to make a proposal in advance which could serve as guidance for the applied toxicologist. We suggest which statistical methods he can rely on and give hints to sources of bias when performing risk assessment for prenatally induced adverse health effects.

Control for typical confounders such as mortality, stage-specificity etc. to avoid bias.
Use per-litter methods for the analysis.
Prefer benchmarks to NOEL's whenever possible.
Incorporate the biological variance of effects into risk extrapolation.

When Bruce Ames (Ames and Gold 1990) wrote his critical remarks on the famous Ames-salmonella test for mutagenicity of substances and gave a critical review of results, he never had in mind that this test is not a good experiment to establish mutagenicity in salmonella. That means, when applying concentration-response analysis for risk assessment we have to restrict ourselves in a first step to the species from which the data came from. Then, using all information available, e.g. pharmacokinetics, metabolism, etc., we may use the animal data for a risk assessment for man. We want to stress that this paper does not go into the problem of risk extrapolation to man. We restrict our discussion to risk assessment within the animal experiment.

There is a difference between the qualitative and the quantitative approach to risk assessment. Whereas the qualitative approach makes only a statement about the potential of a chemical substance or a medicinal product, the quantitative approach gives more detailed information about the potency.

We prefer the term *concentration* and we will use it instead of *dose* since dose-response relationships are heavily dependent on the bioavailability of a substance. The use of the term *concentration* shall stress the importance of *pharmacokinetics* for risk assessment.

To give an overview, we will discuss some general aspects and concepts of concentration-response relationships. Special aspects of some importance for the proper understanding and an appropriate analysis of prenatal toxicity tests will be considered in the second part. The last part will cover some material on the estimation of concentration-response models of prenatally induced adverse health effects.

General Aspects of Concentration-Response Experiments

When looking for possible reasons for adverse reactions we have to consider the following points:

- Adverse reactions, say malformations or carcinogenicity or death of an animal may occur without any treatment. We call this *background* or *spontaneous* effects.

- When adverse reactions occur concentration-dependently under treatment we introduce the *threshold assumption* to explain the observed reaction rates.
- There might be *competing effects* which can make it impossible to observe an induced effect. It can happen then that we observe lower reaction rates under treatment than those originally related to a given concentration of a chemical compound.

Going into detail we will discuss the threshold concept: We assume that the interaction between the organism and the chemical substance is concentration related. For any individual treated we make the assumption - and it is a natural one - that an *individual threshold* exists. Below that threshold no adverse reaction is observed, above that some kind of reaction will be induced. Thresholds may vary for different effects (e.g. may be different for cleft palate and limb malformations). The assumptions on individual thresholds lead to the most simple model for concentration-response curves. It can be written down in statistical terms:

$$\begin{aligned} P\ (\text{reaction} \mid \text{concentration}) &= \\ &= P\ (\text{individual tolerance} < \text{concentration}) \\ &= F\ (\text{concentration}), \end{aligned}$$

where P means *probability* and F denotes the statistical *distribution function* of the individual thresholds.

For prenatally induced effects the existence of a *global threshold* different from zero for all individuals and all effects has often been postulated. However, there is no possibility to establish such a general limit on the basis of experimental data. Therefore, a more promising approach is the use of *tolerance distribution models* to describe the variability of the individual tolerances in a group or a population of experimental animals. We know that we can use different types of tolerance distributions such as the *log-normal-* or the *log-logistic-distribution*, which are the basis for the well-established probit and the logit approaches. But, there are other functions, for example the *Weibull distribution*, which allow a more sophisticated analysis of observed concentration-response curves. We can finally determine characteristics of the concentration-response curve which are not model-dependent and have the same meaning for all possible types of underlying distributions. The median effective concentration called EC50 value, above which a reaction of 50% or more of the animals is expected, is very often used. We can estimate the slope of the curve and transform it into an estimate for the biological variance. A curve with a large value for the slope showing a steep increase in the response probabilities demonstrates a small biological variability among the individual animals. Very often effective concentrations other than the EC50 value are required. We can estimate these ECp values and may calculate statistical confidence limits.

Summarizing, we state that approaches to quantitative analysis of concentration-response experiments can be described in terms of the threshold and an underlying tolerance distribution.

Quantitative Risk Assessment

In quantitative risk assessment we are interested in the *additional risk over background*. The AROB is the probability of an adverse reaction directly related to treatment.

$$\text{AROB} = P(\text{reaction} \mid \text{treatment}) - P(\text{reaction} \mid \text{control})$$

Questions concerning risk assessment can be posed in two different ways. We can ask for the risk at a given concentration, or vice versa, ask for a concentration where an *acceptable* risk is expected. There are different approaches to answer these questions: the more traditional approach is to establish a *no-observed-effect-level* (NOEL) experimentally and then to divide this NOEL by a safety factor of e.g. 10, 30, 100 and call this value a *safe concentration.*

$$\text{Safe concentration} = \text{NOEL} \,/\, \text{safety factor}$$

The assumption behind this approach is that a global threshold exists and the combination of the NOEL and the chosen safety factor is assumed to guarantee that the global threshold is larger than the calculated value.

The more rational approach seems to be the use of *benchmark doses* (see Chen and Kodell 1989) - here called benchmark concentrations - and the application of safety factors. Benchmark concentrations are defined as lower confidence limits for the EC01 or EC05 values, the safety factors are used in a similar way as in the case of the NOEL approach. Chen and Kodell suggested a *linear extrapolation* from the benchmark in their 1989 publication - the safety factor is than dependent on a specified acceptable risk.

Safe concentration = benchmark / safety factor

Although a model has to be specified when estimating the benchmark, different models yield very similar results when the observed data can be fitted sufficiently.

The third possibility is the model-dependent estimation of safe concentrations. Model parameters are estimated and the extrapolation is performed using the model function and the estimated parameters.

Safe concentration = F^{-1} (negligible risk)

For a predetermined, practically negligible risk the *virtual safe dose (VSD) concept* for risk assessment in carcinogenicity testing was introduced by Mantel (1969) using a probit model with an assumed conservative slope of 1 for the extrapolation. The use of parametrical models for the tolerance distribution has been proposed by many authors, see e.g. Krewski and Van Ryzin (1981) or Armitage (1982) for an overview.

The choice of the magnitude of a negligible or acceptable risk is not a scientific decision. This choice is basically a political one and has to be made by the society, i.e. the regulatory authorities.

Specific Aspects: Endpoints and Predictors of Prenatal Toxicity

When transferring known methods to a new problem we always have to consider specific questions related to the subject, otherwise good mathematical methods might be inappropriately applied to the new field. Some important points were made by Neubert et al. in 1987.

We will make an attempt to characterize the specificity of prenatal toxicity tests. One possibility to do this is to look at the *endpoints* used to measure possible effects and to classify the *factors*, which have some influence on the results (Meister et al. 1991).

- Endpoints in prenatal toxicity testing are prenatal mortality, structural or functional anomalies and transplacental carcinogenicity. There are, of course, other effects which can be of interest in these experiments.
- Factors which have an influence on the results may be given by the experimental design and are therefore under the control of the experimenter, these are the selected species, the developmental stage at the time of treatment, naturally the substance used and the dose or concentration applied to the randomly assigned animals of the different treatment groups.
- There are, on the other hand, factors which are not under the control of the experimenter, so-called confounders, which can introduce some difficulties for the analysis. Such confounders are e.g. the litter-effect, sex, maternal toxicity, prenatal mortality, etc.

Prenatal Mortality - Response and Confounder

We will discuss some of the problems associated with possible confounders using an example. If we consider the endpoint "prenatal mortality" we see that this can also be regarded as a confounder. Chahoud (this volume) presented some data on resorption rates which increase with concentration. In that case, prenatal mortality was an important endpoint for the risk assessment. For other endpoints, however, prenatal mortality can be a serious confounder. Often all adverse reactions are combined and then there is of course no problem when the resorption rates increase with the dose. The study of a single and specific effect will be difficult or even impossible in that case. We use mathematical notations:

$$P(A) = P(A|S)\,P(S) + P(A|M)\,P(M),$$

where P denotes the *probability measure*, A means *anomaly*, S *survival* and M *mortality*.

In practice we can estimate P(A|S), but often we cannot decide whether a fetus has carried an anomaly if it died during pregnancy. Therefore, we cannot find an estimator for P(A|M) and, consequently, for P(A). We see from Table 1 the possible effects on the observed abnormality rates when mortality is present. For example, if we have observed 50% abnormalities among the living and we have 50% mortality, then the true underlying abnormality rate can vary between 25% and 75%.

With such results an enormous uncertainty for the risk assessment is present and there is no way to improve this situation with statistical tricks.

My personal conclusion regarding this problem is: We have to avoid prenatal mortality as far as possible in the experiment by the choice of the experimental design if single, specific endpoints are to be used for the determination of prenatal toxicity.

Table 1. Possible true response probabilities P(A) for different rates of mortality P(M) and abnormalities among the living P(A|S). The lower values correspond to the assumption P(A|M) = 0, the higher to P(A|M) = 1

P(M)	P(A|S)					
	0.1		0.25		0.5	
0.1	0.09	0.19	0.225	0.325	0.45	0.55
0.25	0.075	0.325	0.1875	0.4375	0.375	0.625
0.5	0.05	0.55	0.125	0.625	0.25	0.75

This goal is not easily achieved when looking at protocols which demand a treatment over a very long period associated with sometimes rather high mortality rates. When treatment is restricted to a short time interval we can get into other difficulties. The next example shall illustrate this problem.

Developmental Stage - Predictor of Response

In this example we show how the developmental stage can be used as a predictor of the response. We use some data from a large scale experiment performed by Dr. Chahoud and coworkers (personal communication).

It is well known (Neubert and Barrach 1977) that the occurrence of malformations is related to the developmental stage of the animals at the time of treatment. Using some simple assumptions on the developmental process, we can give a good explanation of extremely different response rates. The importance of an adjustment of response rates to the developmental stage is obvious when these data are the basis for risk assessment for man.

We make the following assumptions: similar developmental stages between species do exist. There might be different time to stage relationships: the developmental stages may be reached at different times of pregnancy. The last assumption is that the observed response rates, according to the underlying tolerance distributions, are stage-specific.

Not only between species, but even between different strains of the same species distinct time patterns of the developmental process may exist. The data we show in Table 2 demonstrate a dramatic difference between the response rates for the NMRI strain and the DBA strain if treated with the same dose of a well known teratogenic substance, methylnitrosourea (see Platzek et al. 1988). This difference can be reduced and explained if the developmental stage measured by the mean somite number on the different days of pregnancy is considered. From the parameter estimates given in Table 2 we deduce that somite growth proceeds with a similar rate in these two strains - about one somite every two hours - but the DBA strain starts with approximately half a day delay. If we consider a treatment on day 10 of pregnancy we expect a somite number of (10-7.85) x 11.95 = 24.6 somites for the NMRI strain. This stage will be reached by the DBA strain after 24.6/12.25 + 8.4 = 10.4 days of pregnancy. The good agreement of our prediction with the experimental results can be found in Table 3. When treatment is given at a comparable stage - here at day 10 for the NMRI strain and at day 10.5 for the DBA strain - not only the estimated somite numbers are more comparable but also the response rates increase from 22% for the DBA strain to 65% on day 10.5. Again, we showed that a factor not of primary interest determines the concentration-response relationship.

Table 2. Estimated parameters in the linear growth model for different strains of mice. The rate[1] gives the speed of growth, start[2] is the estimated onset of growth (e.g. E(*som(start)*) = 0)

	Rate	Start	Sample size
DBA	12.25	8.4	600
NMRI	11.95	7.85	1500

1 = Somites/day
2 = Day of gestation

Table 3. Induction of cleft palates by treatment with MNU (5 mg/kg i.p.). The estimated response rate p, the total number of living fetuses n, the number of litters l and the ratio DI = ... of the estimated variance (Gladen, 1979) to that expected under the binomial model are given for different times of treatment (from: *Meister et al. 1991*)

	DBA day 10	DBA day 10.5	NMRI day 9.5	NMRI day 10
p	0.22	0.10	0.65	0.89
n	109	220	27	134
l	30	22	5	10
DI	2.3	3.2	5.2	2.6

Conclusion: The developmental stage should be used as a covariate, or the treatment times should be adjusted to achieve comparable developmental stages if a comparison between species or strains is made. This point seems to be of special importance when performing risk extrapolation to man.

Multiple Endpoints

The next specific point we will discuss is the presence of multiple endpoints. We know that usually more than one anomaly may be induced and observed in one animal (embryo or fetus). How can the analysis deal with this problem? There are some proposals in the literature showing that a combined analysis of multiple endpoints is possible. For the problem of testing this was described by O'Brien (1984) for multiple endpoints in clinical studies. Lefkopoulo, Moore and Ryan (1989) have demonstrated that this approach can also be applied to data from teratological studies. They used the more general *estimation equation method*. For the comparison of several treatment groups this approach seems to be promising. The application to concentration response data, however, is still questionable. In teratology we know (Neubert and Barrach 1977) that reactions are dose and substance specific, but we do not have enough prior knowledge to predict which anomalies will occur at the different concentrations.

Conclusion: Up till now, no sufficient statistical approach to the simultaneous analysis of multiple endpoints in concentration-response experiments is available which could be applied on a routine basis.

Litter Effects - Additional Variance

We now make some remarks on the most prominent problem when analysing prenatal toxicity data: the description and incorporation of litter effects. Litter effects have to be taken into account since the experimental units are the treated female animals and we have multiple outcomes since each pregnant animal has 10 to 15 or more fetuses. For the problem of continuous variables Healy (1972) described litter effects using variance components. There have also been approaches using covariances or covariates. For the case of quantal response several alternatives to the binomial model, which represents the "no-litter-effect" case, have been proposed. There are the generalized binomial models introduced by Williams (1975), Kupper and Hasemann (1978) and Altham (1978). Other alternatives are the weighted least squares- and the quasi-likelihood methods which have been introduced by Kleinmann (1973) and Williams (1982) and the use of resampling techniques as Gladen (1979) and Frangos and Stone (1984).

We won't go into technical details of these approaches but we state that the methods are well described. They can be applied and are computationally feasible.

Conclusion: The problem of incorporating litter effects into the analysis of teratological data, which has received broad discussion in the biometrical literature, can now be handled. However, the biological interpretation and the biological significance of litter effects is still open for discussion.

Estimation of the Concentration-Response Curve

The investigations on the statistical properties of per litter analysis in teratological experiments showed that for a statistically correct concentration-response estimation the litters have to be incorporated (see Piegorsch and Haseman 1991a, 1991b for overview). Therefore, special models have been developed. Rai and Van Ryzin (1985) published a model in which litter effects had been taken into account via a *dose-dependent litter size*. The model has been reanalysed by Williams (1987) and Carr and Portier (1990). Both papers show that the Rai and Van Ryzin model has some drawbacks and cannot be recommended as standard for general use in analysing concentration-response in prenatal toxicity tests. Kupper and his coworkers (1986) showed how a *logistic model without background* rate could be fitted using a beta binomial likelihood. We proposed (Meister 1986) the use of *general tolerance distributions including background.* The approach was based on weighted nonlinear regression where the weights were derived from Gladen's (1979) jackknife variance estimator. Chen and Kodell (1989) used a *Weibull model* as type of the tolerance distribution incorporating *independent background* and using a beta binomial likelihood similar to Kupper et al. (1986). The beta binomial model was taken as control for litter effects.

Some Experiences with Concentration-Response Estimation in Prenatal Toxicity Testing

During the last years we have analysed several concentration-response curves from various substances. We did this always with the aim to improve risk assessment. Some of the most relevant facts observed in this process are reported in the following.

The recently introduced benchmark dose approach can be compared to the NOEL approach. An example will demonstrate this: If we assume that the background rate is zero we take an upper confidence limit for the possible underlying response probability at the NOEL. If there is no litter effect at the NOEL it turns out that for 30 litters with approximately 300 fetuses the 95% upper limit for the true underlying response rate is 1%. In other words, for these data the NOEL is a lower limit for the EC01 and that is exactly the definition of the benchmark concentration. There is no real controversy between the benchmark and the NOEL approach if the upper limit for the risk at the NOEL is taken into account. In general, however, a multiple number of animals will be needed to establish a NOEL compared to a model-based benchmark estimate.

The idea of a possibly existing global threshold is reflected by the observation that often the observed concentration-response relationships in prenatal toxicity testing are "J-shaped". Such a shape implies a low likelihood for linear low-concentration behaviour. Models such as the on-hit, which is linear for low doses, show significant lack of fit for J-shaped data. The observed shape is, of course, in strong relation to the measurement of the concentration at the target tissue. This seems to be a very important approach as we can get more simple models and perhaps less J-shaped curves if an appropriate effective concentration at the target tissue can be determined. Platzek (this volume) showed how such a determination could be performed.

We demonstrated that for steep and J-shaped data moderate safety factors in a magnitude of 30 to 100 applied to the NOEL, may yield lower limits than model-based extrapolation (Meister 1990). There is no general trend in using safety factors towards anticonservative safe concentrations. Therefore, it is an important point for regulatory agencies to be able to compare the model-based and the safety factor approach.

There is another point important for routine experiments where often only concentrations with zero or 100% response are observed. We have recently shown (Meister 1991) how lower confidence limits for the slope and, therefore a benchmark concentration, can still be estimated in this case.

Discussion

We draw some final personal conclusions to stimulate the discussion considering concentration-response models for quantitative risk assessment of prenatally induced adverse health effects.

1. *Model-based estimation of concentration-response curves is now possible for teratological data. Statistical properties of these estimators still need more evaluation.*

2. *Risk assessment based on NOEL and the benchmark concentration is in good agreement if the NOEL is accompanied by an upper confidence limit for the additional risk over background at the NOEL and this is reflected by the safety factor used.*

3. *We have to stress that we are still in no way able to deduce the low concentration behaviour of concentration-response curves from the observed data. The estimation of response rates at concentrations far beyond the experimental range is always speculative. It is also speculative to assume the existence of global thresholds. There is no way to confirm this hypothesis without a better understanding of mechanisms and biologically based models.*

4. *Risk assessment and estimation of virtual safe concentrations should reflect the observed concentration-response relationship and, therefore, all the ap-*

proaches should incorporate information on the observed toxicity, e.g., the NOEL or the benchmark concentration and on the biological variability of the effects. This means taking the information on the slope of the concentration-response model into account.

References

Altham PME (1978) Two generalizations of the binomial distribution. Appl Statist 27: 162-167

Ames BN, Gold LS (1990) Too many rodent carcinogens: Mitogenesis increases mutagenesis. Science 249: 970-971

Armitage P (1982) The assessment of low-dose-carcinogenicity. Biometrics 38 Suppl: *Current Topics in Biostatistics and Epidemiology*, pp 119-129

Carr GJ, Portier C (1990) An evaluation of the Rai and Van Ryzin dose-response model in teratology. in press

Chahoud I, Bochert G, Neubert D (1992) Dose-Response Relationships in Reproductive Toxicology: Importance of Skeletal Variations for Risk Assessment. In: Neubert D, Kavlock RJ, Merker H-J, Klein J, Webb D (eds) Risk Assessment of Prenatally-Induced Adverse Health Effects, Springer-Verlag, Heidelberg, pp

Chen JJ, Kodell RL (1989) Quantitative risk assessment for teratological effects. J Am Statist Assoc 84: 966-971

Frangos CC, Stone M (1984) On jackknife, cross-validatory and classical methods of estimating a proportion with batches of different sizes. Biometrics 71: 361-366

Gladen B (1979) The use of the Jackknife to estimate proportions from toxicological data in the presence of litter effects. J Am Statist Assoc 74: 278-283

Healy MJR (1972) Animal litters as experimental units. Appl Statist 21: 155-159

Kleinmann JC (1973) Proportions with extraneous variance: single and independent samples. J Am Statist Assoc 68: 46-54

Krewski D, Van Ryzin J (1981) Dose-response models for quantal response toxicity data. In: Csorgo M, Dawson D, Rao JNK and Saleh E (eds), Statistics and Related Topics, North-Holland, New York, pp 201-231

Kupper LL, Haseman JK (1978) The use of a correlated binomial model for the analysis of certain toxicological experiments. Biometrics 34: 69-76

Kupper LL, Hogan MD, Portier C, Yamamoto E (1986) The impact of litter effects on dose-response modeling in teratology. Biometrics 42: 85-98

Lefkopoulo M, Moore D, Ryan L (1989) The analysis of multiple correlated binary outcomes: Application to rodent teratology experiments. J Am Statist Assoc 84: 810-815

Mantel N (1969) Some statistical viewpoints in the study of carcinogenesis. In: Progress in Experimental Tumor Research 11, S. Karger, Basel, pp 432-443

Meister R (1986) Zur Analyse von Dosis-Wirkungsbeziehungen in der Embryopharmakologie. In: Proceedings of the Colloquium: Statistische Methoden in der experimentellen Forschung, FU Berlin

Meister R (1990) Safety limits and remaining risk - problems of risk assessment. Biomet Bull, abstract

Meister R, Chahoud I, Jürgens M, Iversen F, Bochert G (1991) Biometrical analysis of strain differences and litter effects. A case study. In: Hothorn L (ed), Lecture Notes in Medical Informatics 43: Statistical Methods in Toxicology, Springer, Berlin, Heidelberg, pp 96-103

Meister R (1991) Interval estimates in quantal response models, when maximum likelihood point estimates do not exist. Biometrics, submitted

Neubert D, Barrach H-J (1977) Organotropic effects and dose-response relationships in teratology. In: Neubert D, Merker HJ, Kwasigroch TE (eds), Methods in Prenatal Toxicology, Thieme, Stuttgart, pp 405-412

Neubert D, Chahoud I, Platzek T, Meister R (1987) Principles and problems in assessing prenatal toxicity. Arch Toxicol 60: 238-245

O'Brien P (1984) Procedures for comparing samples with multiple endpoints. Biometrics 40: 1079-1087

Piegorsch WW, Haseman JK (1991a) Per-litter analyses for studies in developmental toxicity. In: Hothorn L (ed) Statistical Methods in Toxicology. Lecture Notes in Medical Informatics, Vol 43, Springer, Berlin, Heidelberg, pp 88-95

Piegorsch WW, Haseman JK (1991b) Statistical methods for analysing developmental toxicity data. Teratogen Carcinogen Mutagen 11: 115-133

Platzek T, Bochert G, Pauli B, Meister R, Neubert D (1988) Embryotoxicity induced by alkylating agents: 5. Dose-response relationships of teratogenic effects of methylnitrosourea in mice. Arch Toxicol 62: 411-423

Platzek T (1992) Prenatal-toxic risk estimation based on dose-response relationships and molecular dosimetry. (this volume)

Rai K, Van Ryzin J (1985) A dose-response model for teratological experiments involving quantal response. Biometrics 41: 1-9

Williams DA (1975) The analysis of binary response from toxicological experiments involving reproduction and teratogenicity. Biometrics 31: 949-952

Williams DA (1982) Extra-binomial variation in logistic linear models. Appl Statist 31: 144-148

Williams DA (1987) Dose response models for teratological experiments. Biometrics 43: 1013-1016

Discussion of the Presentation

Bolt: You have presented the case where you have just two doses with responses of 0% and 100%, respectively. Is it meaningful to apply statistics in such an extreme case, or can one simply take the 0% response dose and apply some safety factor?

Meister: As I already mentioned, there is no possibility to check the model assumptions in this case. However, a sharp increase from 0% to 100% response at two adjacent doses gives additional information on the slope if an assumption about the underlying dose-response is made. Just using a safety factor means ignoring this information.

Palmer: Risk assessment modelling is largely based on very positive compounds but in testing we mainly get negative or equivocal values.

Meister: I must admit that working with very positive compounds might induce a biased view. We try to avoid this by stating that quantitative risk assessment is difficult for less "positive" compounds.

Palmer: NOELs are different for different variables in a study. These would vary for the same study in different labs.

Talk of mean and variance always worries me as most variables in reproductive toxicity are non normal. Calculation of a mean value followed by subtraction of one SD gives biologically impossible values.

Meister: We have to distinguish between descriptive statistics for observed experimental data, where the use of parametrical methods might be inappropriate and the discussion of biological variance in terms of an assumed underlying tolerance distribution. Here the use of parametrical terms is an essential for any quantitative model.

Neubert: The comment of Dr. Palmer concerning mean and median values is certainly very important, although not really relevant to this point, since no mean values of the data were used here. However, you rightly pointed out that many if not the majority of biological data do not necessarily show a normal distribution. We feel it is important to know this information and therefore, in almost all of our publications we give median values, ranges and Q_1- and Q_3-values.

Dose-Response Relationships in Reproductive Toxicology: Importance of Skeletal Variations for Risk Assessment

Ibrahim Chahoud, Gerd Bochert and Diether Neubert

Introduction

Evaluation of the embryotoxic risk of substances is a two-stage process:

- first the "effect" (or biological endpoint) has to be defined,
- and second, the dose-response for this (or these) endpoint(s) has to be assessed.

The effects, i.e. deviations from the norm (controls), as in the case of the skeletal system induced by xenobiotics, may be classified into two categories:

(1) The first category includes those anomalies which only occur in exposed fetuses. Although not entirely without problems the evaluation of this first category creates the least difficulties, especially when drastic and frequent gross-structural abnormalities are induced.

(2) The second category consists of an increased frequency of anomalies of exposed fetuses also occurring in the controls, i.e. *spontaneous anomalies* (which, for simplicity, will be designated here as *variations*). The toxicological significance of these deviations from the norm are considered very differently by various investigators.

Another attempt at classification would be to distinguish deviations from the norm according to severity of the effect. In this way "major" and "minor" abnormalities have been classified, also with respect to man (Heinonen et al. 1977). However, again all investigators use their own classification and no general agreement has been attempted.

A fourth possibility would be to distinguish anomalies which are "indirectly" inducible through maternal toxicity from "direct" interferences with embryonic development. Although some types of anomalies have been suggested to be induced "indirectly" (Khera 1984), this problem is far from being solved.

There is no general agreement about the necessity to detect and record all of the anomalies which belong to the second category. Anomalies which also occur in the controls, so-called *spontaneous anomalies* or *variations*, are not evaluated and documented in an accurate manner by all laboratories. This fact is obvious when the published different frequencies of such anomalies are compared, but it is also well-known that considerable strain differences exist (Palmer 1977; Chahoud 1988; Fritz 1990).

In order to assess the risk of embryotoxic agents there is a need for quantitative data and dose-response estimates for low exposure levels. Therefore, one is forced to use mathematical procedures to reveal the relationship between dose and effect. Provided sufficient data exist (which is often not the case) such an evaluation is not too difficult in the range of the established data. However, extrapolations are difficult or impossible to very low exposure levels which can never be directly ascertained. There are a number of extrapolation models which fit dose-response data for anomalies and allow some predictions for a possible risk at low levels of exposure.

From the notation that teratogenic substances may increase the frequency of *variations*, it follows that the shape and course of the dose-response curves will be different if an increase of the *variations* in exposed fetuses is recorded and considered as substance-related or not. Consequently, the decision to consider *variations* may be crucial for risk assessment of embryotoxic agents and the strategy to be followed should be clarified in advance (but it is not generally agreed upon).

The purpose of this paper is to provide examples on the relationship between the determination of exposure and the assessed risk (dose-response relationship) for gross-structural abnormalities. Dose-response relationships are determined with or without considering the *variations* which are increased as the result of the exposure.

Mathematical transformations were used to assess the risk, and the data from dose-response relationship studies of three well-known teratogenic substances: 5-fluoro-desoxyuridine (FUDR), hydroxyurea (HU) and 6-mercaptopurine-ribosid (6-MPr) were used for the calculations.

Materials and Methods

Experimental Design

Wistar rats (Bor:Wisw/spf, TNO) were used. Animal maintenance, the mating procedure, as well as the teratological procedure, have been described elsewhere (Chahoud et al. 1988). Dams were treated with a single dose on day 11 of pregnancy with FUDR, HU or 6-MPr, respectively. The dose regimens of all three substances are shown in Table 1. Altogether, 4330 fetuses from 445 dams were evaluated for the three studies on dose-response relationships.

In order to estimate whether the frequency of *variations* in exposed fetuses is increased when compared to the controls, the anomalies of control fetuses were recorded accurately. As a data-base historical controls from our laboratory were used for the comparative assessment.

Table 1. Dosage regimens

Substance	Dose (mg/kg)					
HU	250	300	350	400	450	---
FUDR	3	14	25	35	55	65
6-MPr	3	7	10	14	---	---

Procedures used for Risk Assessment

Based on the data from the dose-response relationship study, risk assessment was performed using the fetus as statistical unit. For statistical analysis SAS software was used (Ray 1985). The independent background model "Gompertz distribution, log10 dose" was employed to estimate the slope as well as the effective doses 50, 10 and 1 for the additional risk over background (AROB).

In some instances also a NOEL (no-observed-effect-level; we deliberately do not state: no-observed-adverse-effect-level in this case) was established. We rather favour the concept of calculating dose-response relationships (curves) and establishing ED_x-values (Bass et al. 1985; Neubert et al. 1987; Platzek et al. 1982).

Up till now, a controversy exists whether *variations* which show a dose-response relationship have to be considered as substance-related effects or not. Therefore, the dose-response relationship was determined proceeding from the assumption that the term "effect" includes either:

(a) all abnormalities including *variations* which increase substance-relatedly, or
(b) all abnormalities excluding these *variations*, or
(c) all abnormalities of one selected skeletal part including its *variations*, or
(d) all abnormalities of one selected skeletal part excluding these *variations*, or
(e) selected gross-structural abnormalities which do not occur in the controls.

Results

Abnormalities in Historical Controls

Variations of the rat strain used in our laboratory are shown in Table 2. The data of 491 untreated dams with 4811 fetuses were evaluated. Only anomalies with a frequency equal or higher than 0.25% were considered. It should be noted that one fetus may have more than one anomaly. Twenty-six different anomalies were observed. The frequency of only four of them was higher than 2%.

Table 2. Anomalies observed in historical controls

Location	Anomaly	Fetuses with anomalies	
		N	%
Forelimbs			
Finger	incompl. oss.	48	0.99
Forepaw	incompl. oss.	42	0.87
Humerus	bent	21	0.43
Humerus	irreg. dim.	19	0.39
Radius	bent	18	0.37
Scapula	bent	17	0.35
Hindlimbs			
Hindpaw	irreg. pos.	35	0.72
Skull			
Back	incompl. oss.	25	0.51
Os frontale	incompl. oss.	16	0.33
Os interpari.	incompl. oss.	68	1.41
Os parietalis	incompl. oss.	76	1.57
Os parietalis	oss. cent. irreg.	21	0.43
Os zygomaticum	fused	528	*10.90*
Os zygomaticum	incompl. oss.	27	0.56
Os zygomaticum	oss. cent. irreg.	21	0.43
Sternum			
	add. oss. cent.	18	0.37
	dislocated	245	*5.09*
	irreg. dim.	18	0.37
	irreg. shape	48	0.99
Thorax			
Ribs	wavy	284	*5.90*
Trunk			
	hematome	35	0.72
	int. bleeding	20	0.41
Vertebral column			
Th. vert.	missing	26	0.54
Th. vert	oss. cent., bicent., sym.	19	0.39
Th. vert.	dumb-bell shaped	98	*2.03*

Number of litters: 491
Number of living fetuses: 4811
Percentage limit: ≥0.25%

Abbreviations: incompl. oss. = incomplete ossification, irreg. dim. = irregular dimension, add. oss. cent. = additional ossification centre, bicent. = bicentric, sym. = symmetrical, Th. vert. = thoracic vertebrae

Dose-Response Relationship Studies

Dose-response relationship studies with 5-fluoro-desoxyuridine, 6-mercapto-purine-ribosid and hydroxyurea were carried out. Tables 3, 4 and 5 show some sectio data (number of dams, number of fetuses and resorption rate) and an example of *variations* which increased dose-relatedly.

In the groups treated with *FUDR* the highest resorption rate (18%) was recorded in the group dosed with 65 mg/kg body wt. The frequency of the control anomaly "thoracic vertebrae dumb-bell shaped" was dose-relatedly increased (Table 3).

The resorption rate in the groups treated with *6-MPr* was lower than 10% in all groups. Compared to the control, the frequency of the anomaly: "os zygomaticum fused" appeared in the control with 13%, and this rate increased to 22% in the lowest treated group (3 mg/kg body wt) and to 44% in the highest treated group (14 mg/kg) (Table 4).

Table 3. Dose-response relationship study with 5-fluoro-desoxyuridine. Selected sectio data. Some *variations* increased dose-dependently. Dams were treated (s.c.) with a single dose of 5-fluoro-desoxyuridin on day 11 of pregnancy

		5-Fluoro-desoxyuridin (mg/kg)					
	Control	3	14	25	35	55	65
Number of dams	47	18	23	21	20	15	15
Number of fetuses	481	176	221	200	207	157	136
Resorptions (%)	3.8	5.9	6.6	2.8	4.6	4.3	18.4
Th. vert. dumb-bell shaped(%)	*3*	*5*	*14*	*27*	*34*	*66*	*68*

Table 4. Dose-response relationship study with 6-mercaptopurine-ribosid. Selected sectio data. Some *variations* increased dose-relatedly. Dams were treated (s.c.) with a single dose of 6-mercaptopurine-ribosid on day 11 of pregnancy

		6-Mercaptopurine-ribosid (mg/kg)			
	Control	3	7	10	14
Number of dams	41	15	17	37	16
Number of fetuses	373	152	177	348	148
Resorptions (%)	8.8	9.4	6.3	7.1	9.7
Os zygomaticum fused (%)	*13*	*22*	*28*	*36*	*34*

Hydroxyurea caused a dose-related increase of two *variations*. The anomaly "os zygomaticum fused" occurred with a frequency of 43% in the highest dose group (450 mg/kg body wt) and the frequency of the anomaly: "thoracic vertebrae dumb-bell shaped" was 1% in the control and raised to 70% in the group dosed with 450 mg/kg body wt; the resorption rate in this group was 26% (Table 5).

Table 5. Dose-response relationship study with hydroxyurea. Selected sectio data. Some *variations* increased dose-related. Dams were treated (s.c.) with single dose hydroxyurea on day 11 of pregnancy

		Hydroxyurea (mg/kg)				
	Control	250	300	350	400	450
Number of dams	53	18	17	21	34	17
Number of fetuses	559	154	188	213	315	125
Resorptions (%)	4.1	12.7	6.0	3.6	13	26.4
Os zygomaticum fused (%)	*9*	*10*	*19*	*20*	*38*	*43*
Th. vert. Dumb-bell shaped (%)	*1*	*14*	*39*	*46*	*65*	*70*

Determination of a NOEL

Tables 6, 7 and 8 present the values of NOEL according to the procedure described above (see Methods: procedures used for risk assessment).

The NOEL for *FUDR* (Table 6) was four times higher when all anomalies or the anomalies affecting the vertebral column were considered *without* taking the incidence of the anomaly "dumb-bell shaped" in account. Regarding the NOEL based on the frequency of gross-structural anomalies, such as "cleft palate", it was 15 times higher (45 mg/kg) than the NOEL for all anomalies (3 mg/kg).

Table 6. Anomalies induced by 5-fluoro-desoxyuridine in rats. Determination of NOEL and extrapolation of safety dose (mg/kg) using the risk factor (10; 100) approach

Location	Anomaly	NOEL/ 1	NOEL/ 10	NOEL/ 100
Fetus	all	3	0.3	0.03
Fetus	all excl. th. vert. dumb-bell shaped	14	1.4	0.14
Vertebr. Column	all	3	0.3	0.03
Vertebr. Column	all excl. th. vert. dumb-bell shaped	14	1.4	0.14
Th. vert	dumb-bell shaped	3	0.3	0.03
Lumbal vert.	*fused*	25	2.5	0.25
Palate	*cleft*	45	4.5	0.45

Printed in italic: clear-cut "malformations"

The frequency of all anomalies in fetuses exposed to the lowest dose of *6-MPr* (3 mg/kg body wt) was still higher than in non-exposed fetuses. Excluding the anomaly: "os zygomaticum fused", the NOEL was 3 mg/kg. According to the gross-structural anomalies, such as "radius short" and "cleft palate", the NOEL was 7 mg/kg (Table 7).

Table 7. Anomalies induced by 6-mercaptopurine (6-MPr) in rats. Determination of NOEL and extrapolation of safety dose (mg/kg) using the risk factor (10; 100) approach

Location	Anomaly	NOEL/ 1	NOEL/ 10	NOEL/ 100
Fetus	all	< 3		
Fetus	all excl. os zygom. fused	3	0.3	0.03
Os zygomaticum	fused	< 3		
Skull	all	< 3		
Radius	*short*	7	0.7	0.07
Palate	*cleft*	7	0.7	0.07

Printed in italic: clear-cut "malformations"

Regarding the gross-structural anomalies (tibia missing, cleft palate) induced by *hydroxyurea* the NOEL was 350 mg/kg. When all anomalies, excluding the anomalies "os zygomaticum fused" and "thoracic vertebrae dumb-bell shaped" were considered, the NOEL was 250 mg/kg (Table 8).

Variables of Dose-Response Curves

The variables of dose-response curves according to different anomalies are shown in Tables 9, 10 and 11. The independent background model "Gompertz distribution, log10 dose" was employed to estimate the slope as well as the effective doses (ED) 50, 10 and 1 for the additional risk over background (AROB).

Table 8. Anomalies induced by hydroxyurea. Determination of NOEL and extrapolation of safety dose (mg/kg) using the risk factor (10; 100) approach

Location	Anomaly	NOEL/ 1	NOEL/ 10	NOEL/ 100
Fetus	all	< 250		
Fetus	all excl. os zygom. fused	< 250		
Fetus	all excl. th. vert. dumb-bell shaped	250	25	2.5
Fetus	all excl. os zygom. fused and th. vert. dumb-bell shaped	250	25	2.5
Skull	all	250	25	2.5
Skull	all excl. os zygom. fused	250	25	2.5
Os zygomaticum	fused	250	25	2.5
Vertebr. Column	all	< 250		
Vertebr. Column	all excl. th. vert. dumb-bell shaped	< 250		
Th. vert.	dumb-bell shaped	< 250		
Os tympanicum	*missing*	250	25	2.5
Tibia	*missing*	350	35	3.5
Palate	*cleft*	350	35	3.5

Printed in italic: clear-cut "malformations"

Table 9. Variables of dose-response curves. Anomalies induced by 5-fluorodesoxyuridine (FUDR) in rats (ED values in mg/kg)

Location	Anomaly	Slope	Extrapolated ED50	ED10	ED1
Th. vert.	dumb-bell shaped	3.5	*45.9*	13.1	*2.7*
Fetus	all	6.0	*32.1*	15.6	6.0
	all excl. th. vert. dumb-bell shaped	8.1	37.7	21.8	*11.1*
Vertebr. Column	all	4.7	34.2	13.5	4.2
	all excl. th. vert. dumb-bell shaped	6.9	44.1	23.6	10.8

Table 10. Variables of dose-response curves. Anomalies induced by 6-mercaptopurine (6-MPr) in rats (ED values in mg/kg)

Location	Anomaly	Slope	ED50	Extrapolated ED10	ED1
Fetus	all	7.5	*9.0*	5.0	2.4
	all excl. os zygom. fused	8.2	9.3	5.5	2.8
Os zygomaticum	fused	1.5	*43.1*	2.3	*0.06*
Skull	all	3.6	16.7	5.0	1.1
Radius	short	13.9	15.6	11.4	*7.7*

Figures 1 to 6 present the courses of the dose-response curves. In the figures the spontaneous response (of the control values) as well as the ED1- and ED10-values for the treated groups are shown. The dose-response curves as well as ED1 and ED10 are related to the frequency of different anomalies, including or excluding the frequency of certain *variations*. The figures are presented as linear-linear plots of observed anomalies and responses to log10(dose) estimated using Gompertz distribution.

The tables as well as the figures clearly indicate that the variables of the dose-response curves depend on the anomalies which are considered. To illustrate this fact the variables of the dose-response curve for hydroxyurea are compiled in Table 11. The table shows the different slopes of the curves for different anomalies (range: 7.3 - 25.4). Concerning the estimated effective doses for different anomalies the change between the different ED1-values is greater than the corresponding change in the ED50-values. For instance, the ED50 ranged between 322 and 472, while the ED1 was calculated to be between 92 and 277.

Table 11. Variables of dose-response curves. Anomalies induced by hydroxyurea (HU) in rats (ED values in mg/kg)

Location	Anomaly	Slope	Extrapolated ED50	ED10	ED1
Fetus	all	10.4	327	215	128
Fetus	all excl. os zygom. fused	10.3	328	215	127
Fetus	all excl. th. vert. dumb-bell shaped	16.8	363	280	203
Fetus	all excl. os zygom. fused and th. vert. dumb-bell shaped	17.5	369	288	212
Skull	all	20.7	395	320	246
Skull	all excl. os zygom. fused	25.4	407	343	*277*
Os zygomaticum	fused	10.1	*472*	307	180
Vertebr. Column	all	7.8	*322*	185	*92*
Vertebr. Column	all excl. th. vert. dumb-bell shaped	13.0	372	267	176
Th. vert.	dumb-bell shaped	7.3	360	198	94
Os tympanicum	*missing*	18.4	435	344	256

Printed in italic: clear-cut malformations

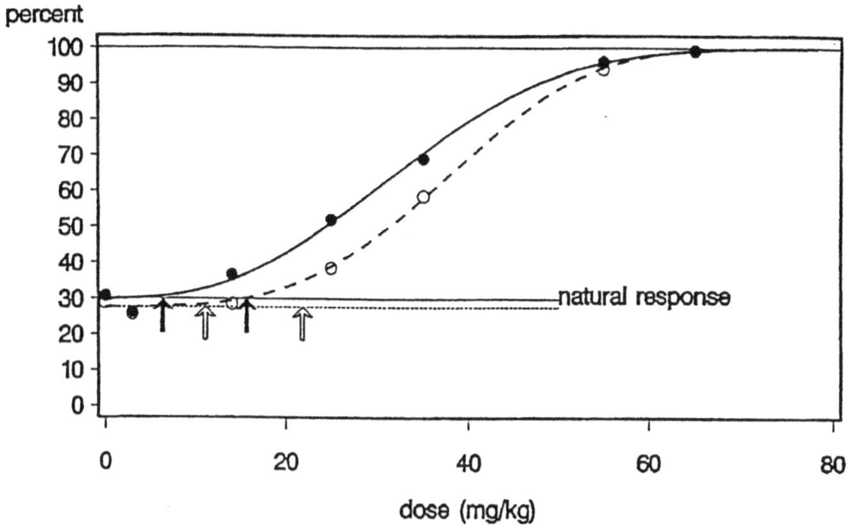

Fig. 1. All anomalies vs. dosage of FUDR. Spontaneous response: frequency of anomalies in the control. *Black circles*: dose-response relationship curve of all anomalies. *Black arrows*: ED1 and ED10 for the black curve. *White circles*: dose-response relationship curve of all anomalies *exclusive* the anomaly "dumb-bell shaped". *White arrows*: ED1 and ED10 for the white curve

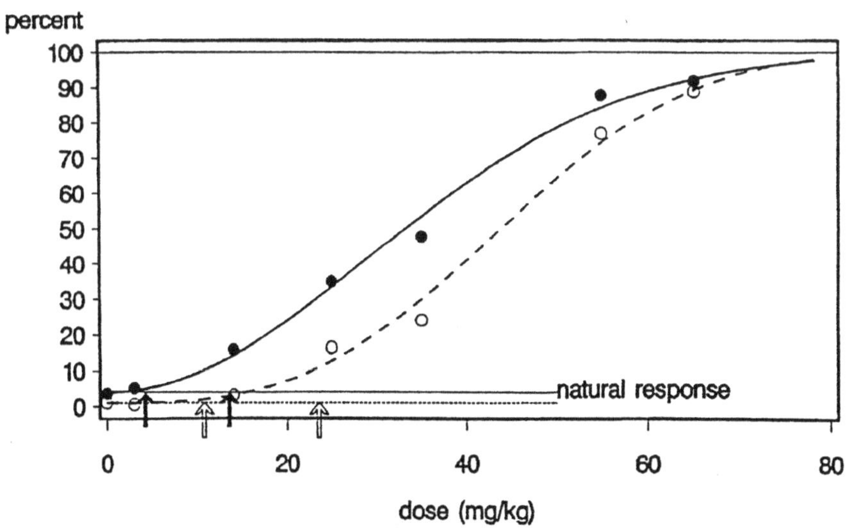

Fig. 2. Anomalies of the vertebral column vs. dosage of FUDR (legend as in Fig. 1)

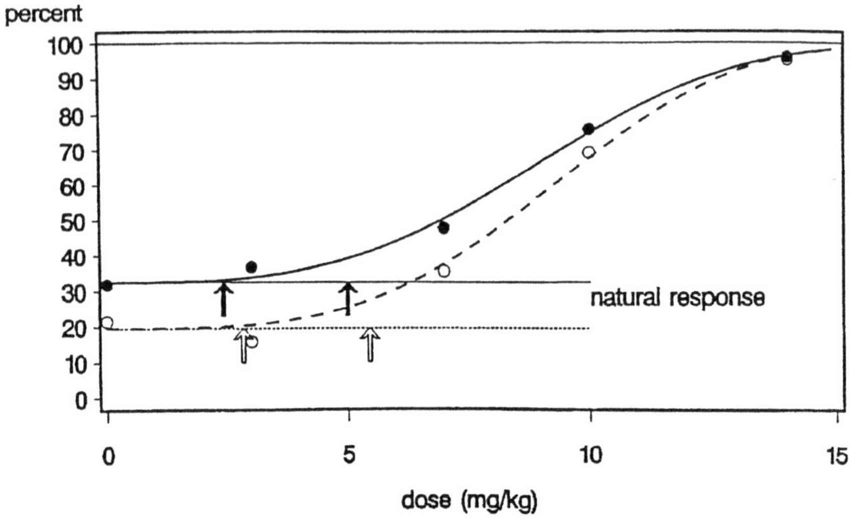

Fig. 3. All anomalies vs. dosage of 6-MPR. Spontaneous response: frequency of anomalies in the control. *Black circles*: dose-response relationship curve of all anomalies. *Black arrows*: ED1 and ED10 for the black curve. *White circles*: dose-response relationship curve of all anomalies *exclusive* the anomaly "os zygomaticum fused". *White arrows*: ED1 and ED10 for the white curve

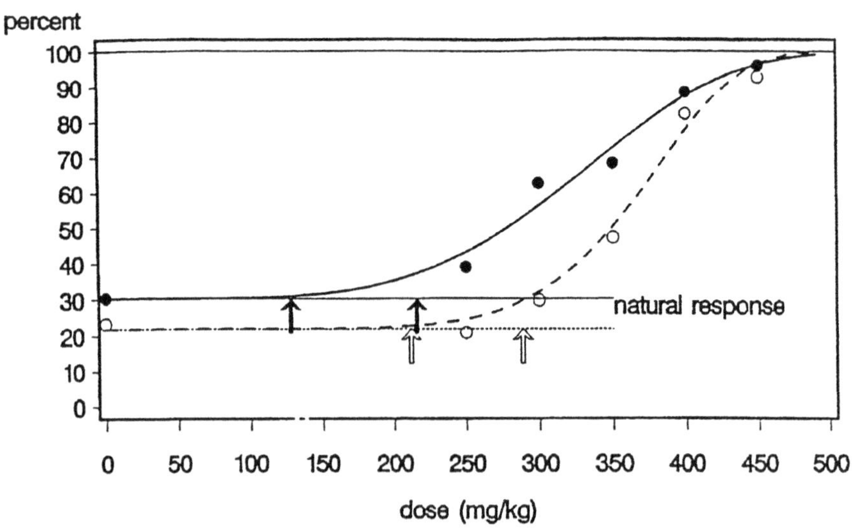

Fig. 4. All anomalies vs. dosage of HU (legend as in Fig. 3)

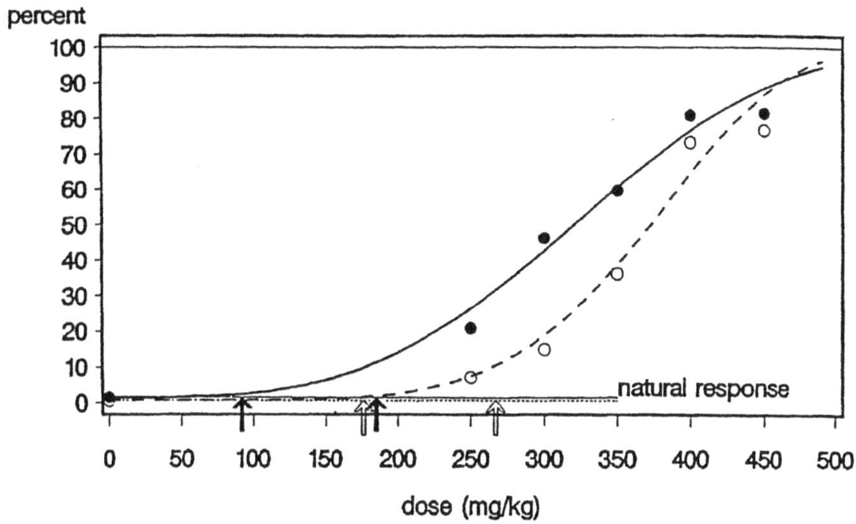

Fig. 5. Anomalies of vertebral column vs. dosage of HU. Spontaneous response: frequency of anomalies in the control. *Black circles*: dose-response relationship curve of all anomalies in the vertebral column. *Black arrows*: ED1 and ED10 for the black curve. *White circles*: dose-response relationship curve of all anomalies in the vertebral column *exclusive* the anomaly "dumb-bell shaped". *White arrows*: ED1 and ED10 for the white curve

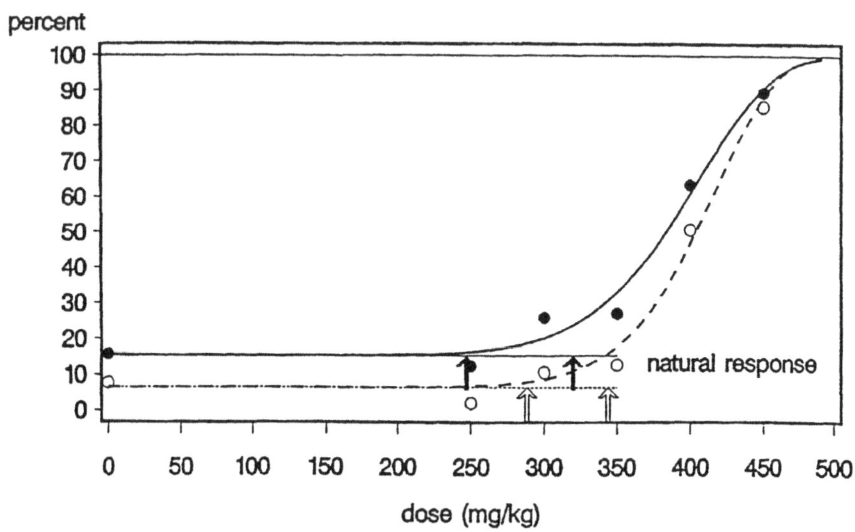

Fig. 6. Anomalies of skull vs. dosage of HU (legend as in Fig. 5)

Discussion

Numerous approaches to assess the risk of xenobiotics are based on the NOEL (Wang 1988; Gaylor and Kodell 1980). This may give the impression that the determination of the "no-observed-*effect*-level" is clear and simple. However, it ignores the fact that the definition of the variable "effect" in different fields of toxicology, in our case assessment of developmental toxicity, is not clear or simple at all.

Furthermore, establishing a NOEL mainly depends on the experimental conditions chosen (e.g. the choice and the number of dose levels), and the assessment may finally rely on a single data point. For this reason, several years ago we suggested (Bass et al. 1985; Neubert et al. 1987; Platzek et al. 1982) calculating ED_x-values from dose-response curves, e.g. ED_{10} (or even ED_1), which also have the advantage of being directly correlated with data on maternal toxicity. This approach has also been used in this presentation.

The data presented in this paper show that the decision whether the increase in the frequency of *variations* should be considered as a substance-related "toxic" manifestation is the first and most important aspect for determining dose-response relationships. Defined values calculated exclusively on the basis of gross-structural abnormalities (to be considered clear-cut "malformations") may be many times higher than values based on all anomalies, including *variations*.

The examples given in this paper indicate that the frequency of certain *variations* in exposed fetuses may increase in the case of substances clearly to be considered as "teratogenic". Such a dose-related increase is certainly a substance-related effect since a dose-dependency in toxicology is often taken as the best proof for a substance-related effect.

However, there is still a considerable controversy whether such increased rates of *variations* or "minor abnormalities" should be considered as relevant for a risk assessment, or whether such an approach is over-conservative. One stand is taken in the guidelines for the Health Assessment of Suspect Developmental Toxicants (1986) stating: "... *if variations are significantly increased in dose-related man-*

ner, these should also be evaluated as a possible indication of developmental toxicity..." However, this interpretation is not shared by many agencies, especially when a possible reproductive hazard of medicinal products is to be evaluated. It is presently difficult to dismiss the possibility that such increased rates of *variations* are induced (dose-dependently) via maternal toxicity, and thereby with *any* agent when using high enough doses and a sufficiently scrutinizing evaluation.

Conclusions

Depending on the biological endpoint to be evaluated the dose-response and defined effective doses (ED_x-values) calculated during the assessment of reproductive toxicity may, under the experimental conditions, differ considerably. The result will depend on the decision whether minor defects (which may even be transient) are given so much weight to consider them clearly "embryotoxic", especially if they occur at dose levels close to maternal toxicity.

The decision which biological endpoint to include in the evaluation is solely a toxicological one, rather than a function of the state of the art of analytical statistics. Unfortunately, presently the scientific basis for making such a decision and for solving such problems is meagre. The question is often put forth, but cannot finally be answered. Many more data are needed to allow meaningful decisions on this topic.

References

Bass R, Neubert D, Stötzer H, Bochert G (1985) Quantitative dose-response models in prenatal toxicology. In: Vouk VB, Butler GC, Hoel DG, Peakall DB (eds) Methods for Estimating Risk of Chemical Injury: Human and Non-human Biota and Ecosystems, SCOPE, pp 437-456

Chahoud I, Stahlmann R, Bochert G, Dillmann I, Neubert D (1988) Gross-structural defects in rats after aciclovir application on day 10 of gestation. Arch Toxicol 62: 8-14

Chahoud I, Bochert G, Fischer B, Neubert D (1988) Substance-induced changes in the frequencies of skeletal anomalies also present in rat control fetuses. Teratology 38: 9A

Environmental Protection Agency (1986) Guidelines for the health assessment of suspect developmental toxicants. Fed Reg 51: No. 185, pp 34028-34040

Fritz H, Giese K (1990) Evaluation of the teratogenic potential of chemical in the rat. Pharmacology 40: 1-28

Gaylor DW, Kodell RL (1980) Linear interpolation algorithm for low dose risk assessment of toxic substances. J Environ Pathol Toxicol 4: 305-312

Heinonen OP, Slone D, Shapiro S (1977) Birth Defects and Drugs in Pregnancy. PSG Publ Co Inc, Littleton Mass, pp 82-108

Khera KS (1984) Maternal toxicity - a possible factor in fetal malformations in mice. Teratology 29: 411-416

Neubert D, Bochert G, Platzek T, Chahoud I, Fischer B, Meister R (1987) Dose-response relationships in prenatal toxicity. Congen Anom 27: 275-302

Palmer AK (1977) Incidence of sporadic malformations, anomalies and variations in random bred laboratory animals. In: Neubert D, Merker HJ, Kwasigroch TE (eds) Methods in Prenatal Toxicology, Georg Thieme Publ, Stuttgart, pp 52-71

Platzek T, Bochert G, Schneider W, Neubert D (1982) Embryotoxicity induced by alkylating agents: 1. Ethylmethansulfonate as a teratogen in mice - a model for dose-response relationships of alkylating agents. Arch Toxicol 51: 1-25

Ray, AA (1985) SAS User's Guide: Statistics, Version 5, SAS Institute Inc., Edition Cary, NC.

Wang GM (1988) Regulatory decision making and the need for and the use of exposure data on pesticides determined to be teratogenic in test animals. Teratogen Carcinogen Mutagen 8: 117-126

Discussion of the Presentation

Buschmann: What was the reason for your idea to exclude just the dose-dependent spontaneous malformations? I, maybe, would have chosen just them as a basis for an assessment of dose-response relationships.

Chahoud: We certainly didn't exclude them. We evaluated once with and once without. The dose-dependent malformation, what we called "spontaneous anomalies" (they also occur in controls), was chosen for two reasons: 1.) The dose-dependent increase of the incidence of any anomaly is the evidence for substance-relationship and 2.) only the dose-dependent increased "spontaneous anomalies" influence the incidence of all anomalies, and therefore the variables of the dose-response curves are different if the incidence of these spontaneous anomalies are included or excluded.

In answer to the second part of your question, I would support such a proposal because it seems that "spontaneous anomalies" are sensitive against teratogens. In our presentation we have demonstrated that, for instance, the NOEL value as calculated according to the anomaly "dumb-bell shaped" was 15 times lower than for the anomaly "cleft palate" in fetuses exposed to FUDR.

Claude: A question to the methology: for a single administration in a segment II study in the rat, is day 11 the most relevant?

Chahoud: Generally, I would suggest day 10 or 11.

Prenatal-Toxic Risk Estimation Based on Dose-Response Relationships and Molecular Dosimetry

Thomas Platzek, Gerd Bochert, Reinhard Meister and Diether Neubert

Introduction

The final purpose of toxicological risk assessment is to determine a human exposure level with acceptable risk to the population. For the great majority of toxic effects the most frequently used approach is the NOAEL risk factor approach: 1. A dose is experimentally determined which induces "no effect" ("no-observed-adverse-effect-level", NOAEL); 2. From this NOAEL a "safe dose" is assessed by division with an arbitrary factor.

In risk assessment of carcinogenic agents much attention was paid, especially by U.S. regulatory agencies, to extrapolation with mathematical models which were derived from dose-response relationships. This procedure is based on the assumption that, in the case of genotoxic agents, no threshold exists and the risk only decreases at lower exposures. With regard to prenatal toxicology there are only few approaches in this direction.

In carcinogenic risk assessment of DNA modifying agents an interesting approach uses molecular dosimetry. Based on experimentally obtained data on DNA adduct rates it was attempted to assess the risk of lower doses (Lutz 1979, 1986, 1987).

The aim of our studies was to assess the prenatal-toxic risk of low doses of alkylating model compounds in NMRI mice by comparative use of different risk assessment procedures. The substances used were the direct acting alkylating agents methylnitrosourea (MNU) and ethylmethanesulfonate (EMS).

This approach was based on previous dose-response studies on teratogenicity which were performed for the three alkylating agents acetoxymethyl-methylnitrosamine (DMN-OAc), EMS and MNU (Platzek et al. 1982; 1983; 1988). Derived from these dose-response studies dose-response relationships were established using probit analysis (Finney 1971). In prenatal toxicity, we have applied this dose-response model already since 1973 (Neubert et al. 1973, Neubert and Barrach 1977), and it was also used later by Biddle (1978). In our approach primarily the probit model and, in addition, further mathematical transformations were used (Neubert et al. 1987a; 1987b).

Furthermore, using these agents DNA adduct formation in the teratogenic dose range was studied in our laboratory. By combining both of these experimental procedures we could provide evidence for our hypothesis that a correlation exists between the initial adduct rate at the O6-guanine site in the DNA of mouse embryos and the teratogenic potency (Bochert et al. 1987; Neubert et al. 1987a; Neubert et al. 1987b; Platzek et al. 1987b). This correlation provides a reasonable basis for using the specific DNA adduct O6-alkylguanine as a molecular dosimeter for prenatal-toxic risk estimation at somewhat lower doses.

Our intention was to compare various procedures of prenatal toxic risk estimation:

1. To assess a virtually safe dose using the NOAEL risk factor approach,
2. to extrapolate the risk of low doses using probit analysis,
3. to extrapolate the risk of low doses by linear extrapolation to zero,
4. to extrapolate the risk of low doses using dosimetry.

It should be stressed that molecular dosimetry provides a measure of the exposure (or internal dose, as the name indicates) and *never* an indication of an effect. For this reason we attempted extrapolations just in the vicinity of measurable effects (not more than two orders of magnitude below the NOAEL).

Material and Methods

The studies were performed on random-bred albino mice (NMRI:Han, Zentralinstitut für Versuchstierkunde, Hannover) kept under spf-conditions and a constant dark-light cycle (dark period from 8:00 p.m. to 9:00 a.m.) at 21 ± 1°C and

50 ± 5% relative humidity. Generally, the animals were acclimatized for two weeks before mating. The animals were fed Altromin[R] 1324 and tap water ad libitum. The mating period was 2 hours (7:00 to 9:00 a.m.). Animals with a body weight between 29 and 35 g were mated. When vaginal plugs were detected the 24 hour period following the mating period was designated day 0 of pregnancy (Chahoud and Kwasigroch 1977).

In the teratogenicity experiments the dams were treated with a single i.p.-dose on day 11 of pregnancy. The animals were sacrificed on day 18 of pregnancy, the fetuses were weighed and inspected macroscopically. After fixation, clearing and staining of the skeleton according to the method of Dawson (1926) all skeletal abnormalities were registered. For the establishment of dose-response relationships only gross-structural abnormalities were used.

DNA alkylation rates of the purine bases 7-methylguanine, O6-methylguanine, and 3-methyladenine in the case of MNU and of the respective ethyl derivatives in the case of EMS in the embryos were determined using a method previously described (Bochert et al. 1978; Platzek et al. 1987a). This method includes the following steps:

(1.) The dams were treated i.p. on day 11 of pregnancy, (2.) 1 h following treatment the embryos were dissected, (3.) nuclear DNA was isolated using a modified phenol method, and (4.) the DNA was hydrolysed to the purine bases (0.1 N HCl, 16 h, 37°). (5.) The non-labelled alkylated bases were added as markers and the natural and alkylated purine bases were separated using cation exchange HPLC. The stationary phase was Dionex DC-6A (Pierce Inc.), mobile phase: ammonium formate (0.2 M). Separation was performed at 80°C oven temperature, the flow was 0.6 ml/min, the pressure was 30 at. A pH gradient was used starting at pH 4.2 and ending at pH 8.5. (6.) The fractions containing the bases were collected and the radioactivity of the fractions was determined by LSC.

The arithmetical and graphical evaluation was performed using a VAX/VMS computer and the SAS software. We fitted the dose-response data to some mathematical models and then used these functions for extrapolation which we confined to the hypothetical incidence 0.1% which, on the one hand, is not too far away from our experimental data on teratogenicity and, on the other hand, in the corresponding dose range we expected to be able to measure DNA adduct rates.

Results and Discussion

Assessment of a "Safe Dose" Using the NOAEL Risk Factor Approach

Table 1 compiles some selected data on the embryotoxicity observed in two experimental series with MNU. In the first two columns some data of our dose-response study are given which show the results of the lowest dose of MNU which was evaluated in that study (2 mg/kg body wt, 9 litters evaluated) and the sum of the controls of the study (58 litters). Treatment with MNU did not result in an increase in the resorption rate, no fetuses bearing gross-structural abnormalities of the extremities were observed and no fetuses with gross-structural abnormalities were found at all in this group. Fetal weights were slightly reduced, but the number of weight-reduced fetuses was not significantly increased (Chi^2-test, Platzek et al. 1988).

In a further experiment this group (2 mg/kg) was increased to 30 litters with 361 fetuses (Table 1, 5th column, NOAEL study). The resorption rate did not exceed the rate of the corresponding controls. The variables "weight-reduced fetuses" and "no. of fetuses with anomalies" were not increased compared to the controls. But, fetuses with gross-structural abnormalities of the extremities were observed (1.9%). These effects were not observed in any of the pooled control fetuses of both studies (1025 fetuses). Thus, a small teratogenic potential of the dose is evident from the occurrence of limb abnormalities.

Although the number of litters was small, in the lower dose group of 1 mg MNU per kg body wt embryotoxicity was not found to exceed the rate of controls, and we designated this dose as NOAEL under our experimental conditions. By use of an arbitrarily chosen risk factor of 30 a "safe dose" of 0.03 mg/kg would be obtained.

Table 1. Selected embryotoxicological data, NMRI mice, single i.p. treatment with MNU on day 11

	Dose-response study		Noel study			
Treatment group (dose mg/kg)	2	Control	1	Control	2	Control
No. of litters	9	58	12	9	30	23
Living fetuses	115	661	150	92	361	272
Fetal weight (g), M ± SD	1.07 ±0.08	1.15 ±0.12	1.14 ±0.12	1.18 ±0.11	1.09 ±0.10	1.16 ±0.12
No. of fetuses with gross-struct. abnormalities (%)	0	8 (1.2)	4 (2.7)	8 (8.7)	11 (3.0)	13 (4.8)
No. of fetuses with gross-struct. limb abnormalities (%)	0	0	0	0 (1.9)	7	0

In the case of EMS the NOAEL was 100 mg per kg body wt and a "virtually safe dose", assessed by dividing by 30, was 3.3 mg EMS per kg body wt (data not shown).

Extrapolation Using Models with Mathematical Transformations

A further possibility for the estimation of the risk of low doses is given by using mathematical models after transformation of the data. This procedure estimates the risk as follows:

Firstly, dose-response studies are performed and some biometrical dose-response models such as probit, logit, or Weibull are fitted to these experimental data.

Figure 1 shows the dose-response relationship (dose vs. percent effect) of EMS for the sum of all gross-structural abnormalities. For comparison in Figure 2 the fitted probit curve of the same data is shown. In Figure 3 the curve was linearized by transformation, the x-axis is now scaled as log dose, the y-axis as probit. In this case the fetus was used as the statistical unit ("per fetus approach"). In Figure 4 probit analysis was based on the response of the litters (percent affected fetuses per litter, "per litter approach") and now the great variability in response of the single litters is obvious. The second significant aspect here is the confidence limit of the curve which, in the case of the "per litter approach", is considerably higher as compared to the "per fetus approach".

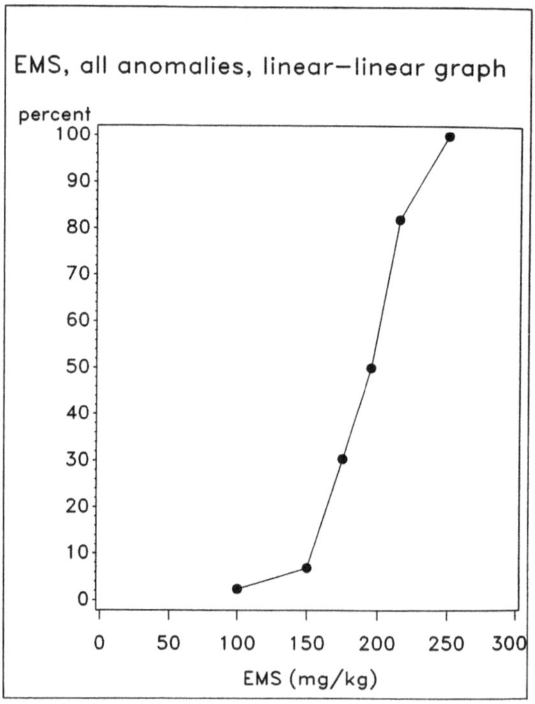

Fig. 1. Graph of the dose-response relationship (dose vs. percent effect), EMS, sum of all gross-structural abnormalities

Secondly, using the calculated parameters of these dose-response functions, the risk of low doses or the doses corresponding to low incidences were assessed. Table 2 compiles the data for the estimated incidence of 0.1% obtained with the models probit, logit and Weibull: these models yielded very similar values. This may be due to two reasons: first, the extrapolation was confined to the incidence

0.1% and secondly, in our experimental example we have steep dose-response relationships (slope 16 [EMS] and 9 [MNU] in the log-probit model). Therefore, only small differences in the results extrapolated with these models were computed.

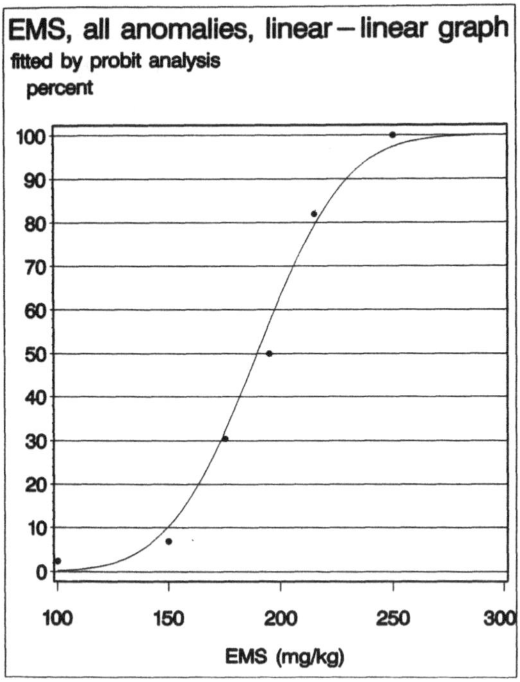

Fig. 2. Graph of the fitted probit curve of the dose-response relationship, EMS, sum of all gross-structural abnormalities

Linear "Extrapolation to Zero"

A further simple procedure for risk assessment is linear extrapolation to zero. Starting from the lowest dose which exhibited an effect in the experiment the risk of lower doses is estimated by linearly extrapolating to zero. This approach is based on the assumption that no "threshold" exists and that a linear relationship exists between risk and dose down to very low exposures. In the case of teratogenicity all existing scientific evidence argues against the assumption that such a linear extrapolation is justified. In contrast to the situation in carcinogenicity, where on theoretical grounds a malignant transformation of a single cell could be speculated to give rise to tumor, an alteration of a single cell in a blastema could not be expected to induce abnormal development. For this reason "extrapolation to zero" over many orders of magnitude is not seriously considered by us for teratological risk assessment.

Recognizing this principle we used this method for comparison. In the case of MNU the dose 2 mg per kg body wt was chosen as lowest effective dose, and as effect the incidence of gross-structural limb abnormalities (1.9%), since this type of effect was not observed in controls. We only estimated the dose corresponding to 0.1% incidence which is not too far below the doses used in the experiment. The extrapolated dose value for an estimated risk of 0.1% was 0.1 mg MNU per kg body wt (Table 2).

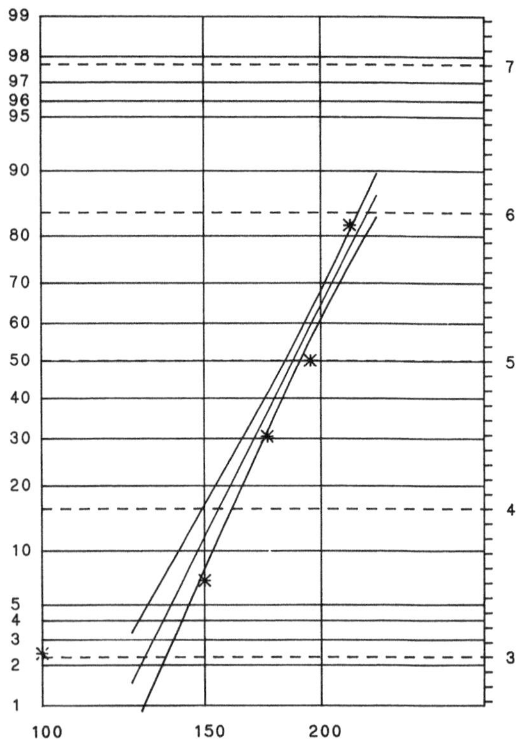

Fig. 3. Graph of the transformed dose-response curve (log dose vs. probit, percent affected fetuses per dose group, "per-fetus-approach"), EMS, all gross-structural abnormalities

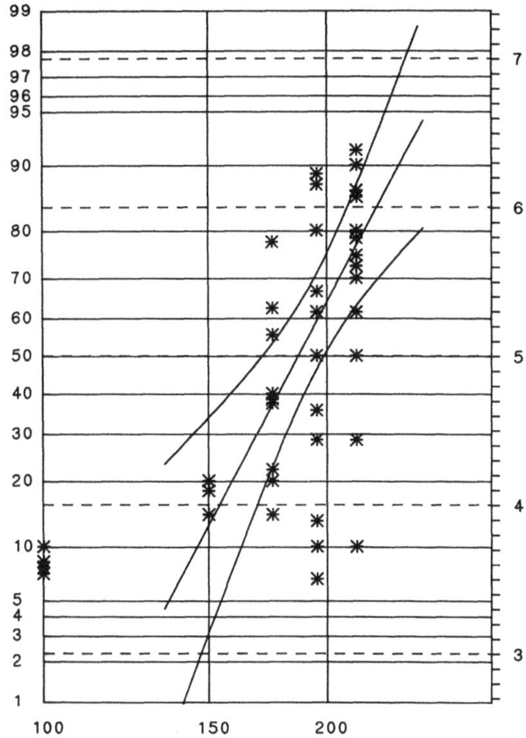

Fig. 4. Probit analysis based on the response of the litters (percent affected fetuses per litter, "per litter approach"), EMS, all gross-structural abnormalities

Table 2. Extrapolation to 0.1% incidence using the models probit, logit and Weibull as well as linear extrapolation based on a dose-response study using NMRI mice. MNU, single i.p. treatment on day 11 of gestation, all anomalies. Values: mg/kg body wt

Model	Probit	Logit	Weibull	Linear
ED-0.1%	1.87	1.64	1.67	0.11
± SD	±0.18	±0.18	±0.24	

Figure 5 shows the graph for this extrapolation as compared to extrapolation using probit analysis. The same comparison was made with EMS (Fig. 6) using limb anomalies as variable. In both figures the great differences between the two models even at 0.1% incidence for both substances is obvious. But, one should keep in mind that "linear extrapolation to zero" ignores the steepness of the dose-response curve and, therefore, is considered over-conservative.

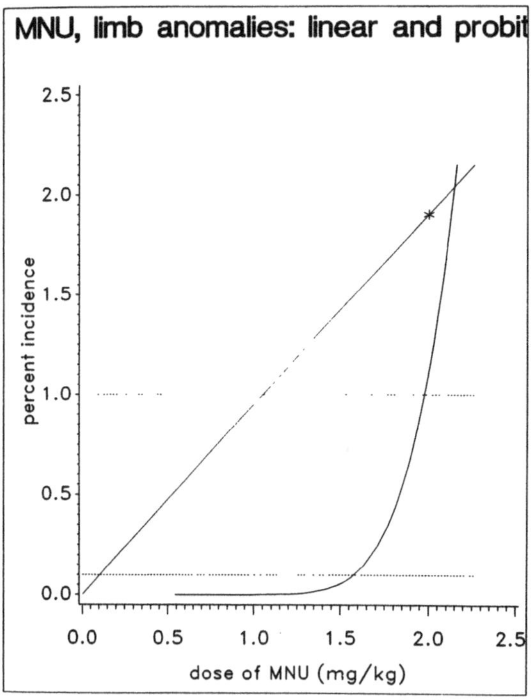

Fig. 5. Extrapolation to low incidences based on limb abnormalities, MNU: comparison of linear extrapolation to zero and probit analysis

Risk Assessment of O6-alkylguanine Using Dosimetry

The last procedure of risk estimation which we used was based on dosimetry and this is, so far as we know, the first approach in the field of prenatal toxicity. In the first step we were able to demonstrate that O6-alkylguanine is a useful dosimeter for assessing the extent of teratogenicity of alkylating agents. For this purpose dose adduct studies were performed using doses which in preceding studies on teratogenicity were found to be effective. Figure 7 shows that dose adduct relationship (O6-alkylguanine) for both MNU and EMS in a log-linear plot. Both functions are very steep, but the substances induce similar adduct rates at very different dose levels.

Table 3 compiles some of the data of the DNA adduct measurements of day-11-embryos. In addition to MNU and EMS the data of DMN-OAc, a further methylating agent, are included. It is obvious that alkylation rates at the effective doses ED-10 are different with regard to 7-alkylation, but practically identical at the O6-guanine site for the three agents. The same holds true at the ED-50 and the ED-90 dose levels.

These data clearly indicate: MNU and EMS are agents of very similar teratogenic efficiency if, for comparison, not the applied dose is used but the target dose, *it est*: the adduct rate of O6-alkylguanine in the DNA of the embryos.

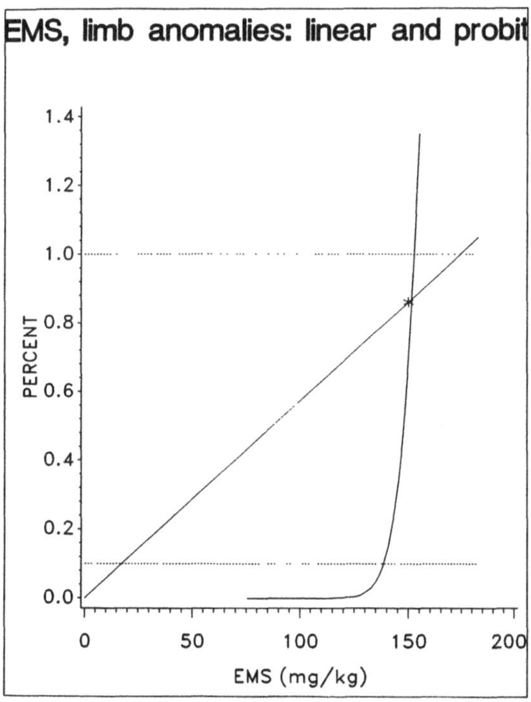

Fig. 6. Extrapolation to low incidences based on limb abnormalities, EMS: comparison of linear extrapolation to zero and probit analysis

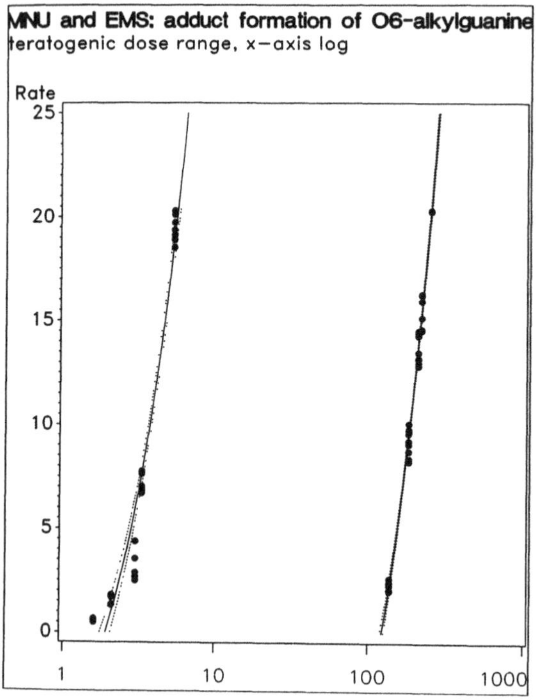

Fig. 7. Log-linear plot of the dose adduct relationship (O6-alkylguanine) in the DNA of day-11-embryos for both MNU and EMS

The data of Table 3 are plotted also as graphs: the adduct rate vs. teratogenicity. In the case of 7-alkylguanine (Fig. 8), the adduct rate exhibits no correlation with the teratogenic effect. In contrast, in the case of O6-alkylguanine, a good correlation between adduct rate and teratogenic effect is obvious (Fig. 9).

For the three alkylating agents a significant parallelism between the teratogenic potency and the initial adduct rate at the O6-guanine site in the DNA of the embryos was found. A possible correlation between initial DNA adduct rate and teratogenicity seems to exist for both methylating and ethylating agents. At this stage of knowledge this correlation is not necessarily interpreted as a causal relationship. It seems to provide, however, a reasonable basis for using this specific DNA adduct (O6-alkylguanine) as a molecular dosimeter in the extrapolation of the risk to lower doses.

Table 3. Interpolated DNA adduct rates of day-11-embryos at the effective doses (teratogenicity) ED-10, ED-50, and ED-90. The data were calculated from dose-DNA adduct relationships of the different alkylated bases. Values: pmoles alkylated bases per μmole guanine

Substance	Alkylated base	Effective dose ED-10	ED-50	ED-90
EMS	7-ethylguanine	97.9	255.4	461.5
	O6-ethylguanine	3.9	10.0	18.0
MNU	7-methylguanine	46.0	90.3	162.9
	O6-methylguanine	5.4	10.8	19.6
DMN-OAc	7-methylguanine	38.6	80.6	156.0
	O6-methylguanine	4.8	9.6	18.2

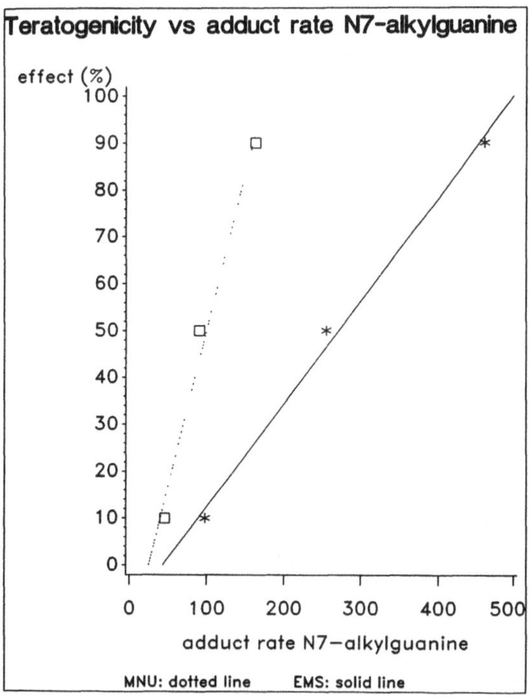

Fig. 8. Adduct rate of 7-alkylguanine induced by MNU and EMS in the DNA of day-11-embryos vs. percent of teratogenicity

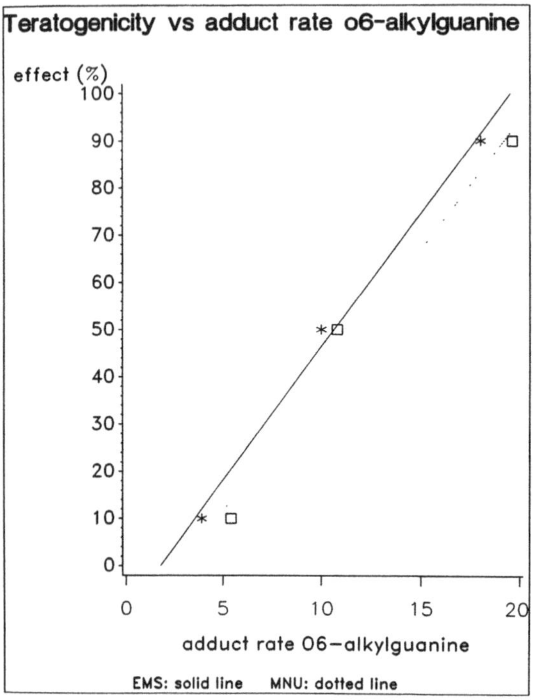

Fig. 9. Adduct rate of O6-alkylguanine induced by MNU and EMS in the DNA of day-11-embryos vs. percent of teratogenicity

As next step we determined DNA adduct rates subsequent to lower doses, following sub-teratogenic doses of MNU (Platzek et al. 1989; Bochert et al. 1991) and EMS (Platzek and Bochert 1990). The lowest doses which were analysed were about 45 mg EMS and 0.2 mg MNU per kg body wt, which means one third of the lowest effective dose of teratogenicity in the case of EMS and one tenth in the case of MNU. Even in this sub-teratogenic dose range measurable DNA adduct rates were found.

Using the SAS NLIN procedure we fitted an exponential function to the complete set of experimental data which is shown for EMS in Figure 10. In the lower (sub-teratogenic) dose range the shape of the dose adduct curve is much flatter when compared to that in the higher (teratogenic) dose range. Additionally, a good fit was obtained using a segmented model (Fig. 11).

If we arbitrarily presuppose that, in the low dose range (not experimentally verified dose range), a similar proportionality may exist between DNA adduct rate and a possible teratogenic effect as demonstrated in the higher dose range, the hypothetical teratogenic risk of these low doses may be deduced directly from the adduct rates of O6-alkylguanine and compared with values obtained with the previously mentioned approaches (Platzek et al. 1989; 1991).

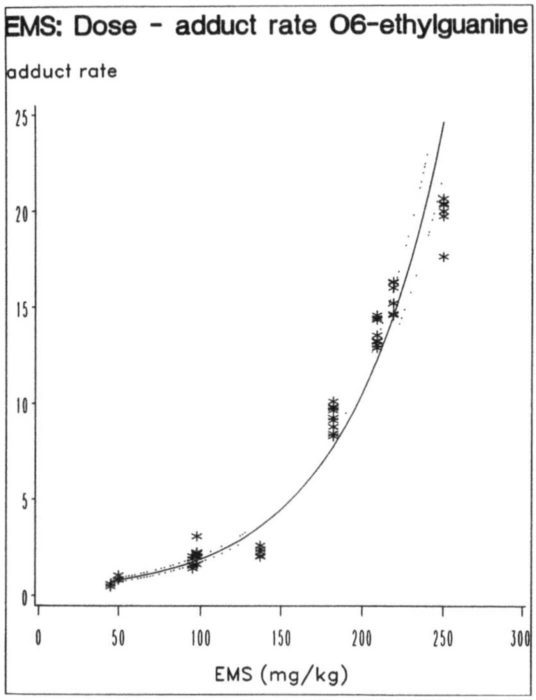

Fig. 10. Exponential fit to the adduct data of O6-ethylguanine induced by EMS in the DNA of day-11-embryos

In Figure 12 the different hypothetical dose to risk functions are shown for EMS: linear extrapolation to zero, probit function and the dosimetry-derived function. Obviously, the different dose to risk function exhibit substantial differences even for the estimated incidence 0.1%.

Table 4 compiles the results of the various methods of risk estimation for MNU and EMS for the hypothetical incidence 0.1%.

When probit and linear extrapolation are compared: For the estimated incidence 0.1% a difference by the factor 16 (MNU) and 35 (EMS) was calculated: this means a great difference between linear extrapolation and the probit model even at this incidence which is not so far away from the teratogenicity experiment.

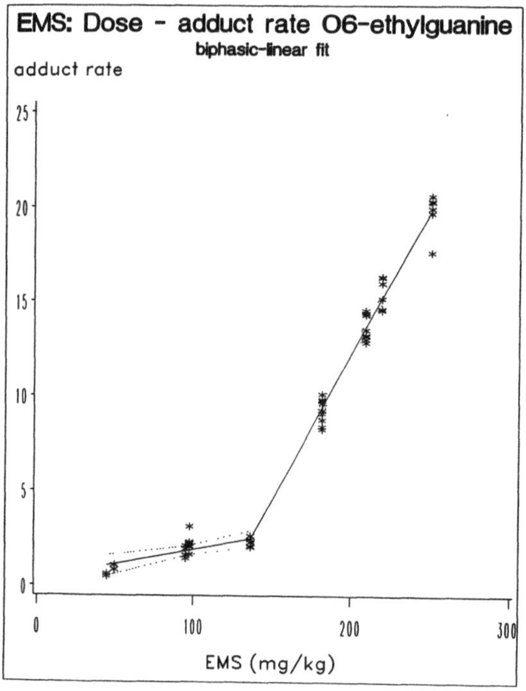

Fig. 11. Biphasic-linear fit to the adduct data of O6-ethylguanine induced by EMS in the DNA of day-11-embryos

When the NOAEL risk factor approach and linear extrapolation are compared: The "virtually safe dose" obtained under our experimental conditions using the safety factor 30 was 0.03 mg/kg (MNU) and 3.3 mg/kg (EMS). Using linear extrapolation hypothetical risks of 0.3% (MNU) and 0.1% (EMS) would be assigned to these doses. In contrast, using probit analysis, these doses have negligible hypothetical risks.

Table 4. Comparison of the various methods of risk assessment. A teratogenic risk of 0.1% was estimated based on linear extrapolation, on dose-response relationships using probit analysis, and derived from DNA adducts. The values are given as mg/kg body wt

Approach	Estimated dose MNU	EMS
Extrapolation based on teratogenicity		
Linear[1]	0.1	3.0
Probit	1.6	106
Extrapolation based on dosimetry		
Exponential fit	0.3	99

[1] See commentary in Results and Dicussion

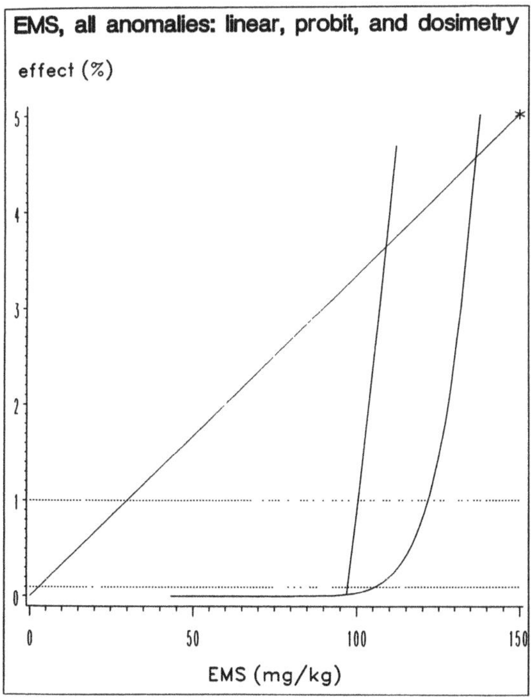

Fig. 12. Comparison of the different dose to risk functions for EMS: linear extrapolation to zero, probit function, and a dosimetry-derived function

When probit and dosimetry are compared: In the case of MNU the dose value for the hypothetical incidence 0.1% obtained by dosimetry was 0.3 mg/kg and that for probit extrapolation 1.6 mg/kg, while linear extrapolation lead to 0.1 mg/kg. In the case of EMS the dosimetry-derived value for the estimated 0.1% incidence (99 mg/kg) was very similar to the probit value (106 mg/kg). In contrast, the value obtained by linear extrapolation (3 mg/kg) was extremely low. Thus, two very similar agents are very different with regard to these aspects.

In conclusion:

- No one knows the real prenatal-toxic risk of low doses (not verified by experimental data).
- Traditional methods such as the NOAEL risk factor approach suffer from the disadvantage of not considering dose-response relationships. But when using factors of 30 to 100 in this approach in the case of teratogenicity conservative "safe doses" are assessed since the dose-response relationships of most teratogenic agents are steep.
- Even when studying alkylating agents we would argue for assuming a threshold for the teratogenic effects.

Low dose risk assessment using mathematical models such as probit, in contrast to linear extrapolation, are based on the steepness of the dose-response relationship and therefore probably more significant. But one should bear in mind that, in any case, the extrapolated dose values are calculated values and are not to be considered as reality.

Dosimetry is a very useful additional instrument especially when used in the vicinity of the experimental range where a biological effect can be verified. Although the extrapolated incidences are based on experimental values, it should be considered that dosimetry gives a measure for the dose and not for the effect. It is not possible to prove the toxicological significance of a certain adduct rate.

Acknowledgments: The support of Bernd Fischer with regard to advise in statistics and program development is gratefully acknowledged.

References

Biddle FG (1978) Use of dose-response relationships to discriminate between the mechanisms of cleft-palate induction by different teratogens: An argument for discussion. Teratology 18: 247-252

Bochert G, Rahm U, Schnieders B (1978) Pharmacokinetics of embryotoxic direct-acting alkylating agents: Comparison of DNA alkylation of various maternal tissues and the embryo during organogenesis. In: Neubert D, Merker H-J, Nau H, Langman J (eds) Role of pharmacokinetics in prenatal and perinatal toxicology, G Thieme Publ Stuttgart, pp 235-251

Bochert G, Platzek T, Neubert D (1987) DNA modification in murine embryos: a primary cause of embryotoxic effects. In: Nau H, Scott WJ (eds) Pharmacokinetics in teratogenesis, Vol II Experimental aspects *in vivo* and *in vitro*, CRC Press, Boca Raton, Florida, pp 73-82

Bochert G, Platzek T, Rahm U, Neubert D (1991) Embryotoxicity induced by alkylating agents: 6. DNA adduct formation induced by methylnitrosourea in mouse embryos. Arch Toxicol, accepted for publication

Chahoud I, Kwasigroch TE (1977) Controlled breeding of laboratory animals. In: Neubert D, Merker H-J, Kwasigroch TE (eds) Methods in prenatal toxicology, G Thieme Publ, Stuttgart, pp 78-91

Dawson AA (1926) A note on the staining of the skeleton of cleared specimen with alizarin red S. Stain Technol 1: 123-124

Finney DJ (1971) Probit analysis. Third edition, Cambridge University Press, London

Lutz WK (1979) *In vivo* covalent binding of organic chemicals to DNA as a quantitative indicator in the process of chemical carcinogenesis. Mutat Res 65: 289-356

Lutz WK (1986) Quantitative evaluation of DNA binding data for risk estimation and for classification of direct and indirect carcinogens. J Cancer Res Clin Oncol 112: 85-91

Lutz WK (1987) Quantitative evaluation of DNA-binding data *in vivo* for low-dose extrapolation. Arch Toxicol [Suppl] 11: 66-74

Neubert D, Zens P, Rothenwallner A, Merker H-J (1973) A survey of the embryotoxic effects of TCDD in mammalian species. Environm Health Perspect 5: 67-79

Neubert D, Barrach HJ (1977) Organotropic effects and dose-response relationships in teratology. In: Neubert D, Merker H-J, Kwasigroch TE (eds) Methods in prenatal toxicology, Thieme Stuttgart, pp 405-412

Neubert D, Chahoud I, Platzek T, Meister R (1987a) Principles and problems in assessing prenatal toxicity. Arch Toxicol 60: 238-245

Neubert D, Bochert G, Platzek T, Chahoud I, Fischer B, Meister R (1987b) Dose-response relationships in prenatal toxicity. Congen Anomal 27: 275-302

Platzek T, Bochert G, Schneider W, Neubert D (1982) Embryotoxicity induced by alkylating agents: 1. Ethylmethanesulfonate as a teratogen in mice - a model for dose-response relationships of alkylating agents. Arch Toxicol 51: 1-25

Platzek T, Bochert G, Rahm U (1983) Embryotoxicity induced by alkylating agents: Teratogenicity of acetoxymethyl-methylnitrosamine - dose-response relationship, application route dependency and phase specificity. Arch Toxicol 52: 45-69

Platzek T, Bochert G, Rahm U (1987a) Embryotoxicity induced by alkylating agents: Some methodological aspects of DNA alkylation studies in murine embryos using ethylmethanesulfonate. Z Naturforsch 42c: 613-626

Platzek T, Meister R, Chahoud I, Bochert G, Krowke R, Neubert D (1987b) Studies on mechanisms and dose-response relationships of prenatal toxicity. In: F. Welsch (ed) Approaches to elucidate mechanisms in teratogenesis, Hemisphere Publ Corp, Washington, New York, pp 59-81

Platzek T, Bochert G, Pauli B, Meister R, Neubert D (1988) Embryotoxicity induced by alkylating agents: 5. Dose-response relationships of teratogenic effects of methylnitrosourea in mice. Arch Toxicol 62: 411-423

Platzek T, Bochert G, Fischer B, Meister R (1989) Low dose prenatal-toxic risk estimation based on dose-response relationships and DNA adducts using MNU as model compound. Naunyn-Schmiedeberg Arch Pharmacol 339 (Suppl): R23
Platzek T, Bochert G (1990) DNA adduct formation induced by ethylmethanesulfonate in mouse embryos. Naunyn-Schmied Arch Pharmacol 341 (Suppl.): R26
Platzek T, Rahm U, Bochert G (1991) Comparative approaches for low dose prenatal toxic risk estimation of the alkylating model compound EMS in mice. Naunyn-Schmiedeberg Arch Pharmacol 343 (Suppl.): R31

Discussion of the Presentation

Scott: First, my congratulations on providing some scientific evidence for risk assessment. At what time following dose administration was the adduct formation measured? How might alterations of this time affect the risk assessment?

Platzek: In our studies the adduct rates were determined at the time point which in previous studies was found to exhibit the maximum value. The persistence and repair of these adducts may, in contrast to carcinogenesis, not play such a major role in prenatal toxicity since the critical period is rather short.

Bolt: Could you comment on the nature of the correlation between malformation rate and extent of O^6-adducts, especially in the low dose range?

Platzek: We arbitrarily presuppose that in the low dose range the same linear correlation between adduct rate and teratogenicity exists as we demonstrated experimentally in the higher dose range.

Weissinger: Which general strategy of risk assessment would you recommend as the consequence of your studies?

Platzek: Generally, I would recommend to obtain information on both dose-response relationships and kinetics. Based upon these data a benchmark dose may be established which can preferably be used as reference dose for the assessment of "virtually safe doses" using uncertainty factors. Additionally, the dose-response data and, if available, the kinetic data should be integrated into the risk assessment procedure.

Neubert: After Dr. Platzek has very critically discussed the various possibilities of risk assessment, the pragmatic problem arises: What shall we do?

I would like to draw the following conclusions:

- Investigators should be encouraged to study more than three dose levels, in order to facilitate establishing dose-response relationships. Furthermore, information on kinetics is almost indispensable.
- Linear extrapolation to zero already has an extremely weak basis in cancer risk assessment, but for me it is completely irrelevant in reproductive toxicology, because there is no theoretical background whatsoever for the assumption that any dose may cause an adverse effect. Any mechanism of teratogenicity would argue against a non-thresholded phenomenon, as Thomas Platzek has mentioned; the question is only where such a "threshold" range would

be. In the majority of cases the dose-response is very steep; apparently a developmental system "flips over" or "breaks down" at a certain critical exposure.

- The NOAEL-safety factor approach also represents some kind of "linear" extrapolation to zero. It certainly is the least scientific procedure, because the NOAEL is ill-defined.
- In my opinion, any approach at a risk assessment taking into account data on dose-response relationships is preferable. From this point of view there seems to be no strong argument against the strategy: probit transformation of the effect vs. log of the dose. Thus we would favour, as we have for many years, using such an assessment. An alternative would certainly be the benchmark approach presented by Dr. Kavlock.
- Evaluation of experimental data should be performed in a range very close to data which could be experimentally verified (e.g. i.e. ED_{10} or even ED_1). Based on such values a small uncertainty factor may be applied.
- Extrapolations should be confined to a limited frequency range, not too far from the experimental data. Choosing an incidence of 0.1% seems reasonable, since rare single abnormalities occur in this range spontaneously. No causal relationship could be established if such a rate would be doubled, and such an incidence is irrelevant from a medical point of view.

I would very much like to learn the opinion of the colleagues here working in regulatory agencies, which approach they would favour or accept in their everyday jobs: Drs. Lingk, Baß, Heger, please.

Lingk: For the Bundesgesundheitsamt, the presented approaches for risk assessment taking into account data on dose-response relationships are welcome. However, for environmental substances (pesticides) the data on reproductive toxicity are often limited and are in most cases not suited for an evaluation on the basis of dose-effect relationships.

Baß: Since most substances exerting a potential risk are excluded from use during pregnancy, quantitative risk assessment is seldomly needed for medicinal products. In such cases, however, pharmacokinetics and pharmacodynamics (mode of action) are applied in the "dose-response relationship" fashion such as you have described.

Heger: From the view of protection of the natural balance in ecosystems another major problem arises: The natural community of organisms comprehends numerous species which are usually very different in nutrition, behaviour, metabolism and biological relation. For this reason the sensitivity of different species to toxic effects of one chemical may differ in several orders of magnitude. On the other hand, the toxicity data submitted for registration of pesticides is usually limited to representatives of six phyla (algae, crustaceans, fish, insect, bird and mammal) with few tests in reproductive toxicity only.

Because of the above outlined problems concerning the extrapolation mode on one side and the heterogeneous data for the ecotoxicity risk assessment on the other side, the Umweltbundesamt favours at present an alternative approach by using experimentally verified no effect concentrations. The ecotoxic risk is assessed on the basis of the lowest available NOEC, which takes into account sublethal effects such as impairment of reproduction, growth, survival and behaviour. If the predicted environmental concentration differs to the NOEC-concentration less than a factor 10, we appraise that adverse effects may occur. This factor is derived from the limited number of animals in the experimental study, the variability of the laboratory tests and the biological variance, but it does *not* consider possible species differences in sensitivity to pesticides of untested species, although being representatives in the ecosystem.

According to the Supreme Administrative Court unjustifiable effects of a pesticide must be almost certainly impossible. This assessment process for ecotoxic effects has proven to stand the weighing up of the German Administrative Court.

This assessment procedure reflects the status of development and discussion in 1991. The problems of extrapolation to low doses, species sensitivity, establishment of a clear causality at low doses, and finally the weighing up of effects needs further discussion.

Kinetic Problems Arising in a Multigeneration Study

Elisabeth Koch, Ibrahim Chahoud and Diether Neubert

Introduction

Multigeneration studies are performed to allow an overall assessment of: gonadal function, estrus cycle, mating behavior, conception, implantation, abortion, embryonic and fetal development, parturition, postnatal survival, lactation, maternal behavior, and postpartum growth. Such an experimental approach is used to evaluate possible reproductive and developmental toxicities of chemicals or environmental xenobiotics, while medicinal products are generally tested with the segment I to III procedure.

Comparing the potency of substances between species is best done on the basis of concentrations in plasma or tissue, preferentially of target-tissue. In the classical three-generation reproduction study the parental generation (F_0) and the offspring of each succeeding generation are exposed to the same daily doses of the test chemical via the feed. During the different stages of development (suckling, phase of rapid growth, pregnancy, lactation) special kinetic situations may arise. Up till now, in such multigeneration studies the multiplicity of kinetic variables has not been taken into account.

We investigated the time courses of tissue concentrations during a multigeneration study using *2,3,7,8-tetrachlorodibenzo-p-dioxin (TCDD)*, a substance with a long elimination half-life. Experimental studies designed to elucidate the toxic effects occurring in rodents during long-term exposure to 2,3,7,8-TCDD have so far mostly been performed using the same daily dose throughout the treatment period (e.g. Rose et al. 1976; Kociba et al. 1978; Murray et al. 1979; Lamb et al. 1981a, b). Since the half-life of TCDD in rats is several weeks, the concentration of the test substance can be predicted to change continuously during such a study: Within the first three months TCDD accumulates within the organism up to a steady-state level. For this reason during a multigeneration study exposure to the maximum dose level would only have been achieved for a short period of the study.

In the course of a study on the reproductive toxicity of TCDD in rats, we intended to achieve rather constant tissue concentrations of TCDD in the animals. For this reason we treated the animals with an initial high loading-dose and attempted to keep a constant exposure level with weekly maintenance-doses. The elimination half-life of the species to be evaluated is already well-established (Abraham et al. 1988), and the loading-dose/maintenance-dose procedure has been proven to be suitable in adult rats (Krowke et al. 1989).

Results and Discussion

Male and female Wistar rats (Bor: spf, TNO) were used. ^{14}C-TCDD was administered with toluene/DMSO (1+2, v/v) as a vehicle, and the substance was given subcutaneously to assure complete absorption (Abraham et al. 1988).

Three dosing regimens were used for these studies, as shown in Table 1. The loading-doses and the resulting maintenance-doses were calculated using established pharmacokinetic estimations (e.g. Dost 1968 or Gladtke and Hattingberg 1973): the maintenance-dose should be about 20% of the loading-dose (based on an elimination half-life of three weeks).

The toxicokinetics of TCDD was studied during the multigeneration study by monitoring the radioactivity of ^{14}C-TCDD. It has been shown before that this procedure measures the original substance only (Rose et al. 1976; Abraham et al. 1988). Three females of the F_0-generation of each group were sacrificed one and three weeks after initial treatment. In every generation the concentration of TCDD in the liver and adipose tissue was measured in three females on day 14 and day 21 of pregnancy and during the lactation period (day 7, 14 and 21). During the suckling period three litters of each group and generation were pooled and the concentrations of TCDD were measured in the liver and adipose tissue on day 7 and 14 of lactation. From 3 weeks to 16 weeks postnatally the concentrations were determined in two male and two female rats at several times during this period.

Tissue Concentrations Obtained in the F_0-Generation

The time course of ^{14}C-TCDD concentration found in samples of liver and adipose tissue of female rats are compiled in Figure 1. Following a single injection of the loading-doses of 50, 120 or 250 ng TCDD/kg body wt, the concentration (pg/g) of ^{14}C-TCDD shortly before application of the first weekly maintenance-dose (one week later) was found to be in liver: 293, 761 and 1,787 pg/g wet wt; in adipose tissue: 216, 379 and 609 pg/g wet wt. The concentration ratio: liver/adipose tissue varied from: 2.9 in the highest dose group (TCDD-250/50) to 1.4 in the lowest dose group (TCDD-50/10). Subsequent to the initial (loading-) dose, the weekly maintenance-doses were capable of maintaining the TCDD levels in the *liver* of the adult female rats. There was a slight decrease (especially at the highest dose) in the concentration of TCDD in the liver during the first three weeks after the loading-dose. During the last third of pregnancy the concentration in the liver declined again. At the same time the concentration of ^{14}C-TCDD in *adipose tissue* slightly but steadily increased in the same animals. Interestingly, there was no decrease of the concentration in this tissue during pregnancy.

During the *lactation period* the concentration of TCDD in maternal liver as well as in adipose tissue declined rapidly. The lipophilic TCDD is excreted from the maternal organism via milk and the exposure of the offspring must be considerable. Surprisingly, at the end of the lactation period (3 weeks) the concentration in maternal liver and adipose tissue of all three treated groups ranged around the same level (Fig. 1), indicating a larger loss of TCDD at the higher doses.

For the investigation of the multigeneration study the changes in the maternal organism during the lactation period were of minor interest, and concentrations in the tissue of the offspring were especially relevant.

Table 1. Dose regimens used in the multigeneration study

Group	Initial dose (ng TCDD/kg body wt)	Maintenance-doses (ng TCDD/kg body wt)
TCDD-50	50	10 (20*)
TCDD-120	120	24 (48*)
TCDD-250	250	50 (100*)

* = Maintenance-doses administered to the F_2- and F_3-generation

Table 2. Variation of mean tissue concentrations in different generations during the rat multigeneration test with TCDD

Generation	Concentrations (pg/g wet weight, ppt) Liver min max	Adipose tissue min max	Male Liver/Fat
TCDD 50/10(20***):			
F0	180- 290	220- 270	640/ 440*
F1	130- 570	190- 540	220/ 330**
F2	180- 570	190- 500	
F3	280- 760	190- 450	
TCDD 120/24(48***):			
F0	450- 760	380- 620	920/ 560*
F1	320- 1,520	300- 740	460/ 540**
F2	320- 1,450	280- 690	
F3	870- 1,690	300- 690	
TCDD 250/50(100***):			
F0	900- 1,790	610- 1,020	2,190/ 780*
F1	710- 2,210	480- 1,050	910/ 690**
F2	720- 2,550	380- 1,190	
F3	1,600- 2,890	410- 1,090	

Values for F_1 to F_3 refer to day 3 postnatally until the end of pregnancy
Lowest and highest mean values are given
* After 14 weeks
** After 20 weeks
*** Loading-dose/maintenance-dose F_0 and F_1 (in F_2 and F_3)

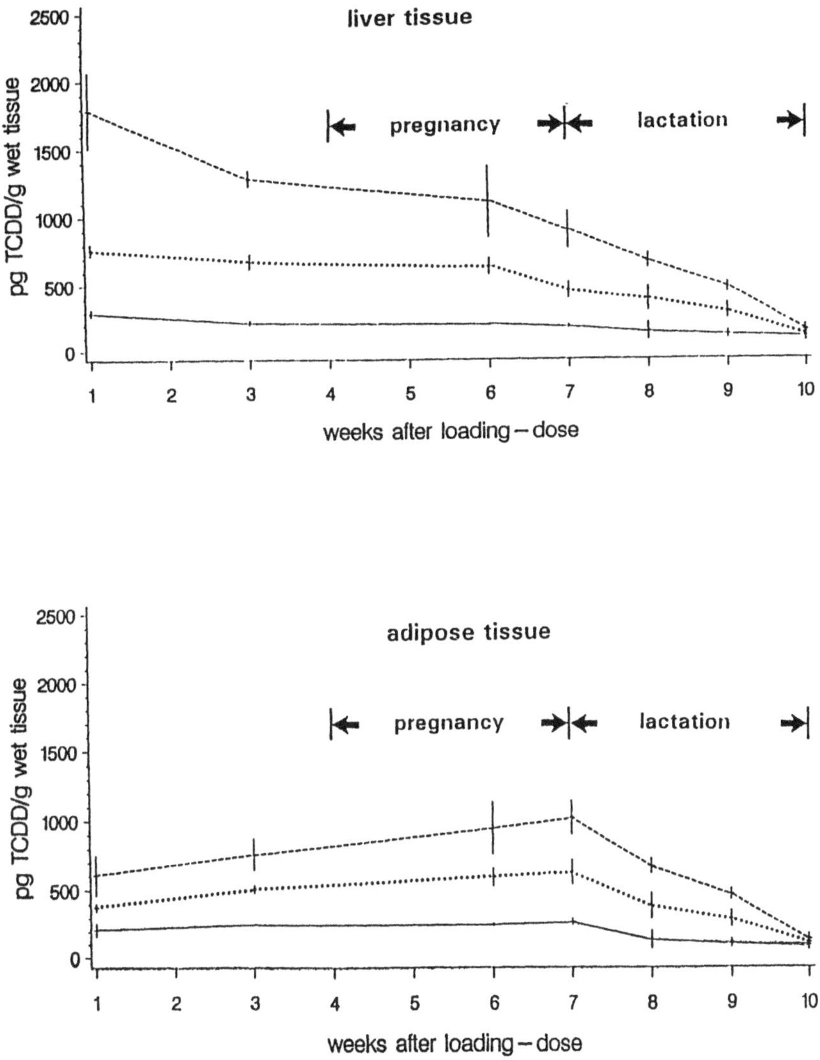

Fig. 1. *F_0-Generation.* Time course of ^{14}C-TCDD concentration in liver and adipose tissue following subcutaneous application of a loading-dose of 50, 120 or 250 ng TCDD/kg body wt (and weekly maintenance-doses of 10, 24 or 50 ng TCDD/kg body wt) to rats of the *F_0-generation. Treatment (loading-dose/ maintenance-dose) ng TCDD/kg body wt*
——— TCDD-50/10; ······· TCDD-120/24; - - - - - - TCDD 250/50

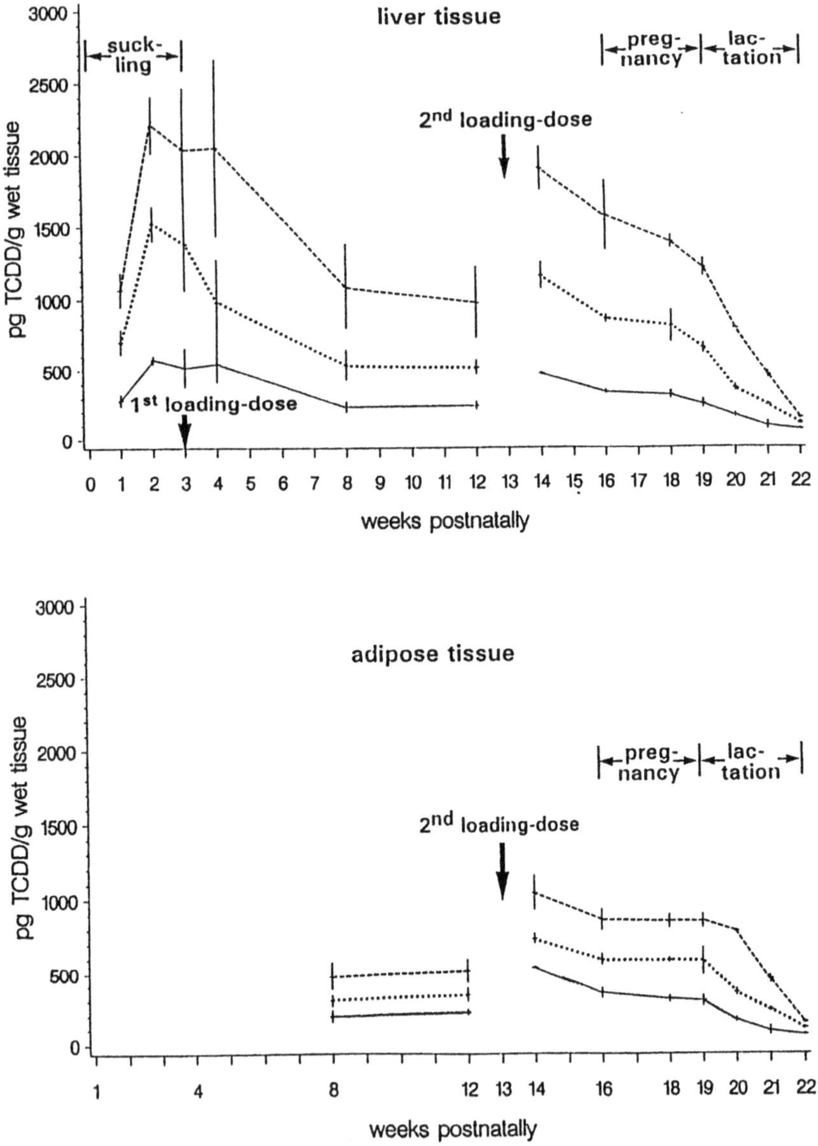

Fig. 2. *F_1-Generation.* Time course of ^{14}C-TCDD concentration in liver and adipose tissue following exposure via milk during the lactation period (+ subcutaneous application of a loading-dose of 50, 120 or 250 ng TCDD/kg body wt in the third and 13th week postnatally and weekly maintenance-doses of 10, 24 or 50 ng TCDD/kg body wt) to rats of the F_1*-generation. Treatment (loading-dose/maintenance-dose) ng TCDD/kg body wt*
——— TCDD-50/10; ········ TCDD-120/24; - - - - - - TCDD 250/50

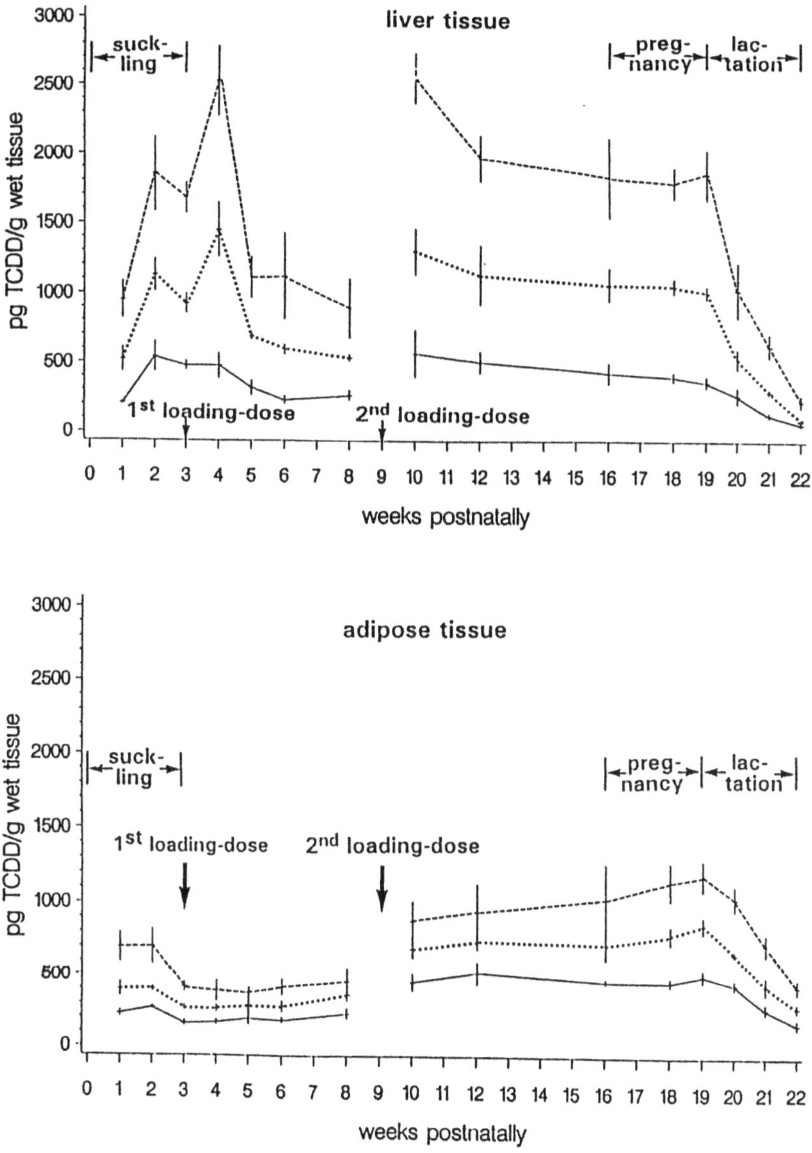

Fig. 3. *F_2-Generation.* Time course of ^{14}C-TCDD concentration in liver and adipose tissue following exposure via milk during the lactation period (+ subcutaneous application of a loading-dose of 50, 120 or 250 ng TCDD/kg body wt in the third and ninth week postnatally and weekly maintenance-doses of 20, 48 or 100 ng TCDD/kg body wt) to rats of the *F_2-generation. Treatment (loading-dose/maintenance-dose) ng TCDD/kg body wt*
——— TCDD-50/20; ······· TCDD-120/48; - - - - - - TCDD 250/100

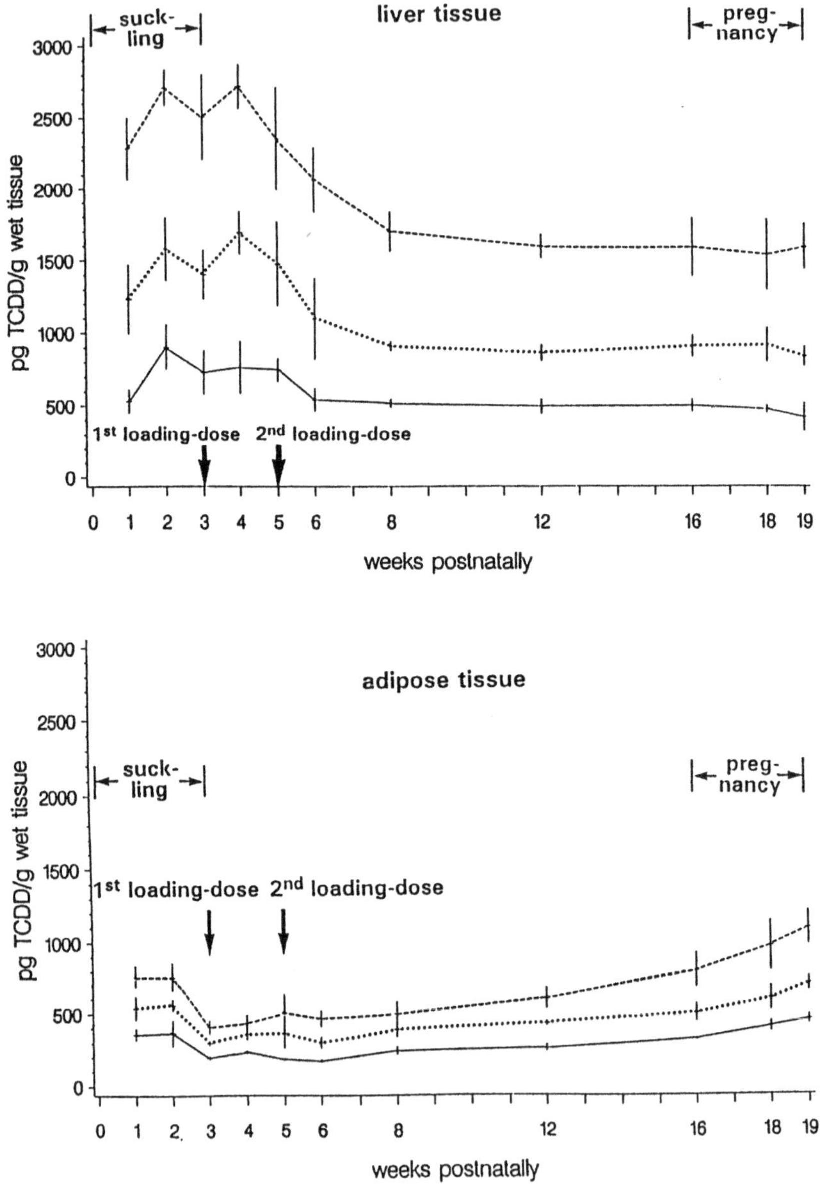

Fig. 4. *F_3-Generation.* Time course of ^{14}C-TCDD concentration in liver and adipose tissue following exposure via milk during the lactation period (+ subcutaneous application of a loading-dose of 50, 120 or 250 ng TCDD/kg body wt in the third and fifth week postnatally and weekly maintenance-doses of 20, 48 or 100 ng TCDD/kg body wt) to rats of the *F_3-generation. Treatment (loading-dose/maintenance-dose) ng TCDD/kg body wt*
——— TCDD-50/20; ········ TCDD-120/48; - - - - - - TCDD 250/100

Tissue Concentrations Obtained in the F_1-Generation

The time course of ^{14}C-TCDD concentration found in samples of liver and adipose tissue of the offspring are compiled in Figure 2. Due to the effective elimination of TCDD from the maternal organism (*see* Fig. 1) with the milk, an increase in the *hepatic* concentration of the suckling offspring was observed. The concentration determined in the liver tissue of the offspring exceeded the hepatic concentration of the dams at the time points investigated during the lactation period. A maximum concentration in the offspring was observed two weeks after birth and the hepatic concentration remained at about this level until weaning (at 3 weeks of age). Because of the rapid growth of the young animals and an expected "dilution" of the TCDD, the F_1 generation was treated at this time with their first loading-dose. Despite this additional dose (and the following weekly maintenance-doses of 20%), the concentration declined to 50% of the concentration measured two weeks postnatally and stayed constant thereafter at this lower level. During the phase of rapid growth (4th to 8th week of age) this dosing schedule proved to be insufficient to keep the hepatic concentrations of TCDD constant. Therefore, we treated the animals with a second loading-dose in the 13th week postnatally. We intended to achieve concentrations in the liver and adipose tissue similar to concentrations measured in the female rats of the F_0-generation a few weeks after the injection of the loading-dose.

The first measurement of the TCDD concentration in the *adipose tissue* of the F_1-generation was performed eight weeks postnatally. At this time the concentrations in the offspring were similar to the concentrations measured in the F_0-generation shortly after the first loading-dose. After application of the second loading-dose the concentrations in the adipose tissue increased and stayed rather constant thereafter at a higher level up to the end of pregnancy. During the lactation period the concentration of TCDD in the liver and the adipose tissue of the females of the F_1-generation declined, as was observed in the pregnant animals of the F_1-generation.

From the experience gained on the toxicokinetics of the F_1-generation we decided to increase the weekly maintenance-doses in the following generation to 40% of the loading-dose.

Tissue Concentration Obtained in the F_2-Generation

As shown in Figure 3 the dosing schedule (weekly maintenance-dose: 40% of the loading-dose) was also insufficient to keep the tissue concentration of TCDD in the liver and the adipose tissue of the F_2-generation constant during the phase of rapid growth. Therefore, we treated the rats nine weeks postnatally with a second loading-dose. From the 12th week postnatally up to the end of pregnancy we achieved high and rather constant concentrations in the liver. On the other hand, the weekly maintenance-doses of 40% of the loading-dose lead to an increase of the concentrations in the adipose tissue. During the lactation period the concentration of TCDD declined in the maternal organism in the usual way.

Tissue Concentration Obtained in the F_3-Generation

In the last generation (F_3-generation) only a slight alteration of the dose regimen used in the F_2-generation was necessary. We gave the second loading-dose directly during the phase of rapid growth (5th week postnatally). The weekly maintenance-doses of 40% of the loading-dose were able to keep rather constant concentrations in the liver tissue (Fig. 4). While the concentration of TCDD in the liver stayed reasonably constant up to the end of pregnancy, the concentration in adipose tissue steadily increased, especially at the highest dosing regimen.

Conclusions

The results of our studies indicate that risk assessment of a multigeneration study may be greatly complicated by the special kinetic situation arising throug the behaviour of substances with long elimination half-lives. Very large variations in the tissue levels are to be expected during the various developmental stages. This has been clearly demonstrated in our studies with TCDD. The variations observed in our studies are compiled in Table 2. While it is possible to control the tissue concentrations in the F_0-generation rather well, a considerably larger variation is observed during the different stages of the F_1-generation. The fluctuations may be reduced by using a complex dosing schedule.

A number of findings are unexpected and deserve attention:

- While the loading-dose/maintenance-dose approach resulted in rather constant tissue levels in adult animals, this was not the case during pregnancy.
- Maternal hepatic concentrations declined during the late stage of pregnancy, especially at the higher doses. Such a decline was not observed in adipose tissue.
- During the lactation period concentrations of the very lipophilic substance declined in both maternal liver and adipose tissue, due to the elimination of TCDD via the milk. At the same time the concentration in the offspring gradually greatly increased, largely exceeding the corresponding levels in the dam.
- The elimination from the maternal organism was more pronounced at the higher doses, thus leading to similar hepatic concentrations at the end of the lactation period.
- Even with rather complex dosing schedules it was impossible to achieve completely constant levels in liver and adipose tissue. This was mainly due to the fact that levels in liver and adipose tissue did not change in the same direction.

These data clearly indicate that attempts at a risk assessment on the basis of a multigeneration test cannot rest on the assumption of defined tissue levels during the study, at least not in the case of a lipophilic substance with a long elimination half-life. In such a case it is also almost impossible to base comparative risk assessments on the assumption that a given dosing schedule will result in the same or even similar tissue levels in various species (including man).

Since pharmacokinetic variables are so difficult to control in a multigeneration study, the segment I to III approach seems to represent a testing strategy during which kinetic variables may be more easily assessed for such substances, unless very complex dosing schedules are used in a multigeneration study.

References

Abraham K, Krowke R, Neubert D (1988) Pharmakokinetics and biological activity of 2,3,7,8-tetrachlorodibenzo-*p*-dioxin: 1. Dose-dependent tissue distribution and induction of hepatic ethoxyresorufin 0-deethylase in rats following a single injection. Arch Toxicol 62: 359-368

Dost F H (1968) Grundlagen der Pharmakokinetik. Georg Thieme Verlag, Stuttgart

Glatke E, Hattingberg HM (1973) Pharmakokinetik. Springer Verlag, Berlin

Kociba RJ, Keyes DG, Beyer JE, Carreon RM, Wade CE, Dittenber DA, Kalnins RP, Frauson LE, Park CN, Barnard SD, Hummel RA, Humiston CG (1978) Results of a two-year chronic toxicity and oncogenicity study of 2,3,7,8-tetrachlorodibenzo-*p*-dioxin in rats. Toxicol Appl Pharmacol 46: 279-303

Krowke R, Chahoud I, Baumann-Wilschke I, Neubert D (1989) Pharmacokinetics and biological activity of 2,3,7,8-tetrachlorodibenzo-*p*-dioxin. 2. Pharmacokinetics in rats using a loading-dose/maintenance-dose regime with high doses. Arch Toxicol 63: 356-360

Lamb JC, Marks TA, Gladen BC, Allen JW, Moore JA (1981a) Male fertility, sister chromatid exchange, and germ cell toxicity following exposure to mixtures of chlorinated phenoxy acids containing 2,3,7,8-tetrachlorodibenzo-*p*-dioxin. J Toxicol Environm Health 8: 825-834

Lamb JC, Moore JA, Marks TA, Haseman JK (1981b) Development and viability of offspring of male mice treated with chlorinated phenoxy acids and 2,3,7,8-tetrachlorodibenzo-*p*-dioxin. J Toxicol Environm Health 8: 835-844

Murray FJ, Smith FA, Nitschke KD, Humiston CG, Kociba RJ, Schwetz BA (1979) Three-generation reproduction study of rats given 2,3,4,8-tetrachlorodibenzo-*p*-dioxin (TCDD) in the diet. Toxicol Appl Pharmacol 50: 241-252

Rose JQ, Ramsey JC, Wentzler TH, Hummel RA, Gehring PJ (1976) The fate of 2,3,7,8-Tetrachlorodibenzo-*p*-dioxin following single and repeated oral doses to the rat. Toxicol Appl Pharmacol 36: 209-226

Application of Mathematical Dose-Response Models vs. Physiological Models in Risk Assessment in Reproductive Toxicology

G.A. de S. Wickramaratne

Introduction

Risk assessment in reproductive toxicology is amenable to both mathematical and physiological modelling techniques. These two methods are not necessarily mutually exclusive and can be complementary in achieving the goals of risk assessment. In this presentation examples of the use of the two approaches with data from studies undertaken in this laboratory are used to exemplify the two approaches.

Mathematical Models

The use of this type of modelling technique is illustrated with a set of data that was generated during the toxicological investigation of a chemical (Wickramaratne and Cook 1990). The initial teratogenicity study elicited an unusual vertebral lesion which had not been seen before in our lab. This led to a series of investigations, the main results of which are summarised in Table 1.

In the first study, which was a conventional teratogenicity study, the lesion was seen in the highest and lowest doses used. Thus, a second study atttempted to determine a clear no-effect-level with lower dose levels. Once again the lowest dose did not achieve a no-effect-level. The third study was designed to investigate the occurrence of the lesion with a view to generating a model system for use in further investigations. This consisted of dosing the rats at higher doses than previously used, but restricting the dosing period to three day intervals of 7-9, 10-12, 13-15, 16-19 during gestation. Thus toxicity to the dams was avoided as well as identifying the critical dosing period, in this case days 10-12 of gestation, which is the data presented. The fourth study was a comparison of the technical material with the pure material and used the model developed, with dosing on each of days 10, 11, 12 of gestation. It included determination of foetal uptake on day 11 with the use of radiolabelled compound. The results presented are for day 11 of gestation, in which study the lesion was seen for the first time in the control group. As radiolabelled compound had been used, the fact that this occurrence was spontaneous and not due to misdosing was unequivocally established by the absence of radiolabel in maternal blood. The fifth study, which overlapped the previous two studies in time used groups of 160 rats to determine a no-effect-level for the low incidence phenomenon.

Table 1. Percentage incidence of lesion (litters)

Dose (mg/kg)	Study 1	Study 2	Study 3	Study 4	Study 5
0	0.0	0.0	0.0	10.0	0.0
0.5					0.6
5		4.5			0.0
10		0.0			0.0
15	4.2				
45	0.0				
135	4.3	8.7			9.6
250			35.0		
400			100.0	25.0	
No.: Litters/Group	24	24	20	10	160

Dosing for the studies 1, 2 and 5 were on day 6-15, study 3 on days 10-12 and study 4 on day 11 of gestation. Ceasarean sections were performed on day 22 (day 1 being the day of successful mating)

The results of all these studies are summarised in Table 2 from which an empirical no-effect-level of 45 mg/kg/day is established, based on the litter incidence. The percentage incidence can be misleading with data of this type, and this is seen in the data set where a percentage incidence of 4.2 is seen at 15 mg/kg/day but is based on a smaller number of litters at this dose level for a single occurrence of the defect.

Table 2. Summary of data from five studies

Dose mg/kg	No.: Fetuses	No.: Litters	Incidence /Litter	Percentage
0	3386	288	1	0.35
0.5	1885	158	1	0.63
5	2188	182	1	0.55
10	2219	184	0	0.00
15	292	24	1	4.20
45	248	22	0	0.00
135	2413	202	18	8.90
250	225	20	7	35.00
400	209	24	18	75.00

This data when fitted to three mathematical models, logit, probit and Weibull (Munro 1981) generate the curves in Figure 1. The 135 mg/kg/day dose was used as the pivotal point in curve fitting as this dose had been replicated in the series of studies with a consistent result. All three curves do show that an apparent threshold for the phenomenon exists and they have similar patterns at the lower dose levels. The exposure levels corresponding to the specified excess risk calculated from these models are in Table 3 and demonstrate that they are reasonably comparable with the probit model being slightly different to the other two. These values are all within the experimental range and a threshold will lie below these values.

Table 3. Tolerance distribution models. Exposure (mg/kg/day) at specified excess risk (95% lower confidence limit in parentheses)

Model/Excess Risk	10^{-2}	10^{-4}	10^{-6}
Logit	67.3 (48.3)	15.5 (7.2)	3.6 (1.1)
Probit	80.8 (62.7)	37.7 (23.6)	21.5 (11.4)
Weibull	57.4 (39.4)	9.3 (4.0)	1.5 (0.4)

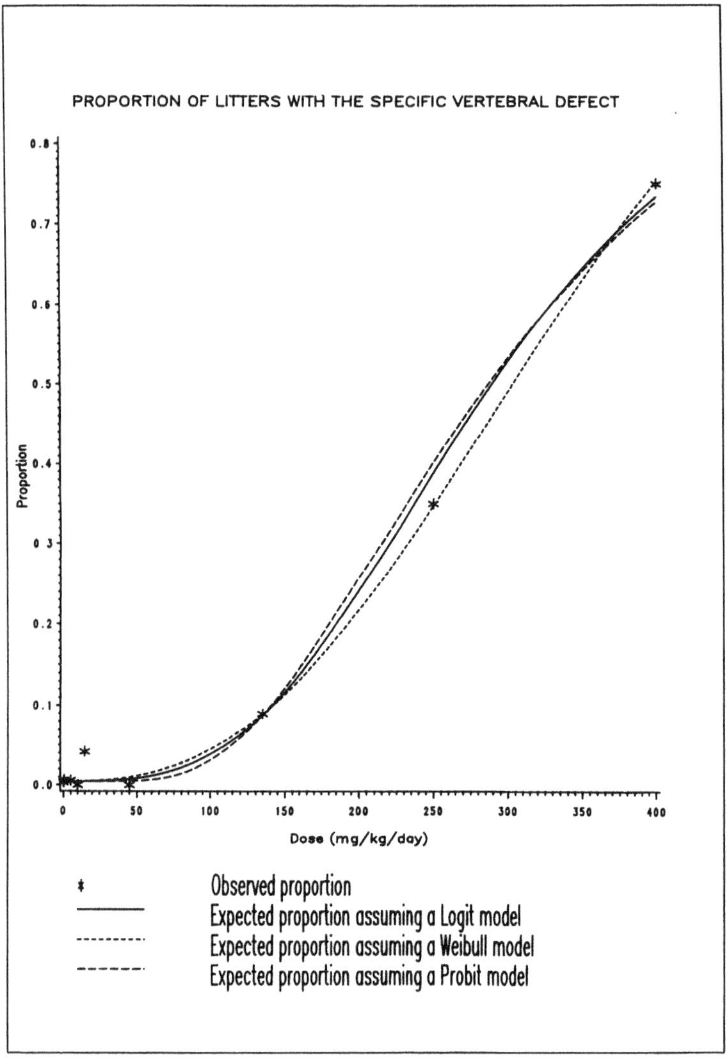

Fig. 1. Tolerance distribution curves

It is thus apparent that for the teratogenicity type of observation, where the incidence of the lesion is usually low in normal studies, a single regulatory type of study is insufficient to provide suitable data for applying the mathematical models. Further studies perhaps with different dosing regimes and at higher doses are necessary to generate sufficient data for modelling.

Physiological Models

This approach is exemplified with a data set from an investigation into eye defects (ECETOC 1989). The initial studies elicited a sporadic distribution of eye defects, Table 4, that was not readily interpretable. This was investigated in a very large study, with examination of the rat foetuses confined to the eye, the results of which are summarised in Table 5. To clarify this situation in terms of risk to man it was decided to move away from the rat model and investigate the likely impact on man more directly. The first step was to use a physiologically based mathematical model to predict the likely blood levels in man from similar exposures as those used in the rat studies. This model predicted that the blood levels in both species would show a rapid uptake and elimination but with peak levels in the rat being twice that in man (approximately 4 mg/l in the rat compared with 2 mg/l in man). This was tested by exposing pregnant rats and determining the blood levels. The result was that the actual levels were half that predicted. The actual levels in man were determined in a human volunteer study (Woolen et al. 1989), and found to be similar to that in the rat. Thus there was no reason from these studies to differentiate between man and the rat model used.

Table 4. Incidence (foetuses/litters) of microphthalmia or anophthalmia in three studies in rats

Exposure mg/M^3	0	350	1050	1750	3500	35000	70000
Study 1	0/21	--	--	--	1A/22	2A/21	--
Study 2	0/34	--	--	1M/33	2A/33	--	1A/35
Study 3	0/38	1M/40	0/35	--	--	2M/34	--
% Incidence	0	2.5	0	3.0	5.4	7.8	2.8

(M): Microphthalmia; (A): Anophthalmia (A)

Table 5. Foetal observations in mega-study

Exposure mg/M^3	Anophthalmia	Microphthalmia	Combined	N (litters)
0	1.65	3.29	4.94	607
350	2.53	10.13	12.66	395
3500	2.56	5.13	7.69	390
17500	15.76*	10.44	26.11**	383

(M): Microphthalmia; (A): Anophthalmia (A)
Statistical significance: * p = 0.5, ** p = 0.1

Discussion

These two examples demonstrate that in general first stage or guideline type studies are inadequate for formulating quantitative risk estimates. Further studies are needed to generate the data on which such estimates can be made. These secondary-stage studies should be of a type that will generate a dose-response curve. Ideally they should be of a multiple dose design, with the dose levels spread across the range, particularly at higher doses where a more substantial response would be expected. The use of such higher dose levels is more readily achieved with developmental toxicity studies than in others as the dosing period can be restricted to the critical window for the effect sought. Particularly for low incidence phenomena this approach is more powerful than the use of large group sizes with a few dose levels. This is particularly so at the lower end of the dose-response curve, where the observations would gradually increase from sporadic zero's and one's to two's and more enabling a good curve fitting exercise. For instance in the first example above, the fifth study used 800 rats distributed in groups of 160 to five dose levels. These same number of rats could have been distributed to 40 groups of 20 each at intervals of 5 mg to generate a very accurate dose-response curve.

Physiological approaches are not an alternative to the above but a complementary way of improving the risk estimates. The generation of useful human information and the use of tested models to predict human handling of xenobiotics is becoming more important with the increase in knowledge and concern over human exposure.

Hence it is my view that the initiative being progressed in the US towards better risk estimates, the use of benchmark doses rather than empirical no-effect-levels and extrapolation to specified residual risk levels is worthy of support. These approaches use the dose-response data more effectively than the empirical one of a NOEL and safety margins. The risk associated with a benchmark dose is specified, whereas that with a NOEL is assumed.

However, to use a benchmark dose to extrapolate in a linear fashion to zero is not at all helpful as it then loses the value of the benchmark dose and also assumes the lack of a threshold which is generally accepted to exist for developmental toxicity. By using appropriate models for low dose extrapolation the exposure associated with a specified excess risk can be determined. Where possible such models should include or be able to accommodate the existence of a threshold below which the occurrence of the effect would be zero.

That these approaches can be successfully used with other reproductive endpoints has been well demonstrated (Pease et al. 1991) by the use of the published data on DBCP to compare four ways of estimating risk. These were the California Proposition 65 method of a NOAEL and 1000 fold safety factor, the USEPA method of a NOAEL and an uncertainty factor based on the quality of data, the benchmark dose of the USEPA and uncertainty factor and quantitative risk estimates based on low dose extrapolation. The first three methods gave reference doses (RfD) without any indication of the risk of infertility at that dose. Only the quantitative method gives this information.

There will however be initial problems in applying these methods from the point of view of the public and its risk perception. This is illustrated by the recent events surrounding 2,4-toluenediamine (TDA) and its putative leakage from breast implants (Park and Brown 1991). The question of its potential to cause birth defects when raised with the USFDA elicited the response that the NOEL was high and that the expected exposure was well below the safety margin based on this. The implication was that at this level there was no risk and this appears to have satisfied the public. However, had the response been that say, one in one million would have been at risk, the public response would probably have been different. Thus it is important that part of the current effort directed towards the development of risk models should be directed towards improving risk communication and the perception of risk in the widest context.

Finally, none of these initiatives are likely to reach fruition unless it is clearly seen that regulatory authorities, not only in the US but elsewhere as well, will accept studies that do not necessarily contain a NOEL, but do contain adequate data for dose-response determinations, that industry is prepared to invest in the type of studies that generate such data and that the toxicologists involved in the provision of expertise in both these fields are committed to more exploratory and less descriptive studies.

References

ECETOC JACC Report No: 9 (1989) European Chemical Industry Ecology And Toxicology Centre, Brussels

Munro IC and Krewski D (1981) Risk assessment and regulatory decision making. Fd Cosmet Toxicol 19: 549-560

Park P, Brown P (1991) Compound from breast implants linked to cancer. New Scientist: 130, No: 1768: P 14

Pease W, Vandenberg J, Hooper K (1991) Comparing alternative approaches to establishing regulatory levels for reproductive toxicants: DBCP as a case study. Environ Health Perspect 91: 141-155

Wickramaratne GA de S, Cook SK (1990) Mathematical modelling and determination of the virtually safe dose from teratogenicity studies and their impact on study design. Teratology 41: 599

Woolen BH, Auton TR, Blain PG, Cocker J, Mahler JD, Makepeace D, Marsh J (1989) Occupational health in the Chemical Industry: Proceedings of the 17th Medichem Congress, Kracow, WHO, Copenhagen 1990, pp 175-180

Discussion of the Presentation

Bolt: I would like to comment on one aspect of incorporating pharmacokinetic models into the dose-response assessment. We now have the situation that we can apply very sophisticated statistics concerning the shape of dose-response curves, but no statistical assessment is incorporated in the kinetic modelling. I think in the future this is a field where also the statistical background should be elaborated.

Wickramaratne: Pharmacokinetic models as with all other model systems will only provide data or information to help in decision making. All the models and numbers derived from the experiments assist in the process which always ends in a judgement made on personal experience and knowledge. There is no system that will provide that final magic number. All they do is increase the "comfort" factor with which that final judgement is made.

Meister: The estimation of pharmacokinetic parameters is a challenge and of similar difficulty as our question on appropriate models. However, if all the doses are transformed the same way there will be no big problems with the statistical variation. Otherwise the variance of the resorbed dose has to be compared to the variance of the estimated concentration.

Neubert: Please, allow me to make two comments:

- The investigations presented certainly cannot be performed routinely. Is the additional information obtained really worth using 1,100 rats in the first example, and 360 rats in the second? I feel strongly that from the point of animal protection it is not justified.
- Why is the new term "physiological model" created. What is performed is in fact comparative pharmacokinetics (*see: Bass et al. SCOPE 1985*).

General Discussion: Use of Biologically Based Models

Bolt: The extensive use of mathematical models is, of course, based on the US point of view which differs from the practice in European countries. I think we agree that much more biological features should be incorporated in the evaluation. However, we are sceptical about the statistical models used so far because they receive too much confidence by politicians and laymen who only see the numerical result and not the biological background. This may lead to a misuse for which we have a recent example in the FRG in the discussion of permitted environmental levels of some six carcinogens.

Anderson: Of course, scientists cannot ensure against the misuse of risk assessment data. However, the merits of incorporating biological data into species to species extrapolations and extrapolation from high dose data in observed responses to estimated low dose responses for humans present a compelling case for further investigation. NOEL and safety factor approaches are clearly arbitrary and do not attempt to take into account differences in mechanisms for reproductive toxicants. Better regulatory results and improved scientific understanding of disease causation should result from biologically based dose-response modeling.

Neubert: If we are going to use biologically based models, which assumptions have to be made and which variables of embryonic development will be relevant? Do we know enough about developmental biology of the embryo to select the proper variables? Which possibilities exist to validate such models?

Anderson: While a completely biologically (mechanistically) based model for reproductive/development risk assessment is still under investigation, interim methods have been developed. In May of 1991 our scientists conducted a workshop for recognized U.S. experts in the fields of reproductive/developmental toxicology, dose-response modeling, and statistics to discuss the biologically-based dose-response models to use to assess risk associated with reproductive/developmental toxicants. The attendees agreed that any biologically-based model should consist of selected key variables or parameters. These variables should adequately describe the following: exposure/dose, the endpoint(s) of interest, threshold/nonthreshold, mechanistic consideration, timing of exposure, and (in animals) litter effects. We have been actively pursuing research in these areas. We have developed in cooperation with the National Center for Toxicological Research (NCTR) three models that incorporate some of these parameters, namely, dose, endpoint of interest (discrete or continuous), threshold, especially inter- and intralitter correlations, and an assessment of the risk for reproductive/developmental toxicity. With regard to dose, physiologically based pharmacokinetic models have been developed that provide an estimate of the appropriate dose metric in the target of interest, including the developing fetus, for the exposure pattern of concern. Such models provide a dose metric that more accurately reflects the actual dose to the fetus and provides a means to extrapolate dose across species, routes, and temporal dosing patterns. Our scientists have developed two software products to aid in this effort. Current efforts are ongoing to more fully use data describing adverse endpoints. Multivariate analyses technique to include multiple endpoints that may occur in sequence or be dependent are under investigation. Mechanism of toxicity is linked to both the timing of the exposure, i.e., the trimester during which exposure occurs, and the dose delivered to the fetus. If the specific target organ for a developmental toxicant is known, then the timing of the exposure would be critical since we have enough knowledge from embryology to understand that certain developing organ systems are more sensitive during certain stages of organogenesis. Typically, mechanism and target organ are unknown and exposure occurs chronically; therefore, timing becomes a function of frequency and duration which can also be assessed with a physiologically based pharmacokinetic model. Hence only the mechanism of action has yet to be fully characterized and incorporated into risk assessment and into biologically based models. However, in lieu of mechanism, biomarkers of exposure and biomarkers of effect could be incorporated into these models. Research in this area between our scientists and other experts in this field are ongoing.

Special Methods and Approaches

Pharmacokinetics and Drug Metabolism in the Design and Interpretation of Developmental Toxicity Studies

Heinz Nau

Introduction

A direct teratogenic effect will depend on the concentration-time relationship of the drug or its active metabolite in the placental-embryonic compartment during the sensitive stages of gestation. Pharmacokinetic studies are therefore important for the interpretation and design of teratogenicity experiments, particularly for the problem of species differences, and for attempts to extrapolate experimental findings to the human ("risk assessment"). Pharmacokinetic studies are also crucial in elucidating whether the parent drug or a metabolite is the proximate or ultimate toxic entity. Finally, it is shown that structure-activity studies can greatly benefit from pharmacokinetic experiments to demonstrate whether the teratogenicity of a substance is related to its intrinsic activity or to the extent of placental transfer or both.

Species Differences

Both the type of malformation produced and the teratogenic dose levels may show drastic species differences (Schardein et al. 1985; Neubert and Chahoud 1985; Nau and Scott 1987a) (Table 1). Most drastic and well known, but still unexplained is the paradigm of thalidomide and its even more potent structural analogue EM-12. These substances produce dramatic limb malformations and other defects in primate species including man (Neubert et al. 1988; Merker et al. 1988), but apparently not to the same degree in the usual laboratory animals, even with several hundred-fold higher doses (Scott et al. 1977; Sterz et al. 1987). This may not be an isolated example, as Table 1 shows. Valproic acid can induce neural tube defects in the human, mouse and hamster, but not prominently in the rat, rabbit and monkey (Nau and Hendrickx 1987; Nau 1986a; Mast et al. 1986; Jäger-Roman et al. 1986; Binkerd et al. 1988). All retinoids tested so far produce a similar pattern of malformations in the species tested (defects of the central nervous system, skeletal structures and cardiovascular system), but effective doses differ widely (Nau et al. 1989; Lammer et al. 1985; Willhite and Shealy 1984; Kistler and Hummler 1985). In the case of isotretinoin, there is a difference in the teratogenic daily doses between man and rabbit by a factor of 20, and between man and mouse/rat by a factor of at least 200 (Table 1). The intrinsic sensitivities of the developing structures, or differences in exposure of the embryo during sensitive stages of gestation may be the two most important reasons for such species differences (Schardein et al. 1985; Neubert and Chahoud 1985; Nau and Scott 1987a; Nau 1988).

In regard to pharmacokinetics, a number of parameters may exhibit significant species differences (Scheme I) (Nau 1986a; Nau 1988). Absorption and distribution phenomena usually do not present great difficulties as these are governed by passive diffusion of exogenous substances through membranes as well as the body composition. However, plasma protein binding and placental transfer, both interrelated, are probably of major importance, although many aspects must still be investigated here (see Scheme II). Of the major pharmacokinetic parameters, drug elimination exhibits the greatest species difference. Small laboratory animals have a relatively high surface area/body weight ratio (Table 2) and thus require high metabolic rates to maintain a physiological body temperature. It is therefore not surprising that the elimination of exogenous substances is also much more rapid in animals than in man. Several approaches have been developed to "scale" results attained in one species to another, particularly to man (Freireich et al. 1966; Bonati et al. 1984-1985; Sawada et al. 1984; Smith 1987). The allometric approaches are most useful for drugs which are excreted unchanged, and here "scaling factors" can be used to compensate for the relatively high elimination rates of laboratory animals. Drugs which are cleared via hepatic metabolism do not follow such a simple species-dependent pattern, as enzymatic processes are involved here which are more specific. Physiological models offer a mechanistic approach which can be applied in some cases. Drug metabolism is much more rapid in animals than man and often result in very different and unexpected metabolic patterns. The significance of differing metabolic pattern in regard to species variations of teratogenic effects has not been clearly established.

The half-life of drugs is therefore often an order of magnitude smaller in experimental animals than in the human. During conventional teratogenicity testing in experimental animals, high and sharp peaks rapidly fall to insignificant levels and "drug holidays" persist between dosing intervals (Nau and Scott 1987a; Nau 1988). The human is usually exposed to more persistent drug concentrations, which still fluctuate, but often much less so than in animals because of the longer drug half-life. This drastic difference between the plasma kinetics of a teratogenic dosing regimen of the antiepileptic drug valproic acid has been demonstrated in the mouse and in man (Nau and Spielmann 1983; Nau 1985b). It is therefore important to establish whether maximal concentrations (as with valproic acid, caffeine, ethanol in mice) or the areas under the concentration-time curves (AUC) (with retinoids, salicylate and cyclophosphamide in mice) are the decisive factors (Table 3). Much less is known about the pharmacokinetics following environmental exposure.

Should inappropriate pharmacokinetic parameters be used for species comparison, misleading answers may be obtained confusing rather than shedding light on this difficult problem. This subject is complicated further by the fact that each specific defect has its own, probably rather short, "window" of sensitivity. This "window" is somewhat "widened" by the considerable variation in development of the embryo/fetus between the animals and also in each litter (Neubert and Chahoud 1985). AUC's and multiple drug peak concentrations within such "windows" are likely to be the most appropriate parameters for interspecies scaling.

Table 1. Species differences in the teratogenicity of selected drugs following oral dosing (once/day)

Drug	Fetal defect observed	Teratogenic dose (mg/kg/day)[b] Mouse	Rat	Rabbit	Monkey	Man
Thalidomide	skeletal	>200	>400	3 x 300	<10	<10
EM-12	skeletal	-[c]	<50	0.1-1	-	
Valproic acid	skeletal	150[a]	200	150-350	200	20-40
	neural tube closure	200-258[a]	low	0	0	20-30
Tretinoin	multiple	4	0.4-2	2	7.5-40	-
Isotretinoin	multiple	100	150	10	2.5	0.4
Etretinate	multiple	0.85; 4	2	2	-	0.2
Retinol (vitamin A)	multiple	50	90	5	-	-
Acitretin	multiple	3	15	0.6	-	-
13-cis-Acitretin	multiple	100	3	3	-	-

[a] s.c. and i.p. administration, which are more effective than oral administration

[b] The lowest teratogenic dose observed is given

[c] i.e. not investigated

Scheme I. Possible species differences in pharmacokinetic processes

DRUG

ABSORPTION — usually not a great problem; can be adjusted by mode of administration/change of drug formulation; first-pass effect often, but not always more extensive in experimental animals than in man (see metabolism).

DISTRIBUTION — usually not major species differences; governed by body composition and passive diffusion of drugs across membranes.

- **PLACENTAL TRANSFER** — probably major species differences, but not well established experimentally; probably related to species differences in placental structure and function; yolk sac placenta is much more prominent during organogenesis in rodents than in man; the transplacental pH gradient is important for the transfer of ionized drugs.

- **PLASMA PROTEIN BINDING** — usually more extensive in man than in experimental animals; this can result in larger tissue concentrations in animals (and possibly increased effects) as compared to man; clearance is therefore also increased in animals because more "free" drug can diffuse into hepatocytes for metabolism or can be filtrated in the glomerulus for renal excretion.

ELIMINATION

- **METABOLISM** — major species differences are possible here; metabolic profiles and rates of biotransformation can differ drastically; the human metabolizes drugs often much slower than experimental animals; prediction and allometric interspecies scaling is difficult because specific enzymes are involved.

- **RENAL EXCRETION** — major species differences are possible here; the rate of elimination is probably related to the body surface area/body weight ratio (Table 2) and thus much higher in small laboratory animals than in man; allometric interspecies scaling is possible.

Scheme II. Parameters defining the placental transfer of drugs

Parameter	Examples
Lipophilicity of drugs	Increase of lipophilicity increases rate of placental transfer (e.g. demonstrated with barbiturates). Too highly lipophilic agents (e.g. TCDD) show low placental transfer because of extensive sequestration into maternal tissue (fat) and the low fat content of the conceptus.
Molecular weight of drugs	Increase of the molecular weight (especially above 1000) decreases placental transfer. Low placental transfer of high molecular weight drugs (digitoxin, erythromycin) and polypeptides/proteins (except when endocytosis is functional).
Ionisation of drugs ("ion trapping")	Accumulation of "free" acidic drugs (valproic and salicylic acid) in the early mammalian embryo (high intracellular pH); accumulation of "free" anaesthetic agents (lidocaine, bupivacaine) in the relatively acidic fetal blood during birth.
Protein binding of drugs	Plasma protein binding of acidic drugs defines the extent of transfer, only the "free" drug equilibrates across the placenta. Extensive binding of all-trans as compound to 13-cis-retinoic acid within the embryo explains the great difference in placental transfer and teratogenicity of these two retinoids (high: all-trans; low: 13-cis).
Active transport	Probably not important for drugs which are usually transferred via passive diffusion. More important for endogenous compounds and vitamins.
Structure of placenta	Probably very important but a dearth of information exists here (see Scheme I).

Table 2. Species differences in body weights, surface areas and dose equivalents

Species	Body Weight (kg)	Surface Area (m^2)	Dose Equivalent/kg (Human = 1.0)
Mouse	0.02	0.0066	12.0
Rat	0.15	0.025	6.0
Dog	8	0.40	1.7
Monkey	3	0.24	3.0
Human child	20	0.80	1.5
Human adult	60	1.60	1.0

Table 3. Correlation of pharmacokinetic parameters of several drugs with the teratogenic potency

Correlation of the teratogenicity with the	
Peak Concentration (C_{max})	Area under the plasma concentration-time curve (AUC)
Valproic acid	Retinoids
Caffeine	Cyclophosphamide
Ethanol	Salicylate

Several compounds, such as arsenate, sodium cyanide, morphine and hydromorphone, were shown to be teratogenic after infusion; an injection-infusion comparison is not available here, so that a direct correlation between the teratogenic effects of these drugs and AUC values has not been established.

The duration of sensitive periods of particular developmental processes are often several-fold shorter in experimental animals than in man (Schardein et al. 1985; Neubert and Chahoud 1985; Nau and Scott 1987a). Thus, for the study of a particular developmental process, a multiple-dosing regimen during a defined period may prove more useful than the conventional daily administration regimen where the time of maximal sensitivity may be missed. The prolonged exposure in the human may also be more accurately reproduced in animals via drug infusion. This is most important with drugs such as retinoids, where the teratogenic effect is related to the AUC reached in the embryo. Such an approach was used recently for the retinoid etretinate and its main metabolite acitretin, which persist for several months after discontinuation of psoriasis therapy because of extensive storage in and slow release from peripheral compartments. We could maintain steady-state levels of these drugs in the mouse to show that very low concentrations in plasma and embryo (low ng/ml or g range) result in retinoid-specific teratogenic effects (Löfberg et al. 1990). Thus, species variations may become less problematic when the teratogenic effects are compared with similar pharmacokinetics (interspecies bioequivalence).

Placental Transfer

Placental transfer depends on several characteristics related to drug structure and placental interface (Scheme II). Many of the important variables have not been completely defined. The widely-known dogma in this regard - drugs with high lipophilicity, molecular weight below 1,000, low protein binding and low ionization will show "good" placental transfer - can be used at the best as initial orientation, as demonstrated by Scheme I (Mirkin and Singh 1976; Nau and Mirkin 1987; Dencker 1987). Passive diffusion across membranes seems to be a highly important transfer mechanism. Such generalizations are not always useful as many exceptions exist: drugs highly ionized at physiological pH often transfer rapidly, because the transferred non-ionized portion may be minor but is rapidly regenerated. Similar factors may also account for highly bound drugs, where the "free fraction" which is able to diffuse across membranes is rapidly re-established. Therefore, not protein binding and ionization are the decisive factors *per se*, but the transplacental protein binding and ionization *gradients* of the drugs. Only if embryonic binding is low does the maternal plasma protein binding govern the placental transfer, as is the case for some acidic substances (salicylic acid, valproic acid): transfer increases when the "free" fraction in plasma increases, either by saturation of plasma proteins via increasing the total drug levels, or by displacing agents such as free fatty acid (Wilson et al. 1977; Itami and Kanoh 1983; Nau et al. 1984; Nau and Krauer 1986).

When the extent of drug binding in embryo or fetus becomes of importance, the placental transfer will depend on the transplacental binding gradient. A good example is the class of local anaesthetic agents (e.g. lidocaine, bupivacaine) which are bound to a much greater degree in maternal plasma than in fetal plasma, because the binding protein for these agents (α_1-acid glycoprotein) exhibits very low levels in the fetus as compared to the mother (Nau 1985a).

A more recent example was established with retinoids to tentatively explain the drastic difference between the placental transfer of all-trans and 13-cis-retinoic acid in the mouse (Creech Kraft et al. 1987; Creech Kraft et al. 1989; Reiners et al. 1988). Both substances are extensively bound to plasma proteins, but only the all-trans isomer is transferred to a great extent to the embryo because of extensive binding to its cellular retinoic acid binding protein which is abundant in the embryo. The 13-cis isomer exhibits little affinity to this protein, is transferred to a very small extent only, and shows very low teratogenic potency as compared to the all-trans isomer.

The pH-gradient across the placenta plays a major role in regard to the placental transfer of acidic substances. It has been shown that the intracellular embryonic pH in the early mammalian embryo is surprisingly high (well above maternal plasma pH). Therefore, acidic substances such as salicylic acid, valproic acid, structural analogues and acidic thalidomide metabolites accumulate during early organogenesis which highlights the special significance of acidic moieties for expression of teratogenic activity (Nau and Scott 1986; Nau and Scott 1987b).

Metabolic Activation (Teratogenic Activation)

The question whether the administered drug and/or a metabolite is responsible for the teratogenic action is often complex, and difficult to answer (Scheme III). In the case of thalidomide and phenytoin, arene oxides were proposed as proximate or ultimate teratogens, which are presumably very reactive and covalently modify crucial macromolecules (DNA, RNA, proteins) in the embryo (Koch 1981; Gordon et al. 1981; Martz et al. 1977). No conclusive results on this question are yet at hand. Other hypotheses (e.g. cooxidation within the prostaglandin cascade) are being followed in the case of phenytoin (Kubow and Wells 1989; Wells et al. 1989).

Valproic acid on the other hand, is teratogenic as parent drug, and the metabolites formed are either not teratogenic, or are active (4-en-VPA), but are produced to a minor degree only (Nau 1986c). Metabolism plays an important role in retinoid teratogenesis, and all-trans retinoic acid appears to be of central importance both in regard to the teratogenicity of 13-cis-retinoic acid (Accutane^{R}) (Creech Kraft et al. 1987; Creech Kraft et al. 1989; Reiners et al. 1988; Creech Kraft et al. 1990) or retinol (vitamin A) (Eckhoff et al. 1989; Eckhoff and Nau 1991; Eckhoff and Nau 1990a). There are large species differences in the metabolism of retinoids. As one example, 13-cis-retinoic acid (isotretinoin) is extensively glucuronidated in the mouse, while in monkey and particularly man the 4-oxo-pathway predominates (Creech Kraft et al. 1990) (Scheme IV). In addition to poor placental transfer of isotretinoin in the mouse, extensive glucuronidation could also play an important role in the low teratogenic potential of the drug in this species because the glucuronide metabolite is transferred to the embryo to a very minor degree only.

On the other hand, all-trans-retinoic acid (tretinoin) is extensively glucuronidated in the monkey, while in the mouse the 4-oxo-pathway predominates (Creech Kraft et al. 1990) (Scheme IV). This may explain why tretinoin is a very potent teratogen in the mouse (the 4-oxo metabolite and the parent drug are equally teratogenic), while the monkey may not be a species particularly sensitive to this drug.

Scheme III. Metabolic activation of drugs: Is the parent drug or a metabolite the proximate / ultimate teratogen?

Drug	**Metabolic Activation**
Thalidomide	unknown; aromatic epoxidation (arene oxides) may be the reactive species, e.g. via intercalation into DNA and covalent binding.
Phenytoin	unknown; arene oxides may also be important here; but also activation via prostaglandin cascade.
Cyclophosphamide	activated by cytochrome P-450 monooxgenase system to phosphoramide mustard and acroleine.
Valproic acid	parent drug is the teratogenic entity; metabolism is complex and extensive, but only defines the crucial parent drug concentration.
Etretinate, motretinide	must be hydrolyzed to the free acid (acitretin).
Isotretinoin	isomerisation to the all-trans retinoic acid (tretinoin) and oxidation to the 4-oxo-metabolite may have major importance.
Retinol (vitamin A)	oxidation to all-trans retinoic acid as well as 13-cis-retinoic acid and 4-oxo-13-cis-retinoic acid metabolite may be crucial.

Scheme IV. Species differences in the metabolism of vitamin A and retinoids between mouse, monkey and man

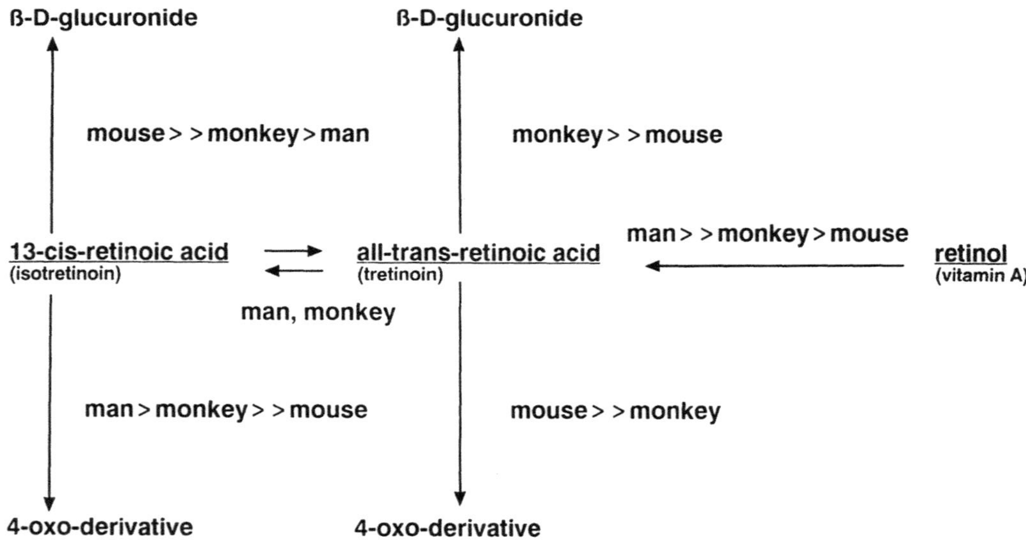

Thus, drug metabolism studies greatly contribute to the interpretation of species differences in retinoid teratogenesis. The appearance of the glucuronide metabolites of tretinoin and isotretinoin in plasma was surprising (Creech et al. 1990; Eckhoff et al. 1990b). It has been shown that the glucuronide of tretinoin exhibits retinoid-like biological activity, reportedly without hydrolysis to the parent drug (Zile et al. 1987; Barua and Olson 1989). The poor placental transfer of this glucuronide may therefore lead the way to developing active retinoids with low teratogenic potential.

Structure-Activity Considerations

The teratogenicity of many compounds is strictly related to the structure in that it shows high structural specificity. In some cases, this may be related to maternal kinetics and embryonic exposure: the low teratogenic activity of 13-cis retinoic acid in the mouse can thus be explained by extensive detoxification (glucuronidation) and low placental transfer in this species (see above). For a number of compounds such as thalidomide (Schumacher 1975; Jackson and Schumacher 1980; Helm et al. 1981) retinoic acid (Nau et al. 1989; Lammer et al. 1985; Willhite and Shealy 1984; Kistler and Hummler 1985) and valproic acid (Nau and Löscher 1984; Löscher and Nau 1985; Nau and Löscher 1986; Hauck and Nau 1989a; Klug et al. 1990) many structural analogues have been synthesized and tested for teratogenicity in order to search for clues towards the mechanism of action, or to develop therapeutically useful agents with low teratogenic potential. Many of these studies have been disappointing. The complexity of the chemical structures render difficult the interpretation of alterations in the teratogenic response obtained by changing the structures of the parent molecules. Also, in most cases it was not shown whether the teratogenicity of a compound is related (a) to its intrinsic (structural) property, or (b) to the extent of placental transfer.

We have been able to show that structural analogues of valproic acid belong to the first class of compounds - all substances reach the embryo, but only a few with very distinct structures are teratogenic. These findings allowed us to develop alternative anticonvulsant agents, 2-en-valproic acid is the first drug of its class which is being developed for clinical use based on our experimental studies (Nau et al. 1984; Löscher et al. 1984; Nau 1986b; Vorhees et al. 1991).

The high structural specificity is best demonstrated by the differences observed in regard to the stereoselective teratogenicity of chiral compounds (Table 4). The enantiomers of some were shown to differ, suggesting that the teratogenic mechanism here includes a stereoselective interaction between these compounds and a chiral molecule, possibly a protein, in the embryo. Experiments with thalidomide were done with rats and mice, presumably "unresponsive" species (see Table 1) (Blaschke et al. 1979). In the rabbit, the teratogenicity of the thalidomide enantiomers did not differ (Fabro et al. 1967). Recently it was shown in the "responsive" species, the monkey, that the teratogenicity of the even more potent thalidomide derivative EM-12 is stereoselective with the S(-) enantiomer being more potent than the R(-) enantiomer (Heger et al. 1988). Extensive racemisation of the enantiomers made a conclusive interpretation difficult, but it appears that the S(-) enantiomer also has a higher intrinsic activity than the R(+) enantiomer (Schmahl et al. 1988; Schmahl et al. 1989).

Table 4. Stereoselectivity of the teratogenic potency of drugs

Drug	Species	Route of application	Teratogenic potency
Thalidomide	Mouse, rat	i.p.	S(-) > R(+)
	rabbit		S(-) = R(+)
EM-12	Monkey	oral	S(-) > R(+)
4-en-Valproic acid	Mouse	i.p.	S(-) > R(+)
4-yn-Valproic acid	Mouse	i.p.	S(-) > R(+)
2-ethylhexanoic acid	Mouse	i.p.	R(-) > S(+)
Indacrinone[a]	Rat	oral	(-) > (+)
2-(4-Chloro-2-methylphenoxy)-propionsäure[b]	Mouse	oral	(+) > (±)
2-(2,4-Dichlorophenoxy)-propionsäure[b]	Mouse	oral	(+) > (±)

[a] Absolute configuration not given; the teratogenic effect of this drug is likely to be related to the saluretic effect and possible electrolytic imbalance at high doses (Robertson et al. 1981)

[b] Absolute configuration not given; the stereoselectivity of the teratogenic response, i.e. (+) vs. (±), was described as low (71)

Some of these questions are more difficult to answer in the case of retinoids, some of which may indeed belong to the second class of compounds. Isotretinoin, and the glucuronides of tretinoin and isotretinoin show only very limited placental transfer, at least in the mouse, and may therefore possess low teratogenic potency. It is not clear whether these compounds have intrinsic teratogenic activity as some *in vitro* studies tend to indicate. These questions can only be answered by extensive measurements of drug and metabolite levels; the mere administration of test substances, both *in vivo* and *in vitro*, followed by biological response evaluations does not suffice.

Recently, we demonstrated that the teratogenicity of valproic acid-related substances is stereoselective (Table 4) (Hauck and Nau 1989b; Hauck et al. 1990). The advantage of this class of compounds for structure-activity studies lies in its structural simplicity and it is surprising that the teratogenicity here is so exquisitely selective. We are now aware that for the expression of significant teratogenic activity (exencephaly in the mouse), the following simple structure is required: a carbon atom bound to a hydrogen atom, to a carboxyl group and to two carbon chains which can be adjusted (e.g. by desaturation) to drastically modify the teratogenic response (Nau and Hendrickx 1987; Klug et al. 1990; Nau et al. 1984; Löscher et al. 1984; Nau 1986b). We hope that further studies in this area will lead the way for the development of "non"-teratogenic compounds, such as 2-en-valproic acid, with therapeutic possibilities (Blaschke et al. 1979; Fabro et al. 1967; Heger et al. 1988; Schmahl et al. 1988). The development of increasingly potent compounds should, on the other hand, provide clues as to the mechanism of the stereoselective action of this class of compounds.

Acknowledgements: Support for the work performed in our laboratory was provided by the Deutsche Forschungsgemeinschaft (SFB 174, C6) and the Free University of Berlin. The manuscript was prepared by A. Zaretzki.

References

Barua AB, Olson JA (1989) Chemical synthesis of all-trans-[11-^{3}H] retinoyl ß-glucuronide and its metabolism in rats *in vivo*. Biochem J 263: 403-409

Binkerd PE, Rowland JM, Nau H, Hendrickx AG (1988) Evaluation of valproic acid (VPA) developmental toxicity and pharmacokinetics in Sprague-Dawley rats. Fund Appl Toxicol 11: 485-493

Blaschke G, Kraft HP, Fickentscher K, Köhler F (1979) Chromatographische Racemattrennung von Thalidomid und teratogene Wirkung. Drug Res 29: 1640-1642

Bonati M, Latini R, Tognoni G, Young JF, Garattini S (1984-85) Interspecies comparison of *in vivo* caffeine pharmacokinetics in man, monkey, rabbit, rat, and mouse. Drug Metab Rev 15: 1355-1383

Creech Kraft J, Kochhar DM, Scott WJ, Nau H (1987) Low teratogenicity of 13-cis-retinoic acid (isotretinoin) in the mouse corresponds to low embryo concentrations during organogenesis: Comparison to the all-trans isomer. Toxicol Appl Pharmacol 87: 474-482

Creech Kraft J, Löfberg B, Chahoud I, Bochert G, Nau H (1989) Teratogenicity and placental transfer of all-trans-, 13-cis-, 4-oxo-all-trans- and 4-oxo-13-cis-retinoic acid after administration of a low oral dose during organogenesis in mice. Toxicol Appl Pharmacol 100: 162-176

Creech Kraft J, Slikker W, Bailey JR, Roberts LG, Fischer B, Wittfoht W, Nau H (1990) Plasma pharmacokinetics and metabolism of 13-cis and all-trans-retinoic acid in the cynomolgus monkey and the identification of 13-cis and all-trans-retinoyl-ß-glucuronides: A comparison to one human case study with isotretinoin. Drug Metab Disp 19: 317-324

Dencker L (1987) Transfer of drugs to the embryo and fetus after placentation. In: Nau H, Scott WJ (eds) Pharmacokinetics in teratogenesis. Vol. I. Interspecies comparison and maternal embryonic fetal drug transfer. CRC Press, Boca Raton, pp 55-70

Eckhoff C, Löfberg B, Chahoud I, Bochert G, Nau H (1989) Transplacental pharmacokinetics and teratogenicity of a single dose of retinol (vitamin A) during organogenesis in the mouse. Toxicol Lett 48: 171-184

Eckhoff C, Nau H (1990a) Identification and quantitation of all-trans- and 13-cis-retinoic acid and 13-cis-4-oxoretinoic acid in human plasma. J Lipid Res 31: 1445-1454

Eckhoff C, Nau H (1991) Vitamin A supplementation increases levels of retinoic acid compounds in human plasma: Possible implications for teratogenesis. Arch Toxicol 64: 502-503

Eckhoff C, Wittfoht W, Nau H, Slikker W (1990b) Characterization of oxidized and glucuronidated metabolites of retinol in monkey plasma by thermospray liquid chromatography/mass spectrometry. Biomed Environm Mass Spectrom 19: 428-433

Fabro S, Smith RL, Williams RT (1967) Toxicity and teratogenicity of optical isomers of thalidomide. Nature 215: 296

Freireich EJ, Gehan EA, Rall DP, Schmidt LH, Skipper HE (1966) Quantitative comparison of toxicity of anticancer agents in mouse, rat, hamster, dog, monkey, and man. Cancer Chemother 50: 219-245

Gordon GB, Spielberg SP, Blake DA, Balasubramanian V (1981) Thalidomide teratogenesis: Evidence for a toxic arene oxide metabolite. Proc Natl Acad Sci USA 78: 2545-2548

Hauck RS, Nau H (1989a) Zu den strukturellen Grundlagen der teratogenene Wirkung des Antiepileptikums Valproinsäure (VPA). Naturwissenschaften 76: 528-529

Hauck RS, Nau H (1989b) Asymmetric synthesis and enantioselective teratogenicity of 2-n-propyl-4-pentenoic acid (4-en-VPA), an active metabolite of the anticonvulsant drug, valproic acid. Toxicol Lett 49: 41-48

Hauck RS, Wegner C, Blumtritt P, Fuhrhop JH, Nau H (1990) Asymmetric synthesis and teratogenic activity of (R)- and (S)-2-ethylhexanoic acid, a metabolite of the plasticizer di-(2-ethylhexyl)phthalate. Life Sci 46: 513-518

Heger W, Klug S, Schmahl HJ, Nau H, Merker HJ, Neubert D (1988) Embryotoxic effects of thalidomide derivatives on the non-human primate *Callithrix jacchus*. 3. Teratogenic potency of the EM 12 enantiomers. Arch Toxicol 62: 205-208

Helm FC, Frankus E, Friderichs E, Graudums I, Flohé L (1981) Comparative teratological investigation of compounds structurally and pharmacologically related to thalidomide. Drug Res 31: 941-949

Itami T, Kanoh S (1983) Studies on the pharmacological basis of fetal toxicity of drugs (IV). Effect of endotoxin and starvation on serum protein binding of salicylic acid in pregnant rats. Japan J Pharmacol 33: 1199-1204

Jackson AJ, Schumacher HJ (1980) Identification of urinary metabolites of EM12 N-(2',6'-Dioxopiperiden-3'-yl)-phthalimidine, a teratogenic analog of thalidomide in rats and rabbits. Can J Pharmaceut Sci 15: 21-23

Jäger-Roman E, Deichl A, Jakob S, Hartmann A-M, Koch S, Rating D, Steldinger R, Nau H, Helge H (1986) Fetal growth, major malformations, and minor anomalies in infants born to women receiving valproic acid. J Pediatr 108: 997-1004

Kistler A, Hummler H (1985) Teratogenesis and reproductive safety evaluation of the retinoid etretin (Ro 10-1670). Arch Toxicol 58: 50-56

Klug S, Lewandowski C, Zappel F, Merker HJ, Nau H, Neubert D (1990) Effects of valproic acid, some of its metabolites and analogues on prenatal development of rats *in vitro* and comparison with effects *in vivo*. Arch Toxicol 64: 545-553

Koch H (1981) Die Arenoxid-Hypothese der Thalidomid-Wirkung. Überlegungen zum molekularen Wirkungsmechanismus des "klassischen" Teratogens. Sci Pharm 49: 67-99

Kubow S, Wells PG (1989) *In vitro* bioactivation of phenytoin to a reactive free radical intermediate by prostaglandin synthetase, horse radish peroxidase, and thyroid peroxidase. Mol Pharmacol 35: 504-511

Lammer EJ, Chen DT, Hoar RM, Agnish ND, Benke PJ, Braun JT, Curry CJ, Fernhoff PM, Grix AW, Lott IT, Richard JM, Sun SC (1985) Retinoic acid embryopathy. New Engl J Med 313: 837-841

Löfberg B, Reiners J, Spielmann H, Nau H (1990) Teratogenicity of steady-state concentrations of etretinate and metabolite acitretin maintained in maternal plasma and embryo by intragastric infusion during organogenesis in the mouse: A possible model for the extended elimination phase in human therapy. Dev Pharmacol Therapeut 15: 45-51

Löscher W, Nau H, Marescaux C, Vergnes M (1984) Comparative evaluation of anticonvulsant and toxic potencies of valproic acid and 2-en-valproic acid in different animal models of epilepsy. Eur J Pharmacol 99: 211-218

Löscher W, Nau H (1985) Pharmacological evaluation of various metabolites and analogues of valproic acid. Neuropharm 24: 427-435

Martz F, Failinger C, Blake DA (1977) Phenytoin teratogenesis: Correlation between embryopathic effect and covalent binding of putative arene oxide metabolite in gestational tissue. J Pharmacol Exp Ther 203: 231-239

Mast TJ, Cukierski MA, Nau H, Hendrickx AG (1986) Predicting the human teratogenic potential of the anticonvulsant, valproic acid, from a non-human primate model. Toxicology 39: 111-119

Merker H-J, Heger W, Sames K, Stürje H, Neubert D (1988) Embryotoxic effects of thalidomide derivatives in the non-human primate *Callithrix jacchus*. I. Effects of 3-(1,3-dihydro-1-oxo-2H-isoindol-2-yl)-2,6-dioxopiperidene (EM 12) on skeletal development. Arch Toxicol 61: 165-179

Mirkin BL, Singh S (1976) Placental transfer of pharmacologically active molecules. In: Mirkin BL (ed) Perinatal pharmacology and therapeutics, Academic Press, New York,San Francisco,London, pp 1-69

Nau H, Creech Kraft J, Eckhoff C, Löfberg B (1989) Interpretation of retinoid teratogenesis by transplacental pharmacokinetics. In: Reichert U, Shroot B (eds) Pharmacology of retinoids in the skin. Vol. 3, Karger, Basel, pp 165-173

Nau H, Helge H, Luck W (1984) Valproic acid in the perinatal period: Decreased maternal serum protein binding results in fetal accumulation and neonatal displacement of the drug and some metabolites. J Pediatr 104: 627-634

Nau H, Hendrickx AG (1987) Valproic acid teratogenesis. ISI Atl Sci Pharmacol pp 52-56

Nau H, Krauer B (1986) Serum protein binding of valproic acid in fetus-mother pairs throughout pregnancy: Correlation with oxytocin administration and albumin and free fatty acid concentrations. J Clin Pharmacol 26: 215-221

Nau H, Löscher W, Schäfer H (1984) Anticonvulsant activity and embryotoxicity of valproic acid. Neurology 34: 400-401

Nau H, Löscher W (1986) Pharmacologic evaluation of various metabolites and analogs of valproic acid: Teratogenic potencies in mice. Fund Appl Toxicol 6: 669-676

Nau H, Löscher W (1984) Valproic acid and metabolites: Pharmacological and toxicological studies. Epilepsia 25(Suppl1): 14-22

Nau H, Mirkin BL (1987) Fetal and maternal clinical pharmacokinetics. In: Avery's drug treatment principles and practice of clinical pharmacology and therapeutics, 3rd Edition, ADIS Press, Auckland, pp 79-117

Nau H, Scott WJ (1987b) Teratogenicity of valproic acid and related substances in the mouse: Drug accumulation and pHi in the embryo during organogenesis and structure-activity considerations. Arch Toxicol(Suppl) 11: 128-139

Nau H, Scott WJ (1986) Weak acids may act as teratogens by accumulating in the basic milieu of the early mammalian embryo. Nature 323: 276-278

Nau H, Scott WJ (1987a) eds. Pharmacokinetics in Teratogenesis. Vol. I & II. Experimental Aspects In Vivo and In Vitro. Boca Raton, CRC Press

Nau H, Spielmann H (1983) Embryotoxicity testing of valproic acid. Lancet i: 763-764

Nau H (1985a) Clinical pharmacokinetics in pregnancy and perinatology. I. Placental transfer and fetal side effects of local anaesthetic agents. Dev Pharmacol Ther 8: 149-181

Nau H (1988) Pharmakokinetische Grundlagen der Teratogenität von Arzneimitteln. Internist 29: 179-192

Nau H (1986a) Species differences in pharmacokinetics and drug teratogenesis. Environ Health Perspect 70: 113-129

Nau H (1985b) Teratogenic valproic acid concentrations: Infusion by implanted minipumps vs conventional injection regimen in the mouse. Toxicol Appl Pharmacol 80: 243-250

Nau H (1986b) Transfer of valproic acid and its main active unsaturated metabolite to the gestational tissue: Correlation with neural tube defect formation in the mouse. Teratology 33: 21-27

Nau H (1986c) Valproic acid teratogenicity in mice after various administration and phenobarbital-pretreatment regimens: The parent drug and not one of the metabolites assayed is implicated as teratogen. Fund Appl Toxicol 6: 662-668

Neubert D, Chahoud I (1985) Significance of species and strain differences in pre- and perinatal toxicology. Acta histochem 31(Suppl): 23-35

Neubert D, Heger W, Merker H-J, Sames K, Meister R (1988) Embryotoxic effects of thalidomide derivatives in the non-human primate *Callithrix jacchus*. II. Elucidation of the susceptible period and of the variablility of embryonic stages. Arch Toxicol 61: 180-191

Roll R, Matthiaschk G (1983) Vergleichende Untersuchungen zur Embryotoxizität von 2-Methyl-4-chlorphenoxyessigsäure, Mecoprop und Dichlorprop bei NMRI-Mäusen. Arzneim-Forsch/Drug Res 33: 1479-1483

Reiners J, Löfberg B, Creech Kraft J, Kochhar DM, Nau H (1988) Transplacental pharmacokinetics of teratogenic doses of etretinate and other aromatic retinoids in mice. Reprod Toxicol 2: 19-29

Robertson RT, Minsker DH, Bokelman DL, Durand G, Conquet P (1981) Potassium loss as a causative factor for skeletal malformations in rats produced by indacrinone: A new investigational loop diuretic. Toxicol Appl Pharmacol 60: 142-150

Sawada Y, Hanono M, Sugiyama Y, Iga T (1984) Prediction of the disposition of ß-lactam antibiotics in humans from pharmacokinetic parameters in animals. J Pharmacokin Biopharmaceut 12: 241-261

Schardein JL, Schwetz BA, Kenel MF (1985) Species sensitivities and prediction of teratogenic potential. Environ Health Perspect 61: 55-67

Schmahl HJ, Heger W, Nau H (1989) The enantiomers of the teratogenic thalidomide analogue EM12. 2. Chemical stability, stereoselectivity of metabolism and renal excretion in the marmoset monkey. Toxicol Lett 45: 23-33

Schmahl HJ, Nau H, Neubert D (1988) The enantiomers of the teratogenic thalidomide analogue EM1. 1. Chiral inversion and plasma pharmacokinetics in the marmoset monkey. Arch Toxicol 62: 200-204

Schumacher HJ (1975) Chemical structure and teratogenic properties. In: Shepard TH, Miller JR, Marois (eds) Methods for detection of environmental agents that produce congenital malformations. Elsevier, Amsterdam, pp 65-77

Scott WJ, Fradkin R, Wilson JG (1977) Non-confirmation of thalidomide induced teratogenesis in rats and mice. Teratology 16: 333-336

Smith C (1987) Pharmacokinetics and toxicity testing: basic principles and pitfalls. In: Nau H, Scott WJ (eds) Pharmacokinetics in Teratogenesis. Vol. I. Interspecies comparison and maternal embryonic fetal drug transfer. CRC Press , Boca Raton, pp 107-122

Sterz H, Nothdurft H, Lexa P, Ockenfels H (1987) Teratologic studies on the Himalayan rabbit: new aspects of thalidomide-induced teratogenesis. Arch Toxicol 60: 376-381

Vorhees CV, Acuff-Smith KD, Weisenburger WP, Minck DR, Berry JS, Setchell KDR, Nau H (1991) Lack of teratogenicity of trans-2-ene-valproic acid compared to valproic acid in rats. Teratology 43: 583-590

Wells PG, Zubovits JT, Wong ST, Molinari LM, Ali S (1989) Modulation of phenytoin teratogenicity and embryonic covalent binding by acetylsalicylic acid, caffeic acid, and α-phenyl-n-t-butylnitrone: Implications for bioactivation by prostaglandin synthetase. Toxicol Appl Pharmacol 97: 192-202

Willhite CC, Shealy YF (1984) Amelioration of embryotoxicity by structural modification of the terminal group of cancer chemopreventive retinoids. J Natl Cancer Inst 72: 689-695

Wilson JG, Ritter EJ, Scott WJ, Fradkin R (1977) Comparative distribution and embryotoxicity of acetylsalicylic acid in pregnant rats and rhesus monkeys. Toxicol Appl Pharmacol 41: 67-78

Zile MH, Cullum ME, Simpson RU, Barua AB, Swartz DA (1987) Induction of differentiation of human promyelocytic leukemia cell line HL-60 by retinoyl glucuronide, a biologically active metabolite of vitamin A. Proc Natl Acad Sci USA 84: 2208-2212

Discussion of the Presentation

Neubert: Simple pharmacokinetic investigations (absorption rate, elimination half-life, etc. mostly from non-pregnant animals) are merely for the planning of the dosing regimen, and such data are also being increasingly considered in routine studies. Data on more sophisticated pharmacokinetics and of assessments of metabolic species differences may be useful, as Dr. Nau has suggested, for the selection of the relevant species. However, this is almost never considered in routine segment or multigeneration studies, and I doubt whether such a selection is feasible and really desirable. The dog will not be chosen for segment II studies because it happens to metabolize a given substance similarly to man.

Nau: It is, of course, best to work with the species which closely reflects the human in regard to pharmacokinetics and metabolism. For many reasons, a species most often is selected which is not ideally suited, e.g. if non-human primates are not available. Pharmacokinetics are of particular importance in this case for the measurement of the active compound(s) in the target tissue and interpretation of the findings.

Manson: You have given an excellent example of the utility of toxicokinetic studies in pregnant animals with agents having potent developmental toxicity. Much less uncertainty is present in estimating risk to humans with potent animal developmental toxicants when toxicokinetic data of the extent you have described are available in laboratory animal species. My question is, however, how to design toxicokinetic studies with agents that lack developmental toxicity. We should also consider the utility of toxicokinetic studies in pregnant animals in general with agents that lack developmental toxicity in biological studies.

The real question is that by the time toxicokinetic studies in pregnant animals are undertaken, complete toxicokinetic data already exist at least in rats with single and repeated dose administration. There is a lot of metabolic data too, so that at the least you know whether the parent compound or a metabolite is the pharmacologically active agent.

Nau: In this case it is important to establish whether the drug is appropriately absorbed and placental transfer/embryonic exposure is extensive. It must also be demonstrated that the lack of effect is not due to rapid detoxification of the drug to inactive metabolites, or to rapid excretion. If any of these possibilities exist, then additional species must be investigated in which it can be established that the lack of effect is *not* due to low exposure of the target tissue.

Stahlmann: Pharmacokinetics in pregnancy is also important for the women themselves, and not only with respect to the embryo/fetus. The kinetics of several drugs is known to change during pregnancy, and the dosing regimen has to be adapted accordingly. Do animal studies provide any relevant information in this respect?

Nau: We do not know, probably not.

Feasibility of Studying Effects on the Immune System in Non-Human Primates

Reinhard Neubert, Ana Cristina Nogueira, Hans Helge,
Ralf Stahlmann, and Diether Neubert

Introduction

The assessment of possible xenobiotic-induced alterations of the immune system is among the most challenging problems of modern toxicology. However, up till now, very few systematic studies have been performed nor has there been any agreement on a testing strategy in this field. The main difficulty rests in the fact that numerous changes may be induced by xenobiotics on various facets of the very complex immune system, and the predictive value of experimental findings for health risk in man is still largely unknown. In recent years some of the prerequisites for such testing in laboratory animals have been improved, and suggestions for testing for xenobiotic-induced alterations in the immune system have been made (Dean et al. 1982; Dean 1987; Luster et al. 1988; Van Loveren and Vos 1989; Neubert et al. 1989).

With respect to the possibility of *prenatally-induced effects* on the immune system much has been speculated, but sound data are scarce. One reason for this is that appropriate systems for testing defined immunological reactions had to be evaluated first in adults, before adequate studies on prenatally-induced effects and on the immune reactions of the developing organism are feasible and meaningful. Up till now, not all of these prerequisites have been established.

Abbreviations:

MABs:	Monoclonal antibodies
Thd :	Thalidomide
TCDD:	2,3,7,8-Tetrachlorodibenzo-*p*-dioxin
PCDDs/PCDFs:	Polychlorinated dibenzo-*p*-dioxins and dibenzofurans

---

Some data presented in this paper are part of a PhD-dissertation of Ana Cristina Nogueira to be submitted at FIOCRUZ, Rio de Janeiro, Brazil, or in Berlin.

Choice of Species for Assessing Effects on the Immune System

Another problem concerns the selection of an appropriate *species*. Mice and guinea pigs, for decades the pets of immunologists, are not the favourite species for toxicological studies. On the other hand, although it has been suggested by many investigators to incorporate immunological evaluations into the frame of routine chronic toxicity studies, the rat used most extensively for such experiments may not exhibit a high susceptibility to immunological alterations induced by typical agents (as we will show in this short presentation). Models for assessing immunotoxic effects in the rat have been established (Van Loveren and Vos 1989), and with some chemicals such as organotin substances (e.g. Krajnc et al. 1984; Vos et al. 1984; Vos et al. 1990) or "dioxins" (e.g. Vos and Moore 1974; Vos et al. 1991) pronounced effects have been reported, at least when using comparatively high doses.

We have, within the last 12 years, gained considerable experience with breeding, maintenance and tackling special toxicological problems using the common *marmoset* (*Callithrix jacchus*), a small, New World monkey weighing about 400 g when mature. A variety of aspects concerning the developmental biology and toxicology of this species has been published before (Neubert et al. 1988b). Presently we maintain a colony in Berlin of altogether about 450 animals, and the colony is self-sustaining without any need for wild-catches (*cf.* D. Neubert et al. 1992b, this book).

The immune system of this non-human primate and effects of several substances on the peripheral lymphocytes of adult animals were assessed. Since the pattern of lymphocyte subsets has been worked out in many details for the adult animal (Neubert et al. 1992f), studies on pre- and perinatally-induced effects are now quite feasible.

In the frame of the risk assessment to be discussed, it is important to stress that restriction to the analysis of the main lymphocyte subpopulations (e.g. total CD4 and total CD8), as it has been routinely done up till now, is *insufficient* for pharmacologic and toxicologic evaluations, as we shall demonstrate here with two examples. This holds for animal studies as well as for studies in humans.

In a number of studies we found that the immune system of the marmoset responds well to the action of small doses of xenobiotics (e.g. "dioxins" and thalidomide), while for the same substances the rat did not serve as a satisfactory experimental model.

For this reason, the marmoset seems to be superior to rodents with respect to risk assessments in this area of research.

Peripheral Lymphocytes of Marmosets as Study Object

There are clear-cut advantages of the marmoset as compared with rodent species when attempting to assess effects on the pattern of lymphocyte subpopulations Neubert et al. 1992f):

(1) The corresponding systems of the marmoset seem to be very similar to those of man (a few differences have also been found),

(2) a good cross-reactivity is found of surface markers on marmoset lymphocytes with monoclonal antibodies (MABs) raised against the corresponding human epitopes, and

(3) a large repertoire of such monoclonal antibodies against human lymphocyte surface receptors is available for very sophisticated analyses of the lymphocyte subtypes and of changes in the pattern of such important surface markers.

A certain disadvantage when using non-human primates, of course, is that there is quite a variability already in the normal data. This fact is exemplified in another presentation in this book (Neubert et al. 1992b). However, the same situation also exists in man.

In this short survey only a few results from our studies can be presented, with the focus on the suggestion of using especially the marmoset as a model for assessing effects on the pattern of lymphocyte subtypes.

Some Data on Effects of Thalidomide (Thd)

For the discussion relevant to this meeting on reproductive toxicity thalidomide (*Thd*) deserves a special place, as the most notorious human teratogen. However, the substance exhibits extremely interesting pharmacologic and toxic characteristics beside the teratogenic potency, as has been clearly demonstrated after the official removal of this substance from the market.

There are indications that the substance is capable of:

- inducing a beneficial effect on the erythema nodosum reaction in lepromatous leprosy (e.g. Sheskin 1965; Iyer et al. 1971; Barnhill and McDougall 1982), and on several other dermatological diseases,
- inhibiting the graft-*vs*-host disease in rats and man (Field et al. 1966; Vogelsang et al. 1986a,b, 1987, 1988; Heney et al. 1990, 1991).

It is common for all of the effects observed empirically in clinical studies with Thd that they must be caused by an interference with inflammatory reactions, and some alterations of the immune system may be expected to take place. However, the exact mechanism of action has not been elucidated.

The studies with Thd provide a good example for the necessity to perform closer analyses of a variety of surface receptors on white blood cells in order to reveal specific effects of xenobiotics on components of the immune system.

We have studied in detail changes in the surface receptors on blood lymphocytes in the marmoset after treatment with Thd or its derivative EM12. The Thd-analogue EM12 exhibits a high teratogenic potency in the marmoset also (Heger et al. 1988; Merker et al. 1988; Neubert et al. 1988a), and its chemical structure is very similar to that of Thd, lacking only one carbonyl-group in the phthalimide ring of Thd. This substance was selected for our studies because it gives rise to much fewer hydrolysis products and metabolites, thus facilitating the interpretation of experimental results.

For studying the surface receptors on lymphocytes in marmosets the substances were given orally once or twice daily for eight days in all studies mentioned in this section, and the receptors on lymphocytes in venous blood were analysed (*a*) before treatment, (*b*) directly at the end, and (*c*) 23 days after discontinuation of the dosing (Neubert et al. 1991c, 1992g,h; Nogueira et al. 1991).

In the case of Thd, "classical" evaluation of the total $CD4^+$ and $CD8^+$ cells did not give impressive results (Tables 1 and 2). At the highest dose of EM12 (50 mg/kg body wt) a slight decrease in total $CD4^+$ cells may have occurred, but some increases were also observed in the various experimental groups. No change was observed in the ratio $CD4^+/CD8^+$.

The data presented in this paper supplement another publication (Neubert et al. 1992h). In all investigations we have evaluated effects induced by xenobiotics as percent of the subfractions, and also with respect to the absolute number of cells (cells/μl blood). Since we feel that the latter information is especially relevant, these absolute data are predominantly given in this presentation.

Effects of Thd on "T Helper Cells"

Closer analysis of the $CD4^+$ subpopulations revealed clear-cut substance-related effects only on the $CD4^+CD45RA^-CDw29^+$ cells in groups treated with the two highest doses of EM12 (10 and 50 mg/kg body wt). This cell population is presently considered as "helper-inducer" or "memory" cells. At a dose of 50 mg EM12/kg the decrease persisted for at least 23 days after discontinuation of the dosing (Table 3).

Table 1. Absolute number (cells/μl blood) of **total CD4 lymphocytes** after treatment with thalidomide (Thd), EM12 or supidimide (Sup). Data measured at the end of the 8-day treatment period

	Control	EM12				Thd			Sup		
mg / kg bw →		1	5	10	50	10	50	100	20	50	100
N =	16	4	9	4	11	4	8	8	3	4	6
Mean	1610	1135	1634	1034	1159	1900	2287	1954	3444	2537	2098
Median	1614	998	1512	959	1237	1857	1928	2014	3268	2305	1913
± SD	500	491	297	590	457	553	881	336	1452	1096	787
Min	849	706	1303	398	319	1294	1469	1481	2088	1512	1365
Max	2554	1840	2228	1821	1652	2593	3813	2366	4976	4026	3506
Q1	1284	769	1442	520	699	1389	1583	1606	2088	1625	1441
Q3	1823	1639	1878	1624	1603	2455	3175	2242	4976	3681	2696
p (Mann-Whitney) =		0.24	0.78	0.14	**0.04↓**	0.42	**0.04↑**	0.09	**0.02↑**	0.07	0.17

Table 2. Absolute number (cells/μl blood) of **total CD8 lymphocytes** after treatment with thalidomide (Thd), EM12 or supidimide (Sup). Data measured at the end of the 8-day treatment period

	Control	EM12				Thd			Sup		
mg / kg bw →		1	5	10	50	10	50	100	20	50	100
N =	16	4	9	4	11	4	8	8	3	4	6
Mean	1240	1009	1230	517	1063	2027	1648	1364	3092	2310	1541
Median	1113	889	1237	337	1136	1836	1544	1305	3406	2023	1452
± SD	550	362	316	387	526	574	800	342	629	962	426
Min	436	720	728	298	185	1584	857	884	2367	1516	1025
Max	2339	1537	1688	1096	1844	2850	3289	1849	3502	3679	2233
Q1	832	754	1012	299	519	1611	1000	1087	2367	1585	1235
Q3	1585	1383	1540	915	1635	2633	1997	1717	3502	3323	1895
p (Mann-Whitney) =		0.48	0.93	**0.03↓**	0.62	**0.03↑**	0.28	0.41	**0.01↑**	**0.03↑**	0.2

Table 3. Absolute number (cells/μl blood) of **CD4 subsets** after treatment with thalidomide (Thd), EM12 or supidimide (Sup). Data of the first row: measured at the end of the 8-day treatment period. Data of the second row (*italics*): measured 23 days after discontinuation of the dosing

	Control		EM12						Thd		Sup	
Dosis in mg / kg bw →			5		10		50		100		100	
N =	8	*8*	5	*5*	4	*4*	7	*7*	8	*8*	4	*4*
CD4+ CD45RA+ CDw29-												
Mean	29	*30*	54	*23*	17	*10*	42	*7*	46	*46*	61	*44*
Median	28	*22*	49	*25*	17	*9*	14	*6*	39	*41*	60	*46*
± SD	11	*23*	35	*18*	11	*6*	56	*7*	22	*18*	23	*24*
Min	16	*8*	10	*0*	5	*3*	2	*1*	28	*30*	34	*13*
Max	46	*72*	105	*44*	30	*18*	156	*17*	87	*84*	90	*70*
Q1	17	*12*	26	*5*	7	*4*	2	*1*	29	*32*	39	*20*
Q3	38	*48*	86	*41*	28	*16*	66	*15*	64	*56*	84	*66*
p (Mann-Whitney) =			0.12	*0.71*	0.13	***0.05↓***	0.60	***0.02↓***	0.10	*0.13*	**0.03↑**	*0.35*
CD4+ CD45RA+ CDw29+												
Mean	170	*267*	259	*140*	210	*199*	268	*260*	410	*412*	184	*304*
Median	155	*289*	255	*149*	166	*204*	263	*239*	365	*286*	188	*267*
± SD	61	*158*	124	*58*	116	*72*	108	*136*	126	*240*	65	*123*
Min	107	*58*	102	*57*	127	*114*	98	*88*	261	*222*	120	*207*
Max	290	*503*	437	*214*	381	*274*	446	*484*	656	*899*	240	*474*
Q1	118	*96*	150	*89*	134	*127*	208	*141*	322	*245*	124	*210*
Q3	203	*390*	370	*187*	331	*266*	339	*344*	490	*574*	240	*434*
p (Mann-Whitney) =			0.27	*0.12*	0.67	*0.45*	0.07	*0.95*	**0.00↑**	*0.37*	0.67	*0.93*
CD4+ CD45RA- CDw29+												
Mean	1378	*1336*	1219	*1236*	710	*629*	743	*594*	1311	*1730*	1935	*972*
Median	1418	*1424*	1108	*1102*	727	*649*	738	*364*	1234	*1633*	1836	*992*
± SD	449	*522*	316	*322*	406	*318*	219	*452*	208	*693*	1012	*645*
Min	628	*540*	936	*899*	201	*244*	401	*221*	1090	*782*	970	*328*
Max	2000	*1931*	1756	*1685*	1185	*976*	968	*1526*	1614	*2751*	3096	*1575*
Q1	1077	*784*	1008	*972*	315	*313*	538	*313*	1123	*1176*	1031	*373*
Q3	1800	*1833*	1484	*1567*	1088	*926*	961	*739*	1541	*2414*	2937	*1551*
p (Mann-Whitney) =			0.42	*0.71*	0.07	***0.05↓***	**0.01↓**	***0.03↓***	0.88	*0.32*	0.55	*0.35*

Effects of Thd on "Cytotoxic T Cells"

Although no pronounced deviations in the *total* number of $CD8^+$ cells were observed, some deviations from the controls were seen when $CD8^+$ *subsets* were evaluated. In the groups treated with the highest doses of EM12 a reduced number of cells carrying the CD45RA marker was observed 23 days after the dosing (Table 4). There was no change in the total $CD8^+$ cells with the CDw29 marker, but it could be shown that the cells reduced in number were all $CD8^+CD45RA^+CDw29^+$ lymphocytes.

Even more pronounced changes were revealed when analysing cells with and without the CD56 marker: a clear-cut reduction in number was observed in the $CD8^+CD56^+$ cells ("*cytotoxic T cells*"), and all the affected cells were $CD8^+$ $CD56^+CDw29^+$. The decrease in these cells occurred at doses of 10 or 50 mg EM12/kg body wt, and in the group treated with 100 mg Thd/kg body wt (Table 5). The effect persisted for at least three weeks after the dosing period.

There was no significant decrease (Table 5) on $CD8^+CD56^-$ cells ("suppressor cells").

General Significance of "Integrin Receptors"

The most important clue on a possible mode of action of Thd on reactions of the immune system came from studies of cells carrying the .*i.CD2 receptor* or related surface markers. The formerly as "rosetting factor" designated surface marker CD2 belongs to the .*i.adhesion receptors*, also called *integrins*.

A large variety of such adhesion receptors or "integrins" with varying specificities have been discovered recently. They represent one of the most interesting classes of cell surface receptors, since they have been recognized as extremely important for many cell/cell- and cell/extracellular matrix-interactions. An excellent review on these receptors was given by Springer (1990). With respect to white blood cells there is evidence that adhesion receptors are participating in extremely important functions, including:

-	antigen presentation	(monocytes	→ T cells)
-	"help"	($CD4^+$ cells	→ B cells)
-	"suppression"	($CD8^+$ cells	→ B cells)
-	"cytotoxicity"	($CD8^+$ cells	→ infected cells)
-	"cytotoxicity"	(NK cells	→ tumour cells)
-	"homing"	(lymphocytes	→ tissue distribution).

However, the function of adhesion factors is by no means restricted to the above mentioned functions nor to lymphocytes and blood cells. Since adhesion receptors often function as pairs, e.g. CD2 → LFA-3, and LFA-1 → ICAM-1 or → ICAM-2, the counterparts are most often located on other cells, such as endothelial cells, and many of these effects may not have been elucidated yet. Adhesion factors (e.g. VLA-1 to VLA-5, etc.) have also been found to be responsible for the interaction of cells with extracellular matrix components, such as fibronectin, laminin, collagens, etc. Even functions of the CNS, like junctional communication, the association of axons with targets, and signals that alter levels of neurotransmitter enzymes (Rutishauser et al. 1988), have been associated with adhesion receptors (NCAM).

Effects of Thd and Derivatives on "Integrin Receptors"

The first effect observed after treatment of marmosets with oral doses of EM12 or Thd was a depletion of cells reacting with mAbs for the CD2 receptor (Table 6). Surprisingly, the number of cells reacting with the antibody for CD2 was smaller than the sum of lymphocytes carrying either the CD4 or the CD8 surface marker (Table 7). This is not the case under normal conditions, since virtually all $CD4^+$ and $CD8^+$ cells are CD2-"positive".

Table 4. Absolute number (cells/μl blood) of **CD8 subsets** with **CD45** and **CDw29** markers after treatment with thalidomide (Thd), EM12 or supidimide (Sup). Data of the first row were measured at the end of the 8-day treatment period. Data of the second row (*italics*) were measured 23 days after discontinuation of the dosing

	Control		EM12						Thd		Sup	
Dosis in mg / kg bw →			5		10		50		100		100	
N =	8	*8*	5	*5*	4	*4*	7	*7*	8	*8*	4	*4*
CD8+ CD45RA+ (total)												
Mean	706	*684*	474	*361*	369	*234*	527	*357*	585	*403*	784	*532*
Median	729	*719*	454	*274*	256	*219*	566	*324*	569	*436*	721	*480*
± SD	178	*240*	159	*241*	287	*73*	182	*173*	97	*215*	375	*160*
Min	361	*234*	304	*212*	170	*164*	202	*172*	490	*112*	438	*407*
Max	918	*982*	726	*790*	795	*336*	712	*700*	777	*643*	1257	*759*
Q1	598	*539*	347	*229*	189	*175*	416	*233*	503	*183*	462	*414*
Q3	848	*857*	612	*536*	663	*310*	699	*432*	637	*602*	1170	*701*
p (Mann-Whitney) =			0.07	*0.07*	0.08	***0.01***↓	0.07	***0.02***↓	0.08	0.05	0.93	*0.20*
CD8+ CDw29+ (total)												
Mean	1406	*1430*	1207	*1224*	693	*842*	1145	*1055*	1410	*1688*	1599	*1141*
Median	1318	*1270*	1105	*1131*	590	*864*	1102	*954*	1365	*1444*	1494	*1154*
± SD	517	*524*	381	*717*	366	*197*	412	*371*	223	*727*	582	*340*
Min	695	*550*	711	*639*	381	*623*	442	*756*	1139	*999*	1086	*797*
Max	2054	*2058*	1661	*2432*	1209	*1019*	1673	*1853*	1824	*3030*	2323	*1461*
Q1	1002	*1170*	879	*685*	410	*650*	998	*813*	1248	*1121*	1108	*823*
Q3	1928	*1987*	1586	*1810*	1077	*1014*	1611	*1107*	1558	*2331*	2196	*1447*
p (Mann-Whitney) =			0.71	*0.34*	0.05	0.05	0.60	0.05	0.88	*0.88*	0.55	*0.55*
CD8+ CD45RA+ CDw29+												
Mean	687	*674*	473	*359*	355	*221*	519	*344*	564	*393*	747	*520*
Median	686	*717*	454	*274*	242	*215*	566	*319*	556	*422*	682	*471*
± SD	178	*240*	160	*242*	282	*64*	181	*169*	107	*216*	340	*160*
Min	346	*223*	300	*212*	162	*149*	192	*162*	452	*107*	438	*393*
Max	901	*976*	726	*790*	774	*304*	697	*679*	773	*623*	1185	*746*
Q1	586	*528*	344	*226*	181	*163*	413	*233*	466	*174*	459	*400*
Q3	843	*846*	611	*536*	642	*284*	682	*408*	612	*595*	1099	*690*
p (Mann-Whitney) =			0.09	*0.07*	0.08	***0.02***↓	0.12	***0.03***↓	0.08	***0.04***↓	1.00	*0.20*

Table 5. Absolute number (cells/μl blood) of **CD8 subsets** with CD56 markers after treatment with thalidomide (Thd), EM12 or supidimide (Sup). Data of the first row were measured at the end of the 8-day treatment period. Data of the second row (*italics*) were measured 23 days after discontinuation of the dosing

	Control		EM12						Thd		Sup	
Dosis in mg / kg bw →			5		10		50		100		100	
N =	8	*8*	5	*5*	4	*4*	7	*7*	8	*8*	4	*4*
CD8+ CD56+ (total)												
Mean	594	*617*	388	*295*	228	*195*	197	*111*	257	*207*	744	*503*
Median	595	*549*	426	*214*	204	*192*	176	*128*	258	*185*	614	*509*
± SD	185	*276*	118	*206*	106	*41*	114	*37*	105	*74*	304	*117*
Min	296	*210*	196	*173*	128	*156*	73	*43*	69	*125*	556	*376*
Max	835	*1039*	508	*661*	375	*241*	380	*144*	405	*320*	1193	*618*
Q1	453	*466*	283	*184*	142	*158*	103	*89*	189	*142*	556	*390*
Q3	778	*903*	476	*448*	338	*236*	310	*142*	347	*275*	1062	*610*
p (Mann-Whitney) =			0.07	*0.07*	**0.01↓**	***0.02↓***	**0.00↓**	***0.00↓***	**0.00↓**	***0.00↓***	0.35	*0.55*
CD8+ CD56+ CDw29+												
Mean	546	*584*	363	*291*	214	*183*	166	*99*	234	*186*	678	*486*
Median	522	*513*	362	*201*	188	*179*	175	*118*	240	*175*	587	*488*
± SD	172	*263*	107	*208*	104	*40*	81	*36*	104	*59*	242	*123*
Min	278	*207*	191	*173*	118	*142*	61	*43*	69	*115*	509	*359*
Max	787	*968*	474	*661*	361	*230*	294	*132*	366	*273*	1029	*611*
Q1	427	*418*	276	*184*	132	*146*	100	*59*	151	*132*	514	*370*
Q3	724	*873*	451	*445*	322	*223*	225	*126*	340	*243*	933	*601*
p (Mann-Whitney) =			0.05	0.05	**0.01↓**	***0.01↓***	**0.00↓**	***0.00↓***	**0.00↓**	***0.00↓***	0.35	*0.55*
CD8+ CD56- CDw29+												
Mean	858	*846*	844	*928*	479	*660*	979	*956*	1175	*1501*	921	*655*
Median	784	*841*	744	*896*	402	*689*	926	*912*	1207	*1289*	916	*646*
± SD	383	*291*	286	*524*	270	*161*	346	*360*	181	*677*	364	*221*
Min	417	*344*	520	*465*	263	*467*	382	*688*	932	*851*	557	*438*
Max	1389	*1247*	1233	*1771*	848	*797*	1448	*1727*	1466	*2757*	1294	*891*
Q1	509	*647*	603	*491*	271	*497*	869	*696*	1003	*973*	584	*453*
Q3	1243	*1098*	1134	*1382*	764	*795*	1318	*989*	1302	*2096*	1262	*867*
p (Mann-Whitney) =			0.94	*1.00*	0.11	*0.20*	0.69	*0.86*	0.13	***0.02↑***	0.80	*0.35*

Table 6. Number of cells with the **CD2** marker after treatment with Thd, EM12 or Sup. Measured at the end of the treatment period (cells/μl blood)

	Control	EM12				Thd			Sup		
mg / kg bw →		1	5	10	50	10	50	100	20	50	100
N =	16	4	9	4	11	4	8	8	3	4	6
Mean	3104	2022	1860	1177	936	3714	3323	1824	7040	4800	3293
Median	3035	1744	1961	955	878	3853	2899	1903	7313	4248	3252
± SD	1094	774	391	744	581	871	1445	638	2036	2043	824
Min	1441	1436	1260	545	182	2628	1906	749	4881	3019	2052
Max	5476	3163	2327	2254	1933	4521	6065	2752	8925	7685	4560
Q1	2426	1511	1457	636	378	2822	2099	1287	4881	3212	2741
Q3	3698	2810	2212	1940	1412	4467	4476	2279	8925	6941	3894
p (Mann-Whitney) =		0.10	**0.00↓**	**0.01↓**	**0.00↓**	0.24	0.93	**0.01↓**	**0.02↑**	0.1	0.6

Table 7. Mathematical evaluation of **CD2** lymphocytes minus all cells being either CD4+ or CD8+: i.e. CD2 minus (CD4+CD8), after treatment with Thd, EM12 or Sup. Measured at the end of the treatment period. Minus values indicate the presence of CD4+ or CD8+ cells lacking the CD2 marker

	Control	EM12				Thd			Sup		
mg / kg bw →		1	5	10	50	10	50	100	20	50	100
N =	16	4	9	4	11	4	8	8	3	4	6
Mean	254	-122	-1003	-375	-1286	-213	-613	-1494	504	-48	-347
Median	229	-186	-975	-342	-1139	-261	-629	-1428	543	-15	-286
± SD	200	168	454	222	676	175	289	626	68	99	466
Min	12	-241	-1612	-663	-2467	-368	-944	-2805	426	-192	-1179
Max	747	124	-396	-152	-235	38	-209	-654	544	31	179
Q1	56	-234	-1501	-603	-1888	-344	-914	-1679	426	-149	-641
Q3	390	54	-568	-179	-840	-34	-306	-1103	544	21	11
p (Mann-Whitney) =		**0.01↑**	**0.00↑**	**0.00↑**	**0.00↑**	**0.01↑**	**0.00↑**	**0.00↑**	**0.03↓**	**0.00↑**	**0.00↑**

Further studies showed that especially pronounced effects of Thd or EM12 could be demonstrated in the marmoset on T lymphocyte receptors that share the common property of being adhesion receptors. These include so far:

- $CD2^+$ the sheep erythrocyte "rosetting" factor (or LFA-2)
- $CD11a^+$ (LFA-1α), the α-chain of Lymphocyte Functional Antigen-1
- $CD18^+$ (LFA-1ß), the ß-chain of LFA-1
- $CD58^+$ (LFA-3)
- ICAM-1

Some of these receptors, e.g. CD2, are present on the lymphocyte surface at different epitope densities, designated as "high density" ($> 5 \times 10^4$ epitopes/cell) or "low density" ($< 5 \times 10^4$ epitopes/cell). This density can also be assessed with the flow cytometer (FACScan). The first effect to be observed after treatment with EM12 or Thd is a shift from high to the low epitope density on the cell surface, and subsequently a disappearance of this receptor from the cells. This is illustrated with the original FACScan dotplots and histogram combinations for CD11a (LFA-1α) and CD18 (LFA-1ß) as shown in Figures 1 and 2.

The mechanism described leads to the appearance of cell subtypes e.g. $CD4^+CD2^-$ or $CD4^+LFA\text{-}1\alpha^-$ cells, which are virtually absent in normal blood of control marmosets or present at a very low percentage. This dose-dependent effect is illustrated in Figure 3 (A and B) for the two polypeptide chains of LFA-1 (α and ß).

The effect on the adhesion receptors is not confined to the T cells, but a disappearance of certain integrin receptors from the cell surface was also observed on B cells. In this case the adhesion receptor: LFA-3 (or CD18) was investigated. There was a slight but significant decrease in this receptor, indicated by the appearance of $CD20^+CD58^-$ cells. However, the effect on B cells was not as pronounced as a corresponding effect on the T cells (Table 8), and it was almost back to normal 23 days after discontinuation of the dosing. In the group treated with 5 mg EM12/kg body wt we detected about 5% of $CD20^+CD58^-$ cells, which were not present before the treatment, and the affect of 50 mg Thd/kg body wt was even less.

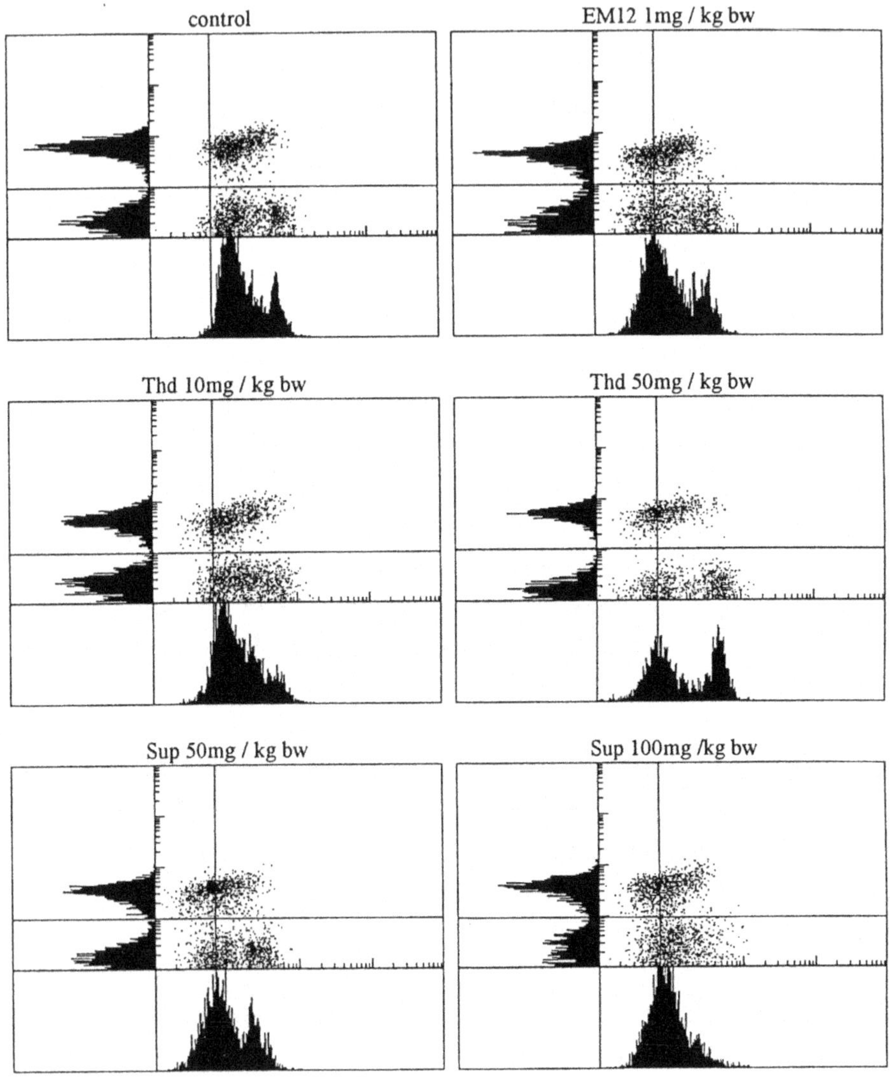

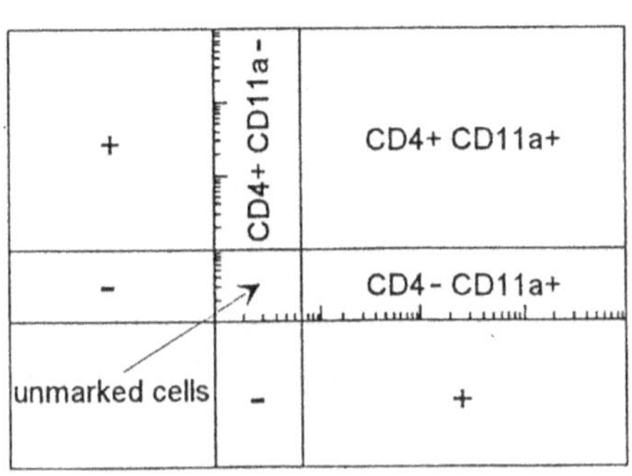

Fig. 1. Original FACScan dotplot and histogram combinations of **CD4** (y-axis) *vs* **CD11a** (LFA-1α) epitopes on lymphocytes after a 8-day treatment with Thd and derivatives as indicated.

◂ An explanation of the four sectors of the dotplot is given on the left side. A loss of CD11α epitopes (shift to the left in the dotplot) is obvious

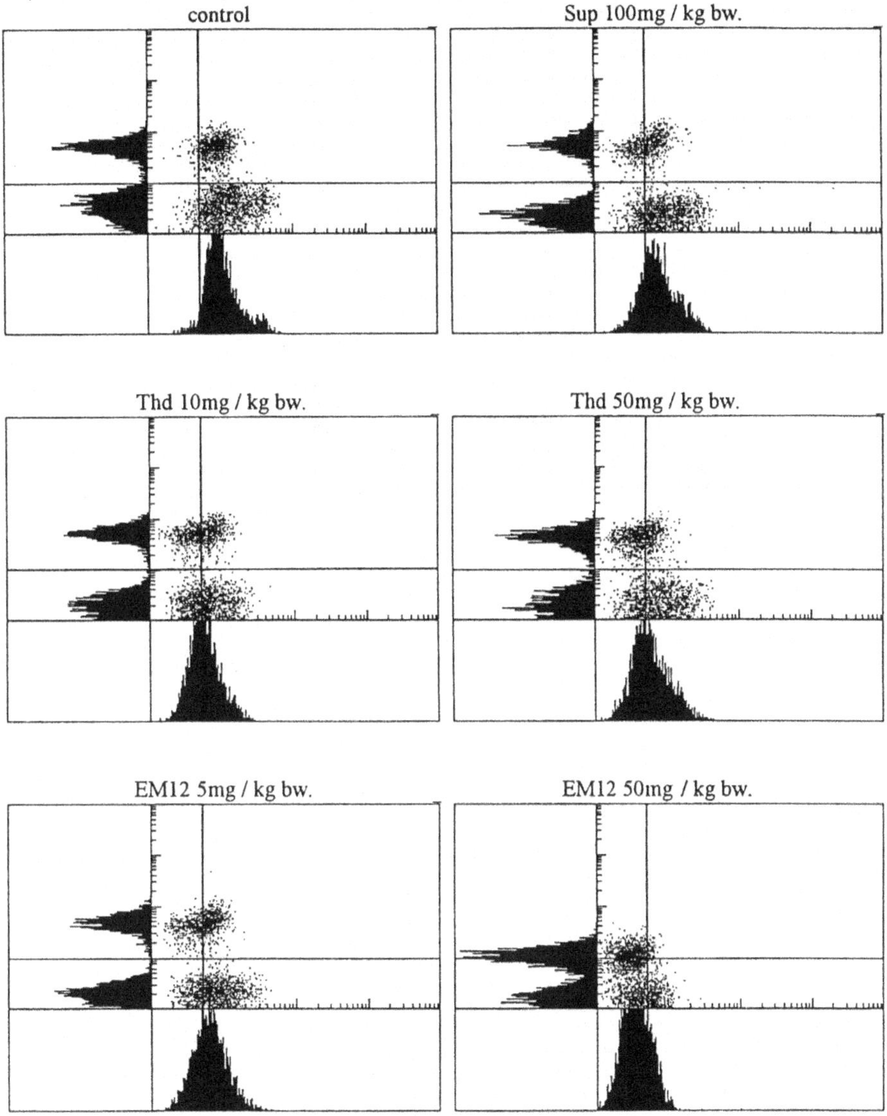

Fig. 2. Original FACScan dotplot and histogram combinations of **CD4** (y-axis) *versus* **CD18** (LFA-1ß) surface markers on lymphocytes (upper two quadrants) after a 8-day treatment with the substances and daily doses indicated. Same blood sample as in the previous figure. Explanation of the dotplot quadrants *see:* Figure 1. Loss of CD18 and CD11a does not completely match. The shift of the cells in the dotplot to the left indicates the loss of the LFA-1 receptors, with concomitant retaining the CD4 marker. The $CD4^-$ cells at the lower two quadrants of the dotplots are mostly $CD8^+$ and B lymphocytes. Some of these cells also lose the CD18 receptor

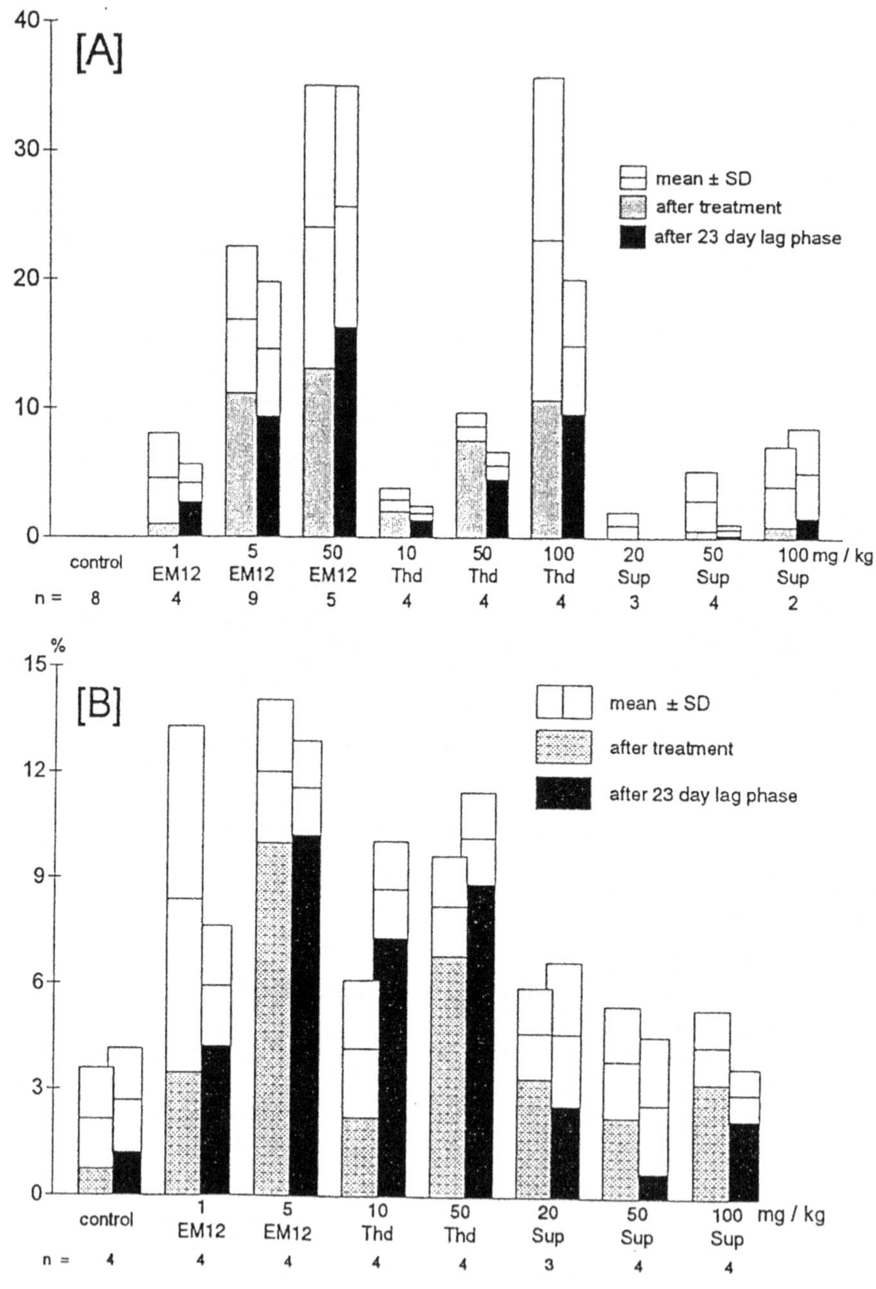

Fig. 3. Appearance of **$CD4^{+}CD2^{-}$** [A] and **$CD4^{+}CD11a^{-}$** [B] lymphocytes directly after a 8-day treatment of marmosets with thalidomide (Thd), EM12, or supidimide (Sup), or 23 days later. These cells are virtually absent or very low under normal conditions. % M ± SD are given

Three weeks after the dosing period there seemed to be a tendency in the EM12-treated groups for an increase in the number of $CD20^+$ cells (Table 9).

Summarizing: The following changes were induced in marmosets by Thd or EM12 in the percentage and in the absolute number (cells/μl blood) of lymphocytes:

- decrease in $CD8^+CD56^+CDw29^+$ cells
- decrease in cells carrying the CD2 marker
- appearance of $CD4^+CD2^-$ lymphocytes
- appearance of $CD4^+CD11a^-$ cells
- appearance of $CD4^+CD58^-$ cells
- slight increase in the number of $CD20^+$ (B) cells, after a lag phase
- appearance of $CD20^+CD58^-$ (B) cells.

Possible Significance of the Effects Observed

In all experiments EM12 was more potent than Thd (apparently at least 5-fold). However, also the derivative supidimide (considered to be "non-teratogenic") was found to affect some of the integrin receptors in our studies, but it was considerably less potent than Thd (possibly about 5-fold).

The interaction of Thd and its derivatives with certain integrin receptors (and possibly with many more adhesion components not yet investigated) would explain the diversity of the effects observed with these agents in clinical trials. The uneven distribution of such receptors within the organism and varying affinities towards the action of Thd would also offer an explanation for organotropic effects seen at various doses of Thd. Not all possible effects would be expected to occur in all the organs and at all the doses.

On the basis of our data we have advanced the hypothesis that the teratogenic effects of Thd may be caused by an interference of Thd with adhesion receptors (Neubert et al. 1992h).

Table 8. **CD20$^+$CD58$^-$(=LFA-3$^-$) cells** occurring after treatment with EM12 or Thd. Data of the first part were measured at the end of the 8-day treatment period. Data of the second part were measured 23 days after discontinuation of dosing (% M ± SD)

	N =	5mg EM12/kg	50mg Thd/kg
Before treatment	8	0.0 ± 0.1	0.0 ± 0.1
After 8 days of treatment	4	**4.9** ± 0.8	**0.7** ± 0.6
23 days after end of treatment	4	0.1 ± 0.1	0.0 ± 0.1

Table 9. Number of **CD20 cells** after treatment with Thd, EM12 or Sup (cells/μl blood). Data of the first part were measured at the end of the 8-day treatment period. Data of the second part were measured 23 days after discontinuation of the dosing

	Control	EM12				Thd			Sup		
mg / kg bw →		1	5	10	50	10	50	100	20	50	100
N =	16	4	9	4	11	4	8	8	3	4	6
At the end of treatment period											
Mean	33.8	35.1	31.3	48.1	40.2	19.4	29.0	41.3	31.1	19.8	36.1
Median	32.7	33.5	31.6	49.5	38.5	19.2	31.1	40.9	29.5	18.4	38.8
± SD	11.0	6.7	8.8	3.7	11.9	3.5	9.9	6.8	5.2	4.4	11.5
Min	19.5	29.7	18.1	42.7	19.7	15.4	14.5	28.9	27.0	16.5	17.4
Max	50.0	43.9	45.9	50.6	57.5	23.8	39.7	51.4	37.0	26.1	46.9
Q1	21.8	29.8	24.2	44.2	32.1	16.2	19.9	38.9	27.0	16.6	25.1
Q3	43.8	42.1	37.8	50.5	51.5	22.8	38.5	46.7	37.0	24.5	46.8
p (Mann-Whitney) =		0.96	0.55	**0.02**↑	0.19	**0.02**↓	0.28	0.26	0.7	**0.02**↓	0.85
23 days after dosing											
N =	12	4	9	4	7	4	4	8	4	4	6
Mean	28.6	43.5	43.4	36.1	49.7	24.8	39.5	37.4	25.4	25.5	39.4
Median	26.7	41.6	38.9	35.5	52.7	26.6	38.9	36.9	25.7	23.8	45.7
± SD	9.5	5.9	8.3	13.3	9.3	5.7	3.3	20.0	2.2	10.5	16.1
min	18.5	38.9	35.6	23.1	38.6	16.6	36.4	8.4	22.5	15.5	9.4
max	40.8	52.0	60.4	50.2	60.3	29.3	43.7	63.4	27.8	39.0	52.7
Q1	19.1	39.2	37.8	24.0	40.6	18.8	36.7	20.9	23.1	16.5	27.5
Q3	39.4	49.8	48.9	48.8	60.1	28.9	42.9	57.1	27.4	36.3	50.2
p (Mann-Whitney) =		**0.03**↑	**0.02**↑	0.20	**0.002**↑	0.59	0.10	0.20	0.95	0.67	0.08

Since cell/cell-interactions as well as cell/extracellular matrix-interactions are well known to play a crucial role in morphogenetic differentiations, such a hypothesis is not without a rationale. In this respect it is worth mentioning that our first data from studies with rats (Korte et al. 1992, unpublished) did not reveal a similar clear-cut action on some integrin receptors on lymphocytes as was observed in the non-human primate: the rat is apparently not a good model for revealing this effect on integrins, and rodents are also known not to be susceptible to the teratogenic action of Thd (Scott et al. 1977).

It is surprising that the many possible effects of Thd on inflammation and immune reactions were not detected during the period of extreme and frequent therapeutic use at the end of the 1950's and beginning 1960's, and it is a tragedy that the most pronounced and dramatic effect of Thd in man was the teratogenic one.

Some Data on Effects of Glucocorticoids

It was also of interest to evaluate whether and in which way the pattern of lymphocyte subpopulations in the marmoset responds to the action of glucocorticoids. This question was of importance since glucocorticoids are therapeutically used during pregnancy for both treatment of the mother or the child. In man it has long been known that hydrocortisone or prednisone may drastically reduce the number of lymphocytes and monocytes (e.g. Fauci and Dale 1974; Yu et al. 1974).

We have assessed whether the adult marmoset represents a good model to study possible effects of these hormones, in order to later extend the studies to the pre- and perinatal period. Dexamethasone (Dmth) was given to the marmosets once and the pattern of lymphocyte subtypes was studied in the blood 24 hr later. Some of the changes which were still obvious after this rather long time are compiled in Table 10.

Table 10. Effect of *Dmth* on lymphocyte subtypes in marmosets

Subtype	0-time	Dexamethasone 50 mg/kg b wt	Dexamethasone 150 mg/kg b wt
Total lymphocytes	5,359-[8,512]-10,966	5,574-[6,910]-7,036	3,840-[**4,204**]-4,568
CD4	1,737-[2,897]-3,908	1,620-[**1,705**]-1,804	983-[**1,108**]-1,234
CD4CDw29	999-[1,300]-1,825	623-[**694**]-727	398-[**484**]-569
Ratio: CD4CDw29 /CD4CD45RA	0.80-[1.06]-1.54	0.63-[**0.66**]-0.69	0.75-[0.80]-0.86
CD8	1,034-[2,087]-2,818	1,467-[1,900]-1,943	961-[**991**]-1,021
CD2HLADR	1,431-[2,444]-3,139	1,388-[**1,396**]-1,669	882-[**911**]-940
CD2IL-2R	94-[208]-361	111-[169]-179	45-[**65**]-85
CD2Transf	253-[433]-684	140-[**151**]-293	101-[**106**]-111
CD20	1,666-[2,295]-4,455	2,019-[2,548]-2,937	1,515-[1,685]-1,854

50 or 150 mg dexamethasone/kg, once s.c. 24 hr before the evaluation
0-time: n = 9; 50 mg/kg: n = 3; 150 mg/kg: n = 2
Values are: minimum- [median] -maximum
Bold and underlined: $P < 0.05$ (Mann-Whitney), compared with controls

The single dose of 50 mg Dmth/kg body wt did not significantly reduce the total number of lymphocytes in venous blood 24 hr after the injection, nor was there a change in the total white blood cell count, but these variables may have recovered at this time. However, there was a significant decrease in the percentage and in the absolute number of $CD4^+$ cells, and especially $CD4^+CDw29^+$ cells ("helper-inducer" cells). Additionally, the ratio of $CD4^+CDw29^+$ to $CD4^+CD45RA^+$ cells ("suppressor-inducer" cells) was significantly reduced. Furthermore, T cells with activation markers ($CD2^+HLADR^+$ and $CD2^+$ cells with the transferin receptor) were clearly reduced in their number. There was no change in the absolute number of $CD8^+$ cells at this dose, and therefore the ratio: CD4/CD8 was found to be clearly reduced. It is noteworthy that no change in the relative and absolute number of B cells ($CD20^+$, or B_1 cells) was observed 24 hr after the dosing.

The effects observed were clearly dose-dependent. Subsequent to the single dose of 150 mg Dmth/kg body wt the number of the total lymphocytes was reduced in the two marmosets examined. The changes found with the lower dose were found to be exaggerated, and additionally the T cells carrying the IL-2 receptor were found to be decreased. At this high dose there was an effect on $CD8^+$ cells, but B cells ($CD20^+$) were also not clearly affected.

These data show that the marmosets reacts to the administration of glucocorticoids with typical and defined changes in the pattern of lymphocyte subpopulations. It will now be interesting to study more closely the changes (possibly also irreversible ones?) inducible during the pre- and perinatal period in this species.

A similar response to Dmth as in man was also found with respect to macrophages in the marmoset (Zwadlo-Klarwasser et al. 1992).

Some Data on Effects of "Dioxins"

As a third example we would like to review some data obtained with a group of *environmental* substances, the "dioxins", i.e. polychlorinated dibenzo-*p*-dioxins and dibenzofurans (PCDDs/PCDFs). Aspects of the general toxicity of PCDDs/PCDFs were summarized recently by us (Neubert 1991; Neubert et al. 1991a, 1992a).

Two of the results obtained with this class of substances are important within the frame of this discussion on risk assessment:

- the rat again was *not* found (Korte et al. 1991) to represent a species to reveal the kind of effects observed in the non-human primate, and
- the data obtained in the marmoset provided important information on general aspects of a *risk assessment* with respect to this pollutant.

There are 210 possible congeners within the class of PCDDs/PCDFs, and most of these can be found in many emissions from combustion processes. Our studies were initiated with the substance exhibiting the highest toxic potency of all the congeners of these classes of chemicals: 2,3,7,8-tetrachlorodibenzo-*p*-dioxin (TCDD).

At comparatively high doses (i.e. in the μg/kg body weight-range) this substance had been shown before to induce a variety of effect on immune reactions in rodents (e.g. Vos and Moore 1974; Vos et al. 1991). Studies in the low-dose range were only published with a complex *in vivo/in vitro* approach in mice (Clark et al. 1981, 1983), and these reports have since not been confirmed. Up till now, there was an apparent lack of a suitable animal model with a sufficient sensitivity.

Extensive studies performed in our laboratory within the last three years using the marmosets as an experimental model revealed a high susceptibility for TCDD-induced changes in the pattern of certain lymphocyte subsets in venous blood. With single doses as low as 10 ng TCDD/kg body wt (Neubert et al. 1990b, 1991b) clear-cut effects were induced (Fig. 4). These included predominantly a decrease in the percentage and in the absolute number (cells/μl blood) of "helper-inducer" cells ($CD4^+CDw29^+$), and also, a reduction in the number of B cells was observed (Fig. 5). In these studies no significant effect was observed with a single dose of 3 or 1 ng TCDD/kg body wt.

Further information was obtained from a repeated-dose study using very small weekly doses (Neubert et al. 1992d). This experimental series extending over a total period of 42 weeks was performed with two dosing schedules used in a sequence (*see:* Table 11): for 23 weeks 300 pg TCDD/kg body wt were given once weekly (period [*a*]), and during this dosing period an accumulation of TCDD was expected to occur at about 84-91% of the steady-state, corresponding to a body burden of 3 ng TCDD/kg body wt. This was followed by a period [*b*] of increased dosing, namely 1.5 ng TCDD/kg body wt given once weekly for 6 weeks, and at the end of this period a body burden of about 11 ng TCDD/kg body wt was expected from the calculation (using for calculation an elimination half-life of 7-9 weeks for TCDD in the marmoset, as reported previously by Neubert et al. 1990a). The last part of the study without additional exposure [*c*] was a recovery period, during which the TCDD body burden could be calculated to drop again to a level similar to that at the end of period [*a*].

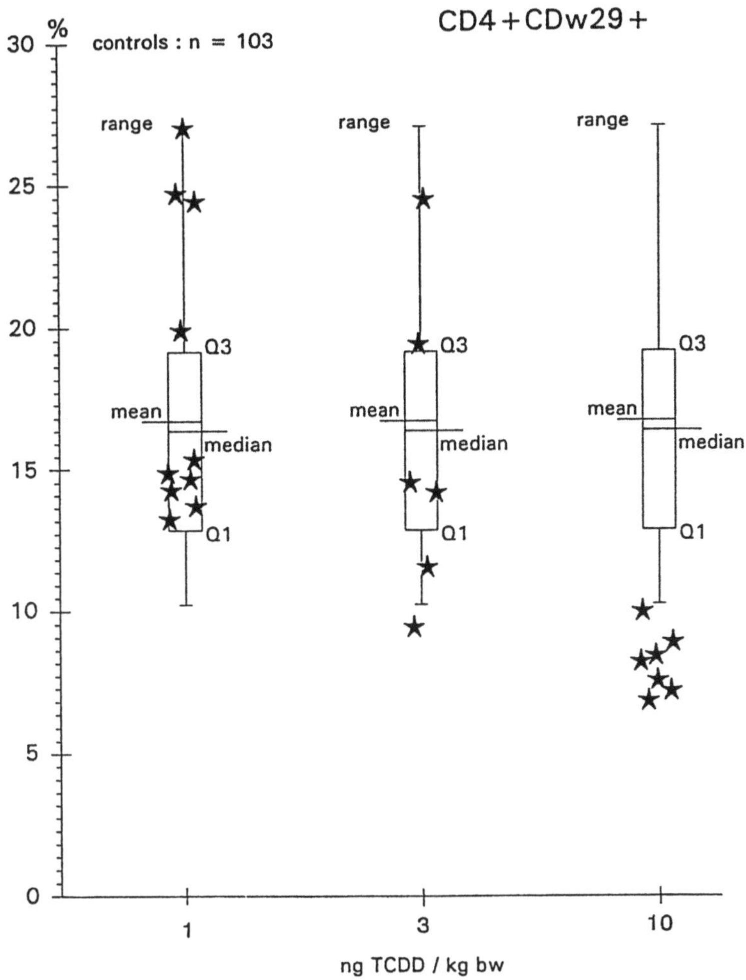

Fig. 4. T lymphocytes in marmoset blood after a single dose of 2,3,7,8-TCDD. "helper-inducer" cells (*CD4+CDw29+*). A single s.c. dose of 10 ng TCDD/kg body wt clearly decreased the percentage and the absolute number of $CD4^+CDw29^+$ and of $CD20^+$ cells. The maximum effect occurred with a lag phase of several weeks. 3 ng TCDD/kg body wt seemed only to exceptionally affect single marmosets, while 1 ng/kg was found to be a no-observed-effect-level (NOEL). Box: median and 1st and 3rd quartile (plus range) of 103 controls. Each star represents one treated marmoset

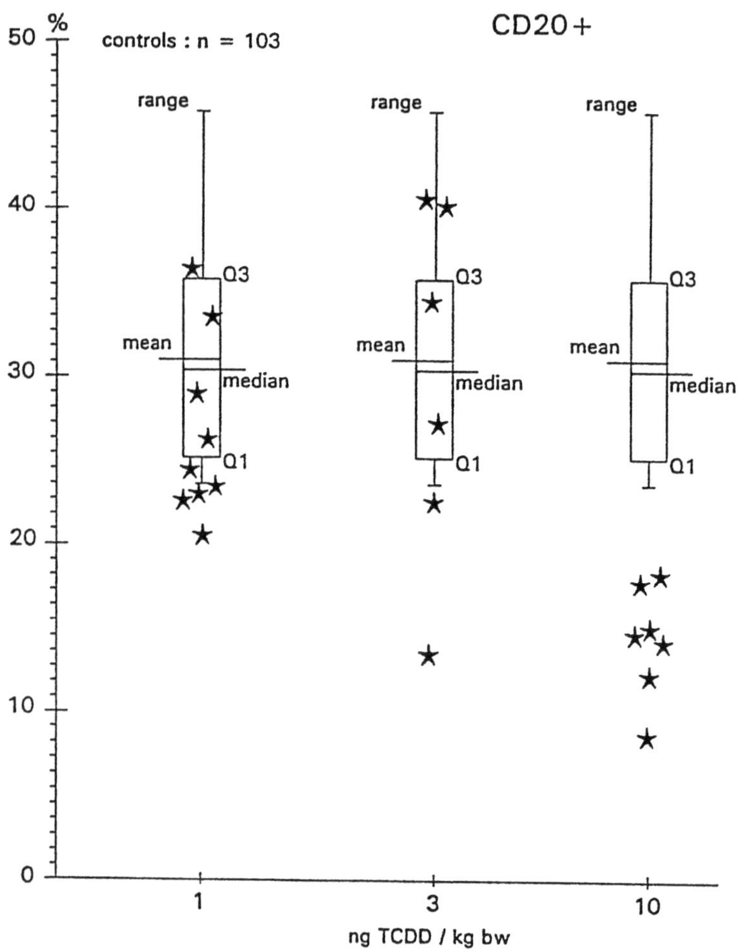

Fig. 5. B lymphocytes in marmoset blood after a single dose of 2,3,7,8-TCDD. B cells ($CD20^+$). Experimental conditions as in Figure 4

Table 11. Dosing schedule used in repeated-dose study with TCDD in marmosets. Indicated are the doses given and the body burdens calculated, also as percent of the steady-state achievable with this dose

Dose (ng/kg body wt, weekly)	After weeks of treatment***	Body burden** (ng/kg body wt)	% of steady-state
0.3	3	0.8-0.8	21-26%
0.3	12	2.2-2.4	60-69%
0.3	19	2.7-3.1	77-85%
0.3	24	2.9-3.4	84-91%
1.5	3(27)	7.0-7.6	21-26%
1.5	6(30)	10.0-10.9	37-45%
1.5	7(31)	10.9-11.8	42-50%
0	5(35)	6.5-8.1	-31-40%*
0	12(42)	3.3-4.6	-61-70%*

Data from: Neubert et al. 1992d and Neubert et al. 1990a
Steady-state concentrations were calculated as: 3-4 ng/kg (dose: 0.3 ng/kg), and 16-20 ng/kg (dose: 1.5 ng/kg)
* % decline from the values at the end of the dosing period
** Calculated for an elimination half-life of 7 (first number) or 9 (second number) weeks
*** Numbers in parentheses give the weeks of the entire study

During both dosing periods the lymphocyte subset predominantly affected was again the "helper-inducer" population (CD4+CDw29+), but in the course of the study the change occurred in different directions (Fig. 6 [*A*]): unexpectedly, during the initial low-dose treatment period (*a*) an opposite effect was observed, namely an *increase* in the percentage and the absolute number of CD4+CDw29+ cells. In order to confirm our previous findings with higher single doses, the weekly dose of TCDD was subsequently raised to 1.5 ng TCDD/kg body wt (period [*b*]. The typical *decrease* in the percentage and absolute number of this lymphocyte subset was again observed, as it had been induced in the previous single-dose studies. However, since in both of the experimental periods (Fig. 6 [*B*]) the "suppressor-inducer" T cells (CD4+CD45RA+) tended to respond in the opposite direction (keeping the total number of CD4+ cells rather constant), the *ratio* of these two lymphocyte subsets (Fig. 6 [*A*] and [*B*] represents an especially sensitive indicator of the effects induced by TCDD in this species (Neubert et al.

1992d,e). The change in this ratio during the course of the repeated-dose study is shown in Table 12. After discontinuing the TCDD-treatment the change in lymphocyte pattern normalized within 12 weeks, indicating the reversibility of the biological change induced by the TCDD.

Table 12. Effect of 2,3,7,8-TCDD on the dose-dependent change in the *ratio* of $CD4^+CDw29^+/CD4^+CD45RA^+$ lymphocytes in blood. Experimental design as shown in Table 11

Weeks of the study	Ratio $CD4^+CDw29^+/CD4^+CD45RA^+$		
0-time	1.6	(0.8 -2.4)	
Low-dose TCDD treatment			
week 3	1.4	(1.1 -2.1)	
week 12	<u>4.2</u>	(1.5 -6.6)	*
week 19	<u>5.4</u>	(1.4-12.6)	*
High-dose TCDD treatment			
week 30	<u>0.7</u>	(0.6 -0.8)	*
week 35	<u>0.7</u>	(0.5 -0.9)	*
Recovery period			
week 42	1.8	(1.0 -3.2)	

Data from: Neubert et al. 1992d
(0-time: n = 12; TCDD-treated: n = 7)
* $P < 0.05$, compared with controls

Figure 6 [A/B] is shown on the next page ►.

Fig. 6 [A/B]. T lymphocyte subsets in marmoset blood during treatment with 2,3,7,8-*TCDD*. **[A]** "helper-inducer" cells ($CD4^+CDw29^+$), **[B]** "suppressor-inducer" cells ($CD4^+CD45RA^+$). Data of each marmoset are connected by a line. Black: treated (n = 6), white: controls (n = 5). Repeated treatments with the higher weekly doses (1.5 ng TCDD/kg body wt) drastically but reversibly decreased the percentage and absolute number of $CD4^+CDw29^+$ cells, but at lower doses (0.3 ng/kg) the effect was reversed: $CD4^+CDw29^+$ cells increased over controls and $CD4^+CD45RA^+$ cells decreased. Total $CD4^+$ cells stayed rather constant during the study

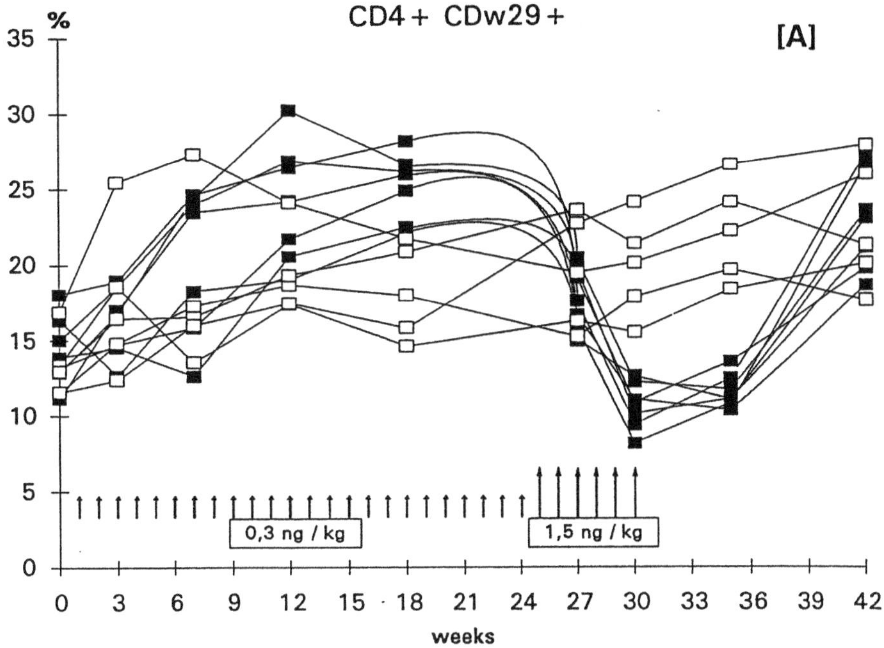
CD4+ CDw29+
[A]
%
0,3 ng / kg
1,5 ng / kg
weeks

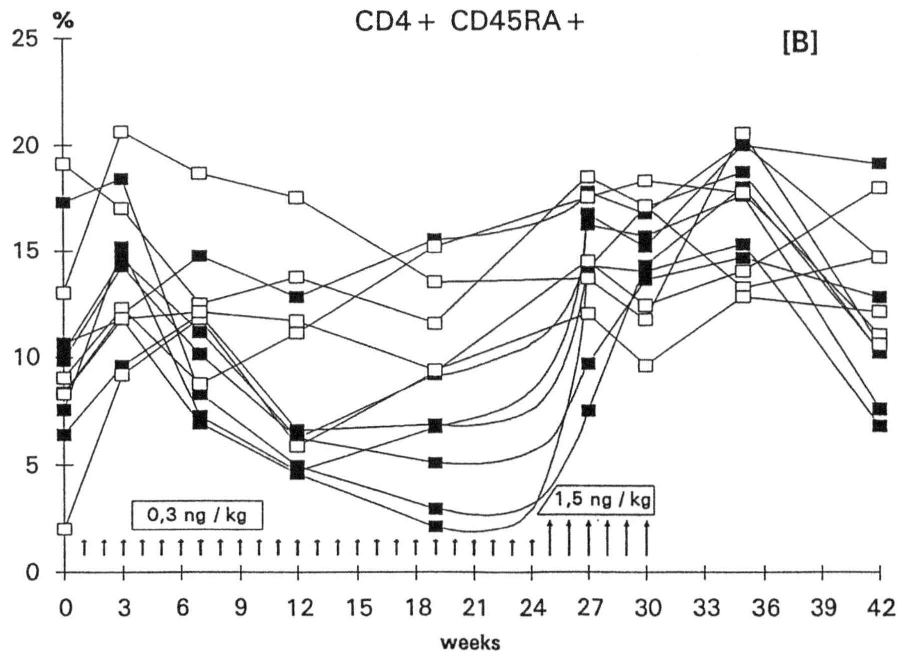
CD4+ CD45RA+
[B]
%
0,3 ng / kg
1,5 ng / kg
weeks

The results of the several independent studies performed during the last three years in our laboratory may be summarized as follows:

- effective doses of > 7 ng TCDD/kg body wt *reduced* the percentage and the absolute number of "helper-inducer" T cells ($CD4^+CDw29^+$) in the blood,
- effective doses of > 2 < 4 ng TCDD/kg body wt *increased* the percentage and the absolute number of "helper-inducer" T cells,
- effective doses of < 2 ng TCDD/kg body wt *had no effect* on the percentage and the absolute number of "helper-inducer" T cells.

These data indicate that the effect of TCDD on $CD4^+CDw29^+$ cells is *reversed* at lower doses (within a rather small window), and it disappears when lowering the dose even further. This result provides evidence for an important aspect:

*it may **not** be meaningful and justified (as is most often done) to extrapolate that an effect observed in experiments at a high dose must also occur at lower exposures (only with a lower incidence).*

A similar biphasic effect of TCDD was reported by Pitot et al. (1987) with respect to the promotion of the occurrence of preneoplastic lesions ("foci" in rat liver).

We have also assessed how much TCDD may be expected to reach the lymphocytes directly *in vivo*. For this purpose ^{3}H-TCDD (40 Ci/mmol) was injected s.c. into rats at a dose of 1,000 ng TCDD/kg body wt, and radioactivity was measured in isolated (Ficoll-hypaque gradient) peripheral lymphocytes of the blood, as well as in isolated lymphocytes of the spleen in three rats each 6, 13 and 20 days after the injection. Table 13 shows the data obtained on day 6 after the application (i.e. at the maximum concentration).

The calculations performed from our studies with ^{3}H-TCDD (assuming Loschmidt's number [6 x 10^{23} molecules/mole]) indicate that in lymphocytes of blood an average of about 100 ± 30 molecules per cell may be expected at maximum absorption (day 6) following a dose of 1,000 ng TCDD/kg body weight. A similar number of molecules (45 ± 11) was found in lymphocytes in the spleen. TCDD radio activity in blood lymphocytes had declined 13 days after the injection to about 59% of the value at day 6, and to 36% on day 20 after the injection. Since placental transfer in the rat is low, much smaller concentrations than calculated here for the adult will be expected in the lymphocytes of a fetus.

The distribution of the ^{3}H-TCDD between the main tissues (hepatic and adipose tissue) agreed very well in this study with corresponding studies performed with ^{14}C-TCDD (Abraham et al. 1988, 1989). Under the experimental conditions chosen, the concentration in the thymus (on a wet weight basis) was about 1/3 that of adipose tissue.

These calculations based on studies with rats do not favour the assumption of a direct effect of TCDD on peripheral lymphocytes at the low single doses (10 ng TCDD/kg body wt) found to be effective by changing the pattern of certain lymphocyte subpopulations in marmosets, although they are compatible with such a direct effect at higher doses (>1 μg/kg body wt).

Up till now, few reliable data are available on possible changes induced by PCDDs/PCDFs on the lymphocyte subtypes of human adults. However, only few human populations at risk have been studied in detail and with modern methods. In one exposed group Webb et al. (1989) did not find any changes in the pattern of surface markers on lymphocytes, but the $CD4^+CDw29^+$ cells were not included in the studies. There is no information on possible effects on the pattern of lymphocyte subpopulations in children after pre- or perinatal exposure to TCDD available. Studies, at least in heavily exposed adults, are urgently needed to judge whether man is as sensitive a species to the action of TCDD on lymphocyte subtypes as the marmoset is. In such studies an assessment of the T cell subpopulations must be included.

With respect to the topic of this meeting it will be important to evaluate whether the effect seen in adult marmosets may also be obtained with prenatal exposure, perhaps even in an exaggerated form. Since we are concerned with the *development* of the immune system, the outcome of such studies with prenatal exposure may be different from that induced in the adult animal.

Table 13. Amount of 2,3,7,8-**TCDD** found in lymphocytes of blood and spleen after application of ^{3}H-TCDD to rats. A dose of 1,000 ng ^{3}H-TCDD (40 Ci/mmol/kg body wt was given s.c. and radioactivity was measured in isolated lymphocytes of blood and spleen after 6 days

	Measured		Calculated (average)		
Rat No.	TCDD (pg)	Cells ($x10^6$)	pg TCDD/cell	fmoles/cell	Molecules /cell
Blood lymphocytes:					
1	0.37	10.0	3.7×10^{-8}	11.5×10^{-8}	**69**
1	0.24	3.8	6.3×10^{-8}	19.6×10^{-8}	**118**
3	0.34	5.2	6.5×10^{-8}	20.2×10^{-8}	**121**
Spleen lymphocytes:					
1	0.38	19.1	2.0×10^{-8}	6.2×10^{-8}	**37**
2	0.53	23.8	2.2×10^{-8}	6.8×10^{-8}	**41**
3	0.29	9.4	3.1×10^{-8}	9.6×10^{-8}	**58**
Liver:					
1	6.3	ng TCDD/g wet wt			
2	8.2	ng TCDD/g wet wt			
3	12.2	ng TCDD/g wet wt			
Adipose Tissue:					
1	1.3	ng TCDD/g wet wt			
2	1.6	ng TCDD/g wet wt			
3	1.5	ng TCDD/g wet wt			
Thymus:					
1	0.4	ng TCDD/g wet wt			
2	0.5	ng TCDD/g wet wt			
3	0.6	ng TCDD/g wet wt			

Summary and Conclusions

The assessment of prenatally-induced lesions that manifest themselves only postnatally (congenital dysfunctions) should not be confined to "behavioural" aspects. Alterations of the immune system may turn out to be of as least as much or even more significance. Although these kinds of dysfunctions are not considered in the frame of the present risk assessment in reproductive and developmental toxicology, this area may be predicted to gain much more importance in the near future.

The marmoset seems to represent a convenient species for assessing complex drug-induced effects on the immune system, especially of changes in the pattern of lymphocyte surface receptors. This New World monkey shows many advantages over Old World monkeys, and it may become the primate species of choice for research in the near future. Studies on possible effects on lymphocyte subpopulations are connected with little distress to the animals because the burden may be limited to the application of the substances at doses with little if any general toxicity, and to a few blood samplings.

Additionally, the marmoset shows the clear-cut advantage that the significance of defined xenobiotic-induced effects on special lymphocyte receptors observed in this monkey may directly be assessed and possibly verified in studies on humans, e.g. in populations at risk. Such a strategy has been found feasible in the case of exposure of infants to PCDDs/PCDFs, and it is possible to judge whether man exhibits a similar susceptibility to a given agent as *Callithrix jacchus*.

Acknowledgement: We thank Dr. E. Frankus (Chemie Grünenthal) for the generous gift of thalidomide, EM12 and supidimide. The reliable, expert help of our animal caretakers, Gerd Leonard, Marion Trapp, Barbara Grote, Monika Meseck, Silvia Rieger, with the studies and their continuous devotion to the marmosets is greatly appreciated.

References

Abraham K, Krowke R, Neubert D (1988) Pharmacokinetics and biological activity of 2,3,7,8-tetrachlorodibenzo-*p*-dioxin. 1. Dose-dependent tissue distribution and induction of hepatic ethoxyresorufin O-deethylase in rats following a single injection. Arch Toxicol 62: 359-368

Abraham K, Weberruß U, Wiesmüller T, Hagenmaier H, Krowke R, Neubert D (1989) Comparative studies on absorption and distribution in the liver and adipose tissue of PCDDs and PCDFs in rats and marmoset monkeys. Chemosphere 19: 887-892

Barnhill RL, McDougall AC (1982) Thalidomide: use and possible mode of action in reactional lepromatous leprosy and in various other conditions. J Am Acad Dermatol 7: 317-323

Clark DA, Gauldie J, Szewczuk MR, Sweeney G (1981) Enhanced suppressor cell activity as a mechanism of immunosuppression by 2,3,7,8-tetrachlorodibenzo-p-dioxin. Proc Soc Exp Biol Med 168: 290-299

Clark DA, Sweeney G, Safe S, Hancock E, Kilburn DG, Gauldie J (1983) Cellular and genetic basis for suppression of cytotoxic T cell generation by haloaromatic hydrocarbons. Immunopharmacol 6: 143-153

Dean JH, Luster MI, Boorman GA, Lauer LD (1982) Procedures available to examine the immunotoxicity of chemicals and drugs. Pharmacol Rev 34: 137-148

Dean JH, Thurmond LM (1987) Immunotoxicology: An overview. Toxicol Pathol 15: 265-271

Fauci AS, Dale DC (1974) The effect of *in vivo* hydrocortisone on subpopulations of human lymphocytes. J Clin Invest 53: 240-246

Field EO, Gibbs JE, Tucker DF, Hellmann K (1966) Effect of thalidomide on the graft-vs-host reaction. Nature 211: 1308-1310

Heger W, Klug S, Schmahl H-J, Nau H, Merker H-J, Neubert D (1988) Embryotoxic effects of thalidomide derivatives on the non-human primate *Callithrix jacchus*; 3. Teratogenic potency of the EM12 enantiomers. Arch Toxicol 62: 205-208

Heney D, Bailey CC, Lewis IJ, (1990) Thalidomide in the treatment of graft-*versus*-host disease. Biomed Pharmacother 44: 199-204

Heney D, Norfolk DR, Wheeldon J, Bailey CC, Lewis IJ, Barnard DL (1991) Thalidomide treatment for chronic graft-*versus*-host disease. Br J Haematol 78: 23-27

Iyer CGS, Languillon J, Ramanujam K, Tarabine-Castellani G, Terencio de las Aguas J, Bechelli LM, Uemura K, Martinez Dominguez V, Sundaresan T (1971) WHO co-ordinated short-term double-blind trial with thalidomide in the treatment of acute lepra reactions in male lepromatous patients. Bull Wrld Hlth Org 45: 719-732

Korte M, Neubert R, Stahlmann R (1991) No effect of 2,3,7,8-tetrachlorodibenzo-*p*-dioxin (TCDD) on leukocytes in peripheral blood of the Wistar rat. Naunyn-Schmiedeberg's Arch Pharmacol 344 (suppl 2): R115(48) [abstract]

Krajnc EI, Wester PW, Loeber JG, Van Leeuwn FXR, Vos JG, Vaessen HAMG, Van der Heijden CA (1984) Toxicity of bis(tri-n-butyltin)oxide in the rat. I. Short-term effects on general parameters and on the endocrine and lymphoid system. Toxicol Appl Pharmacol 75: 363-386

Luster MI, Munson AE, Thomas PT, Holsapple MP, Fenters JD, White KL, Lauer LD, Germolec DI, Rosenthal GJ, Dean JH (1988) Development of a testing battery to assess chemically-induced immunotoxicity: National Toxicology Program's guidelines for immunotoxicity evaluation in mice. Fund Appl Toxicol 10: 2-19

Merker H-J, Heger W, Sames K, Stürje H, Neubert D (1988) Embryotoxic effects of thalidomide derivatives on the non-human primate *Callithrix jacchus*: 1. Effects of 3-(1,3-dihydro-1-oxo-2H-isoindol-2-yl)-2,6-dioxopiperidine (EM12). Arch Toxicol 61: 165-179

Neubert D (1991) Peculiarities of the toxicity of polyhalogenated dibenzo-p-dioxins and dibenzofurans in animals and man. Chemosphere: 23: 1869-1893

Neubert D, Abraham K, Golor G, Krowke R, Krüger N, Nagao T, Neubert R, Schulz-Schalge T, Stahlmann R, Wiesmüller T, Hagenmaier H (1991a) Comparison of the effects of PCDDs and PCDFs on different species taking kinetic variables into account. In: Banbury Report 35, Biological Basis for Risk Assessment of Dioxins and Related Compounds. Cold Spring Harbor Laboratory Press, pp 27-44

Neubert D, Heger W, Merker H-J, Sames K, Meister R (1988a) Embryotoxic effects of thalidomide derivatives on the non-human primate *Callithrix jacchus*: 2. Elucidation of the susceptible period and of the variability of embryonic stages. Arch Toxicol 61: 180-191

Neubert D, Golor G, Stahlmann R, Neubert R, Helge H (1992a) Einige Ausführungen zur Toxizität von polyhalogenierten Dibenzo-*p*-dioxinen und Dibenzofuranen. Organohalogen Compounds 5: in press

Neubert D, Klug S, Golor G, Neubert R, Felies A (1992b) Marmosets (*Callithrix jacchus*) as useful species for toxicological risk assessments. In: Neubert D, Kavlock RJ, Merker H-J, Klein J (eds) Risk Assessment of Prenatally-Induced Adverse Health Effects, Springer-Verlag, Heidelberg, Berlin, New York, this book

Neubert D, Merker H-J, Hendrickx AG (eds. 1988b) Non-Human Primates - Developmental Biology and Toxicology . Ueberreuter Wissenschaft, Wien, Berlin, pp 1-606

Neubert D, Neubert R, Stahlmann R, Helge H (1989) Immunotoxicology and -pharmacology. Brazilian J Med Biol Res 22: 1457-1473

Neubert D, Wiesmüller T, Abraham K, Krowke R, Hagenmaier H (1990a) Persistence of various polychlorinated dibenzo-*p*-dioxins and dibenzofurans (PCDDs and PCDFs) in hepatic and adipose tissue of marmoset monkeys. Arch Toxicol 64: 431-442

Neubert R, Delgado I, Dudenhausen JW, Neubert D (1992c) Some data on lymphocyte pattern in blood during the perinatal period. In: Neubert D, Kavlock RJ, Merker H-J, Klein J (eds) Risk Assessment of Prenatally-Induced Adverse Health Effects, Springer-Verlag, Heidelberg, Berlin, New York, this book

Neubert R, Golor G, Stahlmann R, Helge H, Neubert D (1992d) Polyhalogenated dibenzo-*p*-dioxins and dibenzofurans and the immune system. 4. Effects of multiple-dose treatment with 2,3,7,8-tetrachlorodibenzo-*p*-dioxin (TCDD) on peripheral lymphocyte subpopulations of a non-human primate (*Callithrix jacchus*). Arch Toxicol 66: 250-259

Neubert R, Golor G, Stahlmann R, Helge H, Neubert D (1992e) Dose-dependent TCDD-induced increase or decrease in a T-lymphocyte subset in the blood of New World monkeys (*Callithrix jacchus*). Chemosphere: in press

Neubert R, Helge H, Golor G, Neubert D (1992f) Characterization of lymphocyte subpopulations in the venous blood of marmosets (*Callithrix jacchus*). J Med Primatol : submitted

Neubert R, Helge H, Stahlmann R, Neubert D (1991b) Einige Wirkungen von 2,3,7,8-Tetrachlorodibenzo-*p*-dioxin (T4CDD) und von 2,3,4,7,8-Pentachlorodibenzofuran (P5CDF) auf periphere Lymphozyten von Primaten *in-vivo* und *in-vitro*. Allergologie 14: 360-371

Neubert R, Jacob-Müller U, Stahlmann R, Helge H, Neubert D (1990b) Polyhalogenated dibenzo-*p*-dioxins and dibenzofurans and the immune system. 1. Effects on peripheral lymphocyte subpopulations of a non-human primate (*Callithrix jacchus*) after treatment with 2,3,7,8-tetrachlorodibenzo-*p*-dioxin (TCDD). Arch Toxicol 64: 345-359

Neubert R, Nogueira AC, Neubert D (1991c) Effect of thalidomide-derivatives on the pattern of lymphocyte subpopulations in marmoset blood. Naunyn-Schmiedeberg's Arch Pharmacol 344 (suppl 2): R123(77) [abstract]

Neubert R, Nogueira AC, Neubert D (1992g) Thalidomide derivatives alter the expression of lymphocyte function-associated antigens in marmoset blood. Naunyn-Schmiedeberg's Arch Pharmacol 345 (suppl): R94(375) [abstract]

Neubert R, Nogueira AC, Neubert D (1992h) Thalidomide derivatives and the immune system. 1. Changes in the pattern of integrin receptors and other surface markers on T lymphocyte subpopulations of marmoset blood. Arch Toxicol: submitted

Nogueira AC, Neubert R, Neubert D (1991) Thalidomide-induced alterations in marmoset lymphocytes in culture. Naunyn-Schmiedeberg's Arch Pharmacol 344 (suppl 2): R124(82) [abstract]

Pitot HC, Goldworthy TL, Moran S, Kennan W, Glauert HP, Maronpot RR, Campbell HA (1987) A method to quantitate the relative initiating and promoting potencies of hepatocarcinogenic agents in their dose-response relationship to altered hepatic foci. Carcinogenesis 8: 1491-1499

Rutishauser U, Acheson A, Hall AK, Mann DM, Sunshine J (1988) The neural cell adhesion molecule (NCAM) as a regulator of cell-cell interactions. Science 240: 53-57

Scott WJ, Fradkin R, Wilson JG (1977) Non-confirmation of thalidomide induced teratogenesis in rats and mice. Teratology 16: 333-336

Sheskin J (1965) Thalidomide in the treatment of lepra reactions. Clin Pharmacol Ther 6: 303-306

Springer TA (1990) Adhesion receptors of the immune system. Nature 346: 425-434

Van Loveren H, Vos JG (1989) Immunotoxicological considerations: a practical approach to immunotoxicity testing in the rat. In: AD Dayan, AJ Paine, eds, Advances in Applied Toxicology, Taylor & Francis Ltd, London, pp 143-163

Vogelsang GB, Hess AD, Gordon G, Brundrette R, Santos GW (1987) Thalidomide induction of bone marrow transplantation tolerance. Transplant Proc 19: 2658-2661

Vogelsang GB, Hess AD, Gordon G, Santos GW (1986a) Treatment and prevention of acute graft-*versus*-host disease with thalidomide in a rat model. Transplantation 41: 644-647

Vogelsang GB, Hess AD, Santos GW (1988) Thalidomide for treatment of graft-*versus*-host disease. Bone Marrow Transplant 3: 393-398

Vogelsang GB, Taylor S, Gordon G, Hess AD (1986b) Thalidomide, a potent agent for the treatment of graft-*versus*-host disease. Transplant Proc 18: 904-906

Vos JG, DeKlerk A, Krajnc EI, Kruizinga W, Van Ommen B, Rozing J (1984) Toxicity of bis(tri-n-butyltin)oxide in the rat. II. Suppression of thymus-dependent immune responses and of parameters of non-specific resistance after short-term exposure. Toxicol Appl Pharmacol 75: 387-408

Vos JG, Krajnc EI, Van Loveren H (1990) Immunotoxicity and altered host resistance in rats: tributyltin oxide as an example. In: GN Volans, J Sims, FM Sullivan, P Turner, eds, Basic Science in Toxicology. V Internat Congr Toxicol, Taylor & Francis, London, NY, Philadelphia, pp 354-362

Vos JG, Moore JA (1974) Suppression of cellular immunity in rats and mice by maternal treatment with 2,3,7,8-tetrachlorodibenzo-*p*-dioxin. Int Arch Allergy Appl Immunol 47: 777-794

Vos JG, Van Loveren H, Schuurman H-J (1991) Immunotoxicity of dioxins: immune function and host resistance in laboratory animals and humans. In: Banbury Report 35, Biological Basis for Risk Assessment of Dioxins and Related Compounds. Cold Spring Harbor Laboratory Press, pp 79-88

Webb KB, Evans RG, Knutsen AP, Roodman ST, Roberts DW, Schramm WF, Gibson BB, Andrews JS, Needham LL, Patterson DG (1989) Medical evaluation of subjects with known body levels of 2,3,7,8-tetrachlorodibenzo-*p*-dioxin. J Toxicol Envir Health 28: 183-193

Yu DTY, Clements PJ, Paulus HE, Peter JB, Levy J, Barnett EV (1974) Human lymphocyte subpopulations. effects of corticosteroids. J Clin Invest 53: 565-571

Zwadlo-Klarwasser G, Neubert R, Stahlmann R, Schmutzler W (1992) Influence of dexamethasone on the RM3/1-positive macrophages in the peripheral blood and tissues of a New World monkey (the marmoset, *Callithrix jacchus*). Int Arch Allergy Immunol 97: 178-180

Marmosets (*Callithrix jacchus*) as Useful Species for Toxicological Risk Assessments

Diether Neubert, Stephan Klug, Georg Golor, Reinhard Neubert and Annegret Felies

Introduction

There are some problems in toxicological research, including prenatal toxicology, for the solving of which the use of non-human primates cannot be avoided:

- The main reason is that certain toxic effects only occur in primates and not in other experimentally used species, and therefore, a possible risk can only be assessed in these species.
- Furthermore, non-human primates are used as a third experimental species when equivocal results have been obtained with the usual rodents and lagomorpha, e.g. in the case of a therapeutically important substance to be used in a large human population.

So far, Old World monkeys have almost been exclusively used for these purposes (rhesus and cynomolgus) in experimental medicine. However, for many reasons the use of these monkey species is hampered with serious limitations. These include animal preservation and the necessity for wild catches, difficulties in breeding (the rhesus, e.g., is a seasonal breeder), handling of large monkeys, cost and the necessity for large breeding and maintenance facilities, etc. More recently, the possibility of using New World monkeys for certain areas of research has been explored and many advantages have become obvious.

In this brief overview a few examples of our studies will be presented which illustrate the suitability of the marmoset as an experimental model for toxicological investigations:

- Studies with medicinal products (effects induced by thalidomide-type substances, or by biotechnology products), and
- studies with environmental substances (e.g. "dioxins").

Advantages of Using Marmosets for Research, Prerequisites and Own Experience

Over the last 15 years we have gained considerable experience with the breeding, maintenance and the application of the common marmoset (*Callithrix jacchus*), a New World monkey from the Brazilian jungle, for tackling special toxicological problems (e.g. Neubert et al. 1978). Using this experience we have built up a successful colony of marmosets in Berlin. The overall size of the colony has ranged between 350 and 500 animals over the last eight years (Fig. 1).

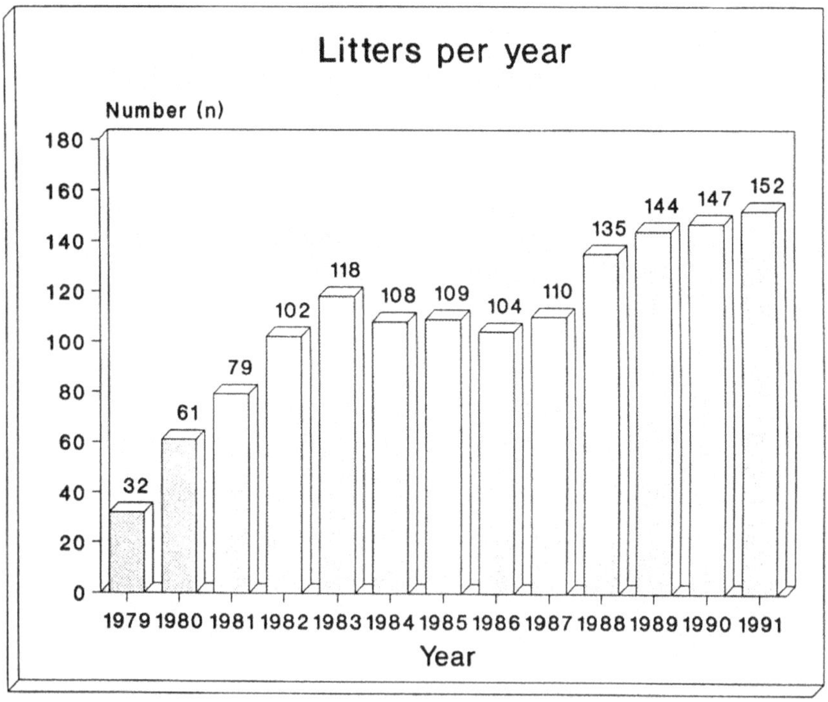

Fig. 1. Number of viable litters born in our colony during the last 13 years. This number is now rather stable. The percentage of (visible) prenatal losses is about 15-30% of the live born; it is (like in man) > 50% when considering the losses from implantation onward. The stillbirths number about 4-6% of the live born (not considering one of the non-viable triplets or two of the quadruplets). The number of litters born in the colony also depends on the number of Caesarian sections performed during the year (generally: 20 to 50)

The annual breeding capacity of our colony in Berlin was increased from about 30 litters in 1979 to about 100 litters/year within three years, and has now been kept at approximately 140 litters/year for the last four years (Fig. 2). From these newborn, about 150 to 200 animals per year are raised to adulthood (Fig. 3). Regularly, some of the older breeding pairs are replaced by younger ones and the rest are available for studies. In this way the colony is self-sustaining without any need for wild catches (Fig. 4). In fact, the breeding capacity within the colony is deliberately kept at a suboptimal level so that there are not too many animals which would overcrowd the colony. The size of the colony is limited only by the space available. Inbreeding is avoided by having the data for all animals in the computer for at least the last three generations.

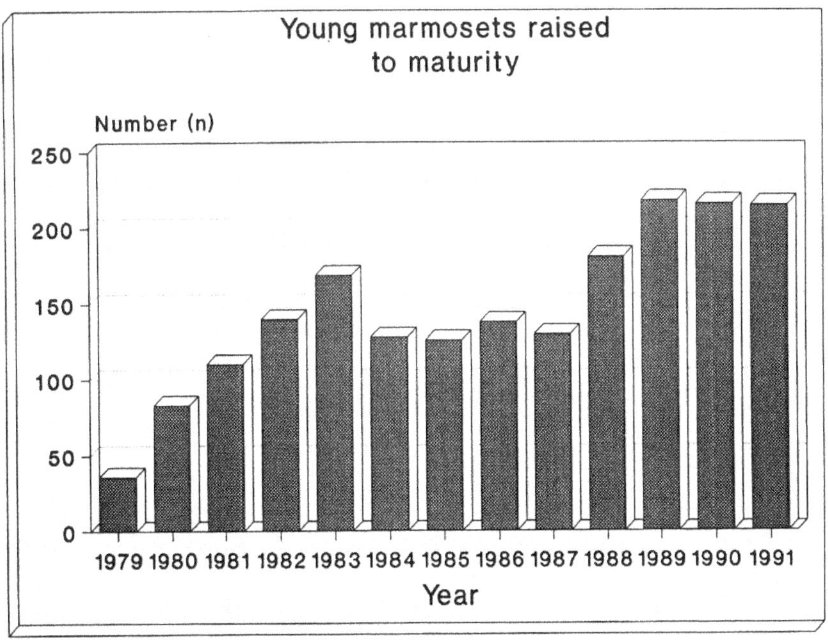

Fig. 2. Number of marmosets raised to maturity per year. This number is adapted to the scientific needs and the space available; it is by no means the maximum number possible for this colony

The marmoset was suggested as an experimental animal by Poswillo and coworkers already in 1972. We were introduced to this non-human primate by the colleagues Siddall and Hiddleston of ICI, England, where the late Hiddleston bred about 1,000 animals per year in the middle 1970's (e.g. Hiddleston 1976; Hiddleston and Siddall 1978).

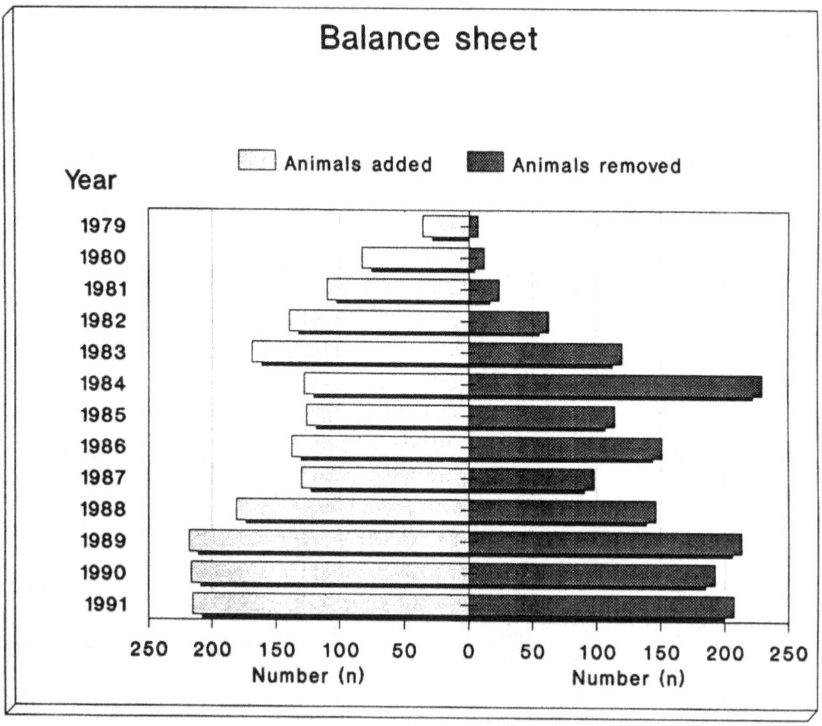

Fig. 3. Balance sheet of the colony, additions and removals from the colony during the last 13 years. Within the last five years the colony has been kept rather stable

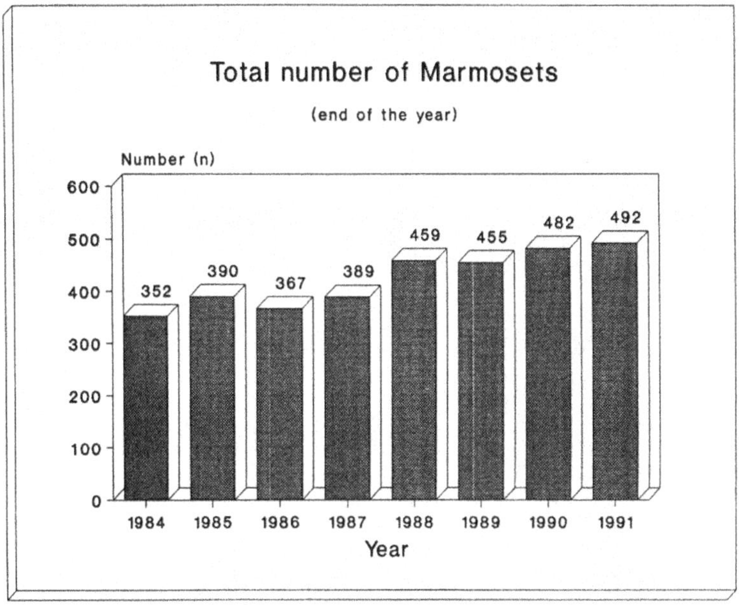

Fig. 4. Total number of animals (including young) in our colony during the last 13 years. The colony is now stable at 450 to 500 marmosets (all animals included)

Advantages and Limitations when Using Marmosets

All animals in our colony are hand-tame (Picture 1); a considerable advantage for studies and from the point of view of animal welfare. Some of the advantages and limitations when using marmosets as an experimental species are listed in Table 1. The advantages when compared with Old World monkeys are quite convincing. One of the main advantages of using marmosets for risk assessment with respect to reproductive toxicity is the fact that they generally carry two to three young. The duration of pregnancy in our colony was found (Heger et al. 1988b) to be 143 days (median), with a range of 129 to 149 days (minimum and maximum of 108 pregnancies).

Picture 1. Marmosets and an animal caretaker from our colony. All marmosets are hand-tame and can be handled without any stress. The animals shown have almost reached full size

When we started our colony the percentage of twins was about 65% and triplets were about 25% (Fig. 5). Soon the percentage of triplets increased to about 40%. Under favourable laboratory conditions the contribution of triplets is now 40-55%, and only about 35-45% are twins. Singles are carried, in general, to < 10%, and 1-5% of the litters may be quadruplets.

Marmosets are only able to raise twins, therefore from a litter of triplets one offspring will always die within the first few days. Since we are interested in effects on the perinatal period, one of the offspring can be sacrificed and the organs used for research since the animal would die anyway.

Table 1. Advantages and limitations of marmosets as experimental primate species

Advantages:	*Marmosets:*
	are small, can be kept hand-tame, can be bred comparatively easily, have a rather short generation time, carry twins and triplets, allow timing of ovulation and pregnancies, embryology has been worked out well, respond well to e.g. thalidomide, immunology has been worked out rather well, are rather inexpensive to breed, can be handled easily and without distress for the animals.
Limitations:	
	Monkeys are not human !
	Metabolism of xenobiotics may deviate from that of man,
	i.v.-infusions may be difficult because of the small size, only studies with limited numbers of animals are feasible.

As in all primates (including man), prenatal mortality is rather high. From implantation to birth the overall postimplantation loss was found to be about 20% (median) in 255 pregnancies (Heger et al. 1988b), with an additional failure to implant after ovulation of about 15% (median). Since, like in man, abortions are often clustered in females with repeated abortions, the rate after selection in a well-organized breeding colony may be smaller than in wild animals.

Some data on the frequency of stillbirths are compiled in Table 2 (from the year 1987). Interestingly, there were one to three stillborn in all six litters with quadruplets. Stillborn were observed in only 19 of the 49 litters with triplets (39%), and this incidence was only 8% (3/39 births) in litters with twins; no stillborn were seen in the 16 litters with single newborn.

When performing Caesarian sections an average of 2.2 fetuses can be expected, compared with only one offspring in Old World monkeys. This is certainly a considerable advantage. Another advantage is the relatively low cost of raising and maintaining these animals (Table 3).

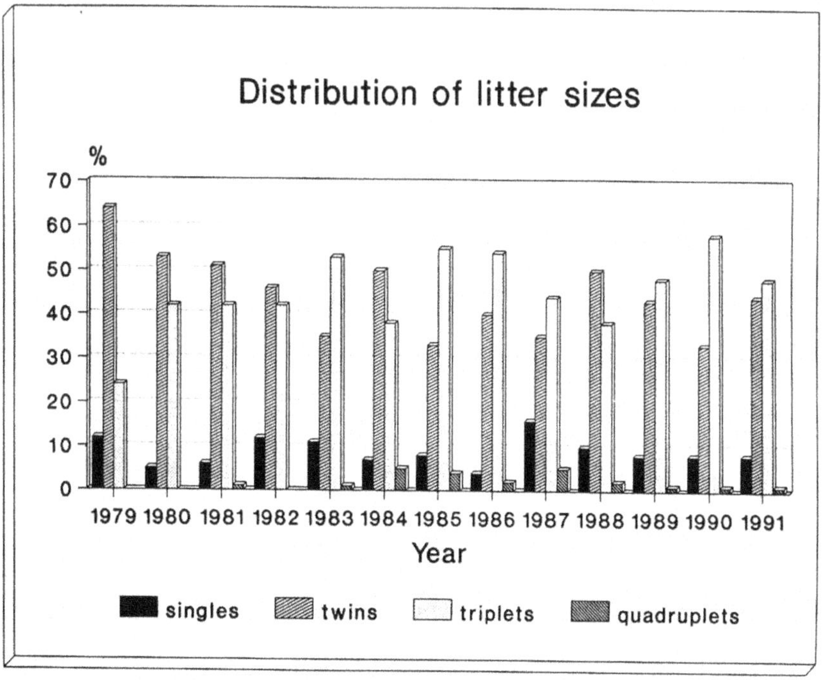

Fig. 5. Percentage of the litter sizes obtained in the various years. The number of triplets increased from the initiation of the colony, it is now at about 40-50%. Marmosets are only able to raise two young, the third is bound to die.

Variability of "Normal Values" Observed in Marmosets

There is one phenomenon which occurs with *Callithrix jacchus* in the same way as in other primates (including man): due to the heterogeneity in genetic background the "normal deviation" of laboratory values and other physiological and biochemical variables may be somewhat greater than in a defined or even inbred rodent strain. This indicates that homeostasis of the organism is achievable within a wide range of variables and is not confined to the narrow range of "normal values", as may be falsely deduced from rodent data.

Table 2. Correlation of stillbirths with litter size in marmosets. Only stillbirths were evaluated and no postnatal mortality of viable newborn were considered. Quadruplets seem to be especially prone to mortality shortly before or during birth

Female #	Litter size	Still-born	Further mortalities
329.1	4	3	
388.1	4	2	
196.1	4	1	+1 died
624.1	4	2	+2 died
207.1	4	2	
510.1	4	2	
395.1	3	1	
522.3	3	1	+2 died
249.1	3	1	+2 died
56.1	3	1	+1 died
361.1	3	1	*
121.3	3	1	
328.1	3	1	inbred female
504.1	3	1	+1 died*
377.1	3	1	+2 died
161.3	3	1	+1 died
315.3	3	1	
391.1	3	2	+1 died
305.1	3	1	
526.1	3	1	inbred female
552.1	3	3	
537.1	3	2	+1 died
521.1	3	1	
536.1	3	1	+2 died
410.1	3	1	
306.1	2	1	+1 died*
213.1	2	2	
220.1	2	2	

Data: from year 1987
* One or more previous abortions

This creates an interesting "philosophical" problem since experimental research on laboratory animals has, up till now, been primed at detecting small changes between experimental groups. While such small differences may be statistically significant it is questionable whether they are of biological or medical relevance in species with an intrinsicly large spontaneous variability. On the other hand, a high spontaneous variability almost excludes the chance to detect small deviations, e.g., induced by substances, especially if relatively small groups of non-human primates are evaluated. For the same reason larger variations in the response to an agent will be expected.

Table 3. Rough cost calculation for 1 fetus, Rhesus vs. Marmoset

Prerequisite:			
Rhesus:	1	fetus at C-section	
Marmoset	2.5	fetuses at C-section	
Cost: (rough calculation)			
1 Rhesus or cynomolgus:			5,000 DM
1 Marmoset			700 DM
(not considering the much more expensive housing and feed for the rhesus)			
	15 Rhesus-litters:	15 fetuses:	75,000 DM
	15 Marmoset litters:	37 fetuses:	10,500 DM
	1 Rhesus fetus:		*5,000 DM*
	1 Marmoset fetus:		*384 DM*

In *Figures 6 to 8* the variability found in our studies is exemplified for some haematological variables: the number of leukocytes, lymphocytes and of erythrocytes per μl blood. It is obvious from the graphs that there is no normal distribution of the data, and the curves show a "tailing" with respect to the higher values. From this rather large data set, representing more than 500 animals, it can be concluded that also in the marmoset many "normal" values of apparently healthy animals cluster around a median value, and the central 50% of the data (Q_1 to Q_3) do not deviate by more than $\pm$ 30% from this value in the case of leukocytes and lymphocytes, and the deviation for the central 50% values is only $\pm$ 8% in the case of erythrocytes. However, especially when studying large numbers of individuals there are also always exceptions which deviate considerably from the median value. While it is almost impossible to decide whether such individuals are "normal" (no signs of impaired health may be obvious), it is advisable to exclude them from an experiment if the deviation can be monitored before the onset of the study. In all of our studies we have attempted to achieve this goal. This will only exceptionally be possible in human epidemiological studies, indicating one *a priori* limitation of such data in man. Another possibility of evaluating data is to set an arbitrary cut-off point, and to state the number of individuals in the various experimental groups with values exceeding this cut-off point (cf. Neubert et al. 1991a). This procedure was originally practised in epidemiological studies in man (e.g. Mocarelli et al. 1986).

Examples of deviations found for some variables in the same individual over a period of more than ½ year are given in *Figures 9 and 10.* In these figures, data on the main T-cell populations ($CD4^+$ and $CD8^+$) in the marmoset are compiled. Although all the data stay within certain limits, it is obvious that the values for one given individual may at one time point be at the lower and at another time point at the upper limit of this "control range". If not properly controlled, such deviations might be taken as evidence for "effects" induced by an exposure if this individual should happen to belong to an exposed group.

Capacity for Drug Metabolism in Marmosets

One of the differences between rodents and primates (including man) is the fact that the hepatic, cytochrome P450-dependent monooxygenase systems responsible for the metabolism of many xenobiotics develop mostly after birth in the rodent, but in primates some of the isozymes (certainly not all) exhibit a substantial capacity during the fetal period (*cf.* e.g. Nau and Neubert 1978). The consequence of this is that for some types of studies and some types of substances the rodent may not represent a suitable experimental model. In such cases, primates may be clearly superior.

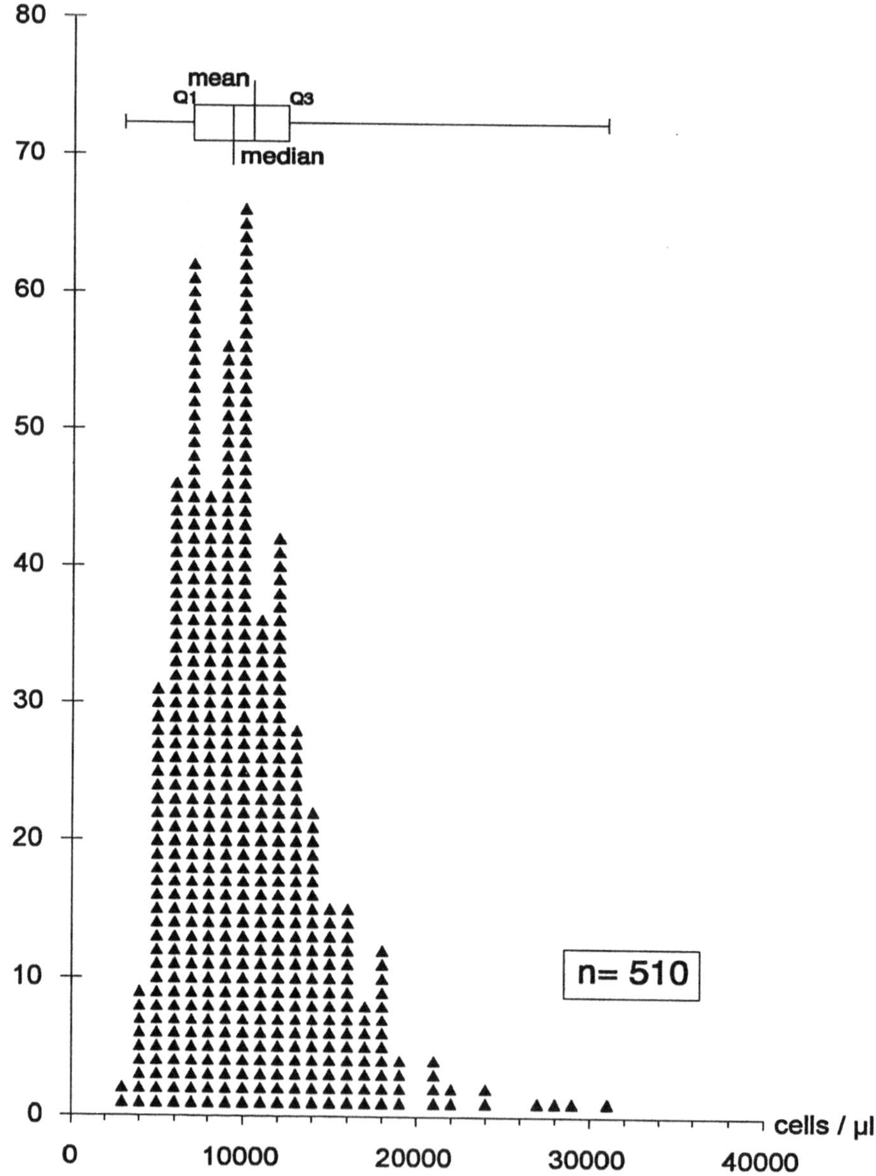

Fig. 6. Distribution of the number of *leukocytes* in the blood of 510 marmosets. The number of leukocytes per μl venous blood (obtained from the femoral vein plexus) was measured with the Sysmex F800 Microcellcounter (TOA Medical Electronics, Hamburg FRG). No obvious signs of infection were observed in the individuals with the comparatively high number of leukocytes

Mean ± SD:	10121 ± 4172	Q_1-Q_3:	7000 - 12100
median:	9500	min-max:	2900 - 30900

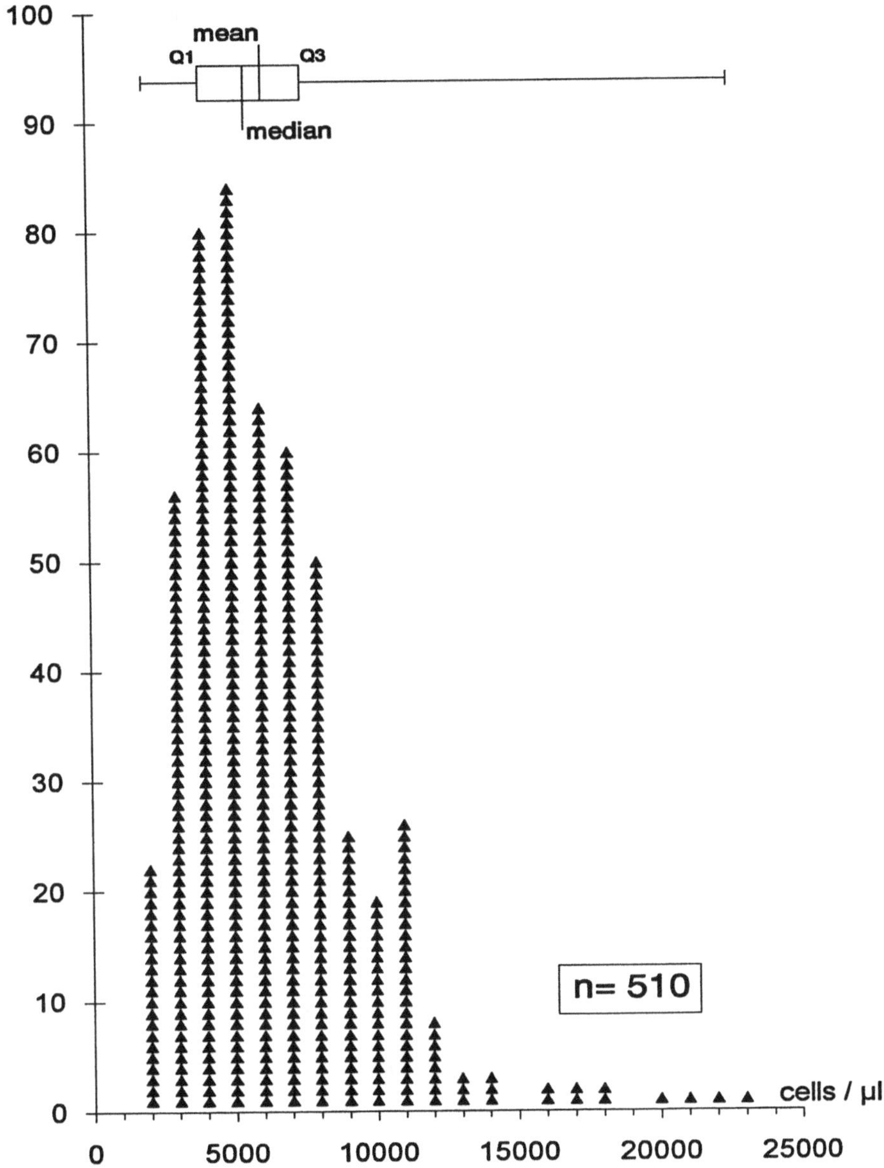

Fig. 7. Distribution of the number of *total lymphocytes* in the blood of 510 marmosets. The number of lymphocytes per μl venous blood (obtained from the femoral vein plexus) was calculated from measurements with the Sysmex F800 Microcellcounter (TOA Medical Electronics, Hamburg FRG) and differential white blood cell counts. No obvious signs of infection were observed in the individuals with the comparatively low or high number of lymphocytes

Mean ± SD:	6279 ± 3171	Q_1-Q_3:	4000 - 7600
median:	5700	min-max:	1500 - 22600

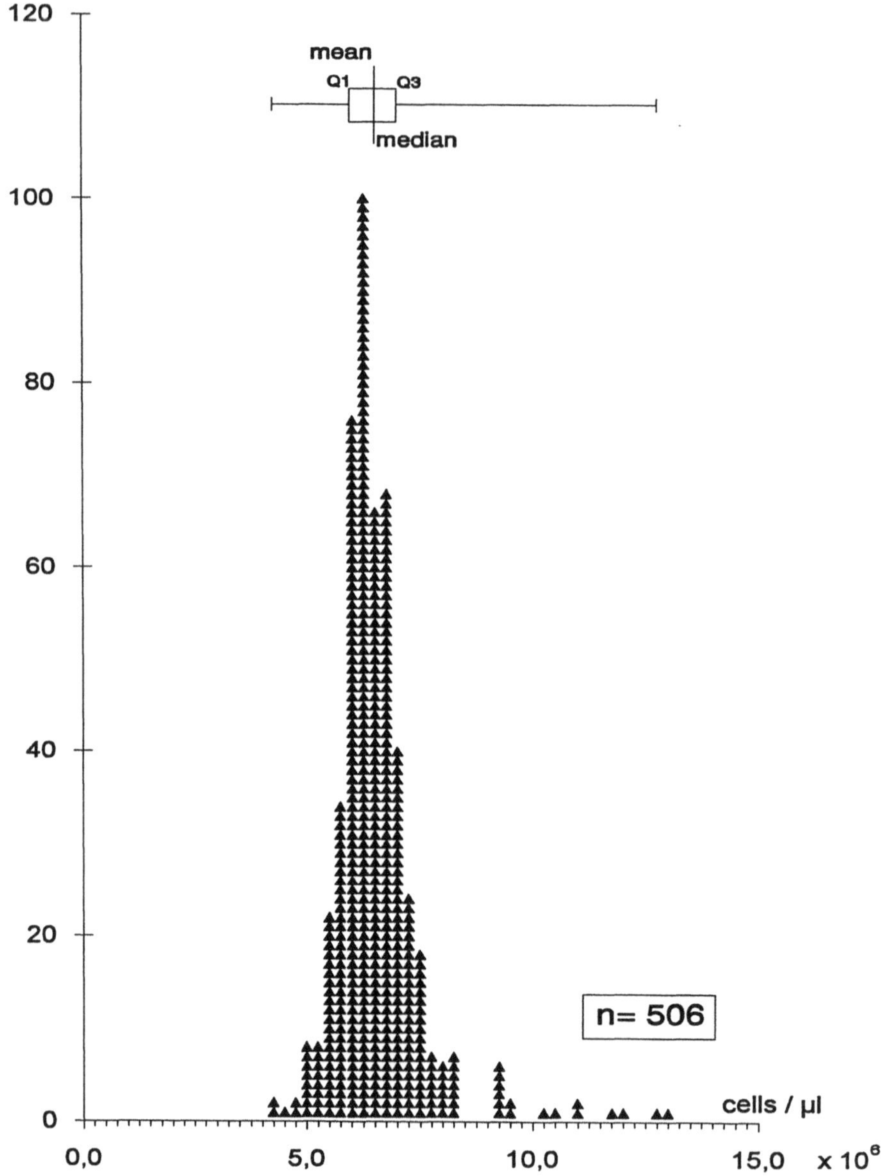

Fig. 8. Distribution of the number of *erythrocytes* in the blood of 506 marmosets. The number of erythrocytes per μl venous blood (obtained from the femoral vein plexus) was measured with the Sysmex F800 Microcellcounter (TOA Medical Electronics, Hamburg FRG) and differential white blood cell counts

Mean ± SD: $6.52 \times 10^6 \pm 1.05 \times 10^6$ Q_1-Q_3: $6.06 \times 10^6 - 6.83 \times 10^6$
median: 6.37×10^6 min-max: $0.3 \times 10^6 - 12.88 \times 10^6$

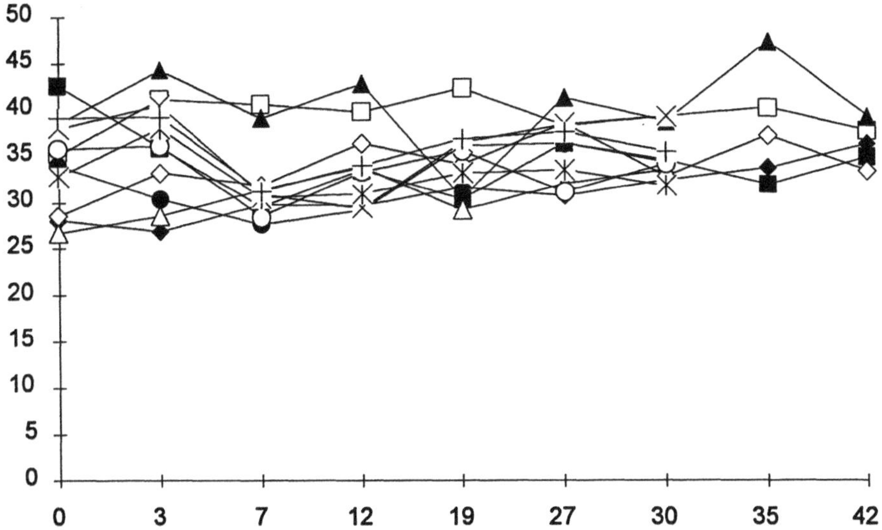

Fig. 9. Variability of *CD4$^+$ cells* in the same marmosets over a period of 30-42 weeks. Total *"helper"* ($CD4^+$) cells in venous blood (obtained from the femoral vein plexus) were measured with monoclonal antibodies and flow cytometry (FACScan). Values represented by the same symbol belong to the same individual

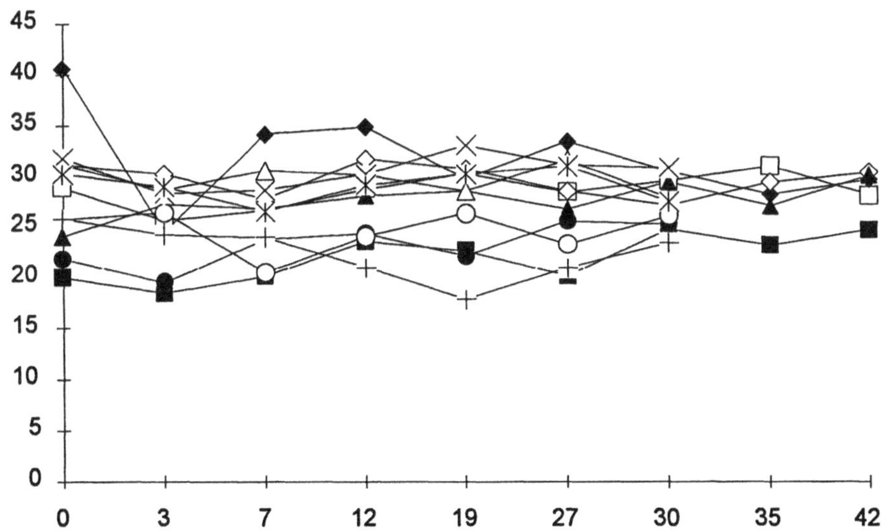

Fig. 10. Variability of *CD8$^+$ cells* in the same marmosets over a period of 30-42 weeks. Total *"suppressor/cytotoxic"* ($CD8^+$) cells in venous blood (obtained from the femoral vein plexus) were measured with monoclonal antibodies and flow cytometry (FACScan). Values represented by the same symbol belong to the same individual

For this reason we have studied some aspects of drug metabolism in liver microsomes of marmosets (e.g. Neubert et al. 1978; Neubert and Tapken 1988), and a number of comparative studies (marmosets vs. rats) were also performed (e.g. Schulz and Neubert 1988; Schulz-Schalge et al. 1990, 1991). Using the technology of a breath-test we have also been able to assess the early postnatal capacity for metabolic conversion of caffeine, methacetin and phenacetin *in vivo* in comparative studies of marmosets and rats (Krüger et al. 1991a).

These studies have shown that the marmoset is also a convenient species for these types of metabolic studies, and that drug-metabolism during the pre- and perinatal period also does not deviate extensively from that in other primates (including man).

Prerequisites for Using Marmosets in Risk Assessment of Prenatal Toxicity

In order to perform studies on reproductive and developmental toxicology there were a number of prerequisites to be met. The most important ones are listed in Table 4.

After being able to breed enough animals, the next challenge was to be able to time ovulation (and thereby pregnancy) accurately in this species. This primate does not show a menstruation nor cyclic changes of the vaginal epithelium. However, they exhibit the typical hormonal changes during the ovulatory cycle. The problem of the timing of ovulation was solved by Dr. Wolfgang Heger in our department by daily measurements of hormone excretion (estrogens and a progesterone metabolite) in small samples of urine. This is now a routine procedure in our colony. There is, in contrast to daily blood samplings performed for this purpose in other laboratories, no distress for the animals connected with our procedure. In this way ovulation can be routinely monitored to $\pm$ 1 day exactly by evaluating the beginning of the peak in hydroxypregnanolone excretion (Heger and Neubert 1983, 1987). It was revealed that the marmoset excretes a unique progesterone metabolite not found in other primates: 5α-pregnane-3α,7α-diol-20-one (Heger et al. 1988a). Since a new pregnancy generally starts in these animals within two to four weeks after delivery, urine sampling can be confined to this period only. Nidation and the continuation of pregnancy may be monitored by measuring estrogens in the small urine samples (about 200 μl).

Table 4. Prerequisites for choosing a species to assess prenatal toxicity

It is essential to:

- *breed* a sufficient number of animals
- *time* the *onset* of pregnancy (e.g. ovulation)
- monitor *implantation* and *maintenance* of pregnancy
- *define* different developmental *stages* during pregnancy
- know the extent of *prenatal wasting*
- have sufficient *historical data* on controls

The embryology of this species has also been worked out in detail by us. This task was predominantly carried out by Prof. Dr. Hans-Joachim Merker in our institute (cf. Neubert et al. 1988b). Caesarian sections can easily be performed under aseptic conditions in isofluran narcosis, with only little distress to the animals. In general, the fetuses are evaluated around day 100 of gestation (length of pregnancy is: 142 $\pm$ 15 days). The wounds heal well in this species and hardly a scar is seen after a few weeks. We routinely keep these animals in the breeding colony after a Caesarian section and they give normal birth to many more offspring.

Examples of Studies with Medicinal Products

Of a number of medicinal products tested in our laboratory only some results obtained with a thalidomide derivative shall be briefly mentioned. Important findings on adverse effects induced with quinolones, important substances in the chemotherapy of bacterial infections, will be presented by Stahlmann et al. in this book.

Studies on the Teratogenicity of Thalidomide-type Substances

Thalidomide (Thd) is the human teratogen *par excellence*. It was marketed as a sedative (or was it the first tranquillizer?) in the middle of the 1950's, and was widely accepted by the public and prescribed and (ab)used in many countries. A pronounced increase in the frequency of hypoplastic and aplastic abnormalities of the limbs was noticed in Germany in 1960/61 (Wiedemann 1961), and the drug was withdrawn from the market in Germany starting at the end of November 1961, but not before 1962 in several other countries. Several thousand babies with severe defects were born within this period of intensive therapeutic use. The typical malformation epidemic abruptly stopped after the drug was no longer available to the public.

The teratogenic action of thalidomide is very special in many respects and it deviates from the developmental toxicity observed with other agents studied experimentally since. Such peculiarities include:

- the typical pattern of malformations can only be induced by Thd in primates, and e.g., rodents are virtually resistant to this toxic effect (despite other claims by some amateurs). Although certain strains of rabbits may exhibit some structural abnormalities (e.g. Lehmann and Niggeschulze 1971; Sterz et al. 1987), these occur only at high doses close to maternal toxicity, not at a high incidence, and the pattern is quite different from that observed in man.
- The teratogenic effect of Thd occurs in primates at doses which are far (orders of magnitude) from those which induce toxic signs in the dam.
- Typical severe defects induced are amelia or phocomelia, which are difficult to induce experimentally by other agents and which are very rare structural abnormalities in man.
- While it is typical for almost all other teratogenic agents tested experimentally so far to exhibit a steep dose-response relationship (e.g. Neubert et al. 1987), the teratogenic effect of thalidomide or its teratogenic derivatives such as EM12 is inducible (of course with a changing frequency) within a dose-range of several orders of magnitude, i.e. 0.1 to 10 mg EM12/kg body wt (e.g. Heger et al. 1988c).

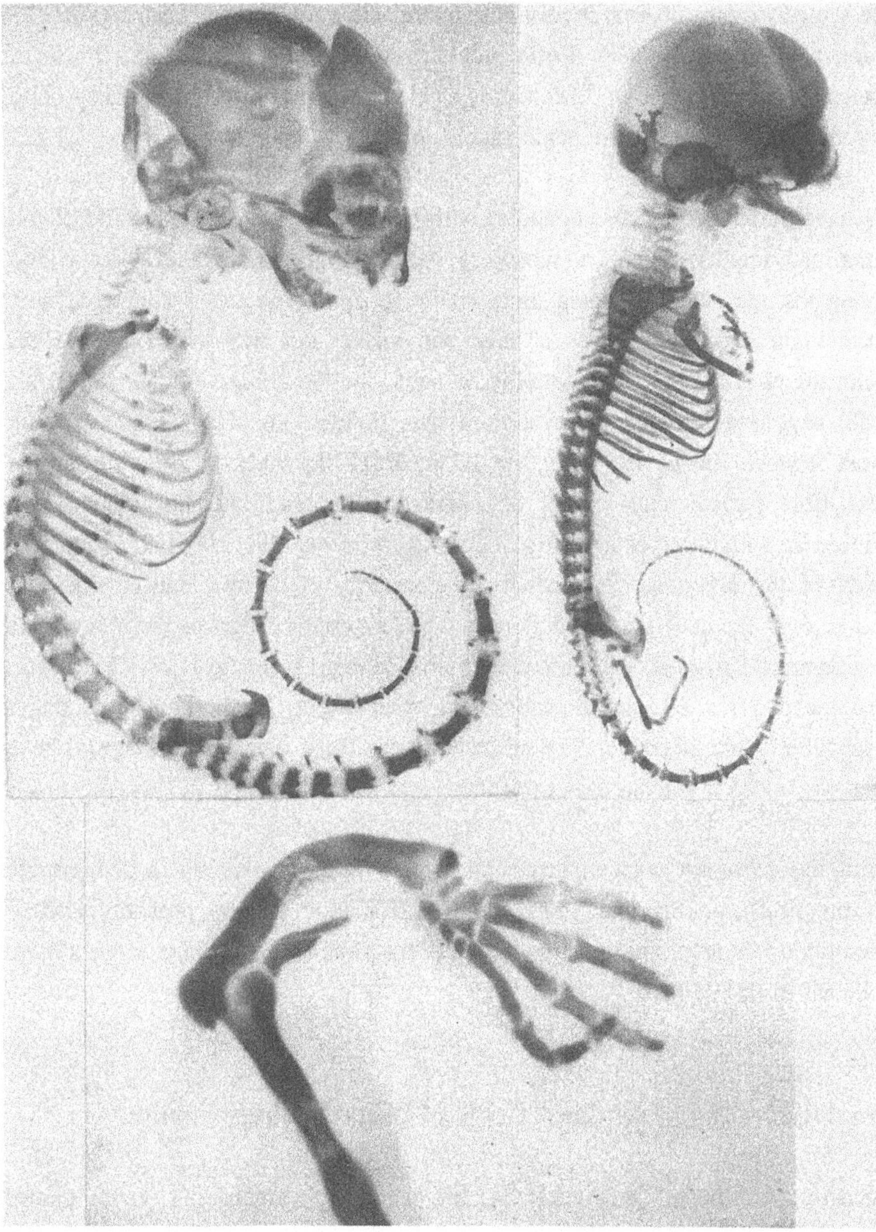

Fig. 11. Example of typical gross-structural abnormalities induced by the thalidomide derivative EM12 in marmosets. The severe skeletal defects are obvious. The substance was given orally (sugar solution) once daily during the days 50 to 57 post ovulation at a dose of 5 mg EM12/kg body wt. The typical picture shown can be induced with a 80-100% certainty under these experimental conditions. These doses do not cause any adverse signs in the pregnant animals. Caesarian section was performed on day 100 of pregnancy. The fetuses were cleared and the skeleton (bone anlagen) stained with alizarin red

The typical pattern of limb malformations inducible by Thd has been reported to occur in all species of Old World monkeys tested so far. Therefore, it was of interest to verify whether *Callithrix jacchus*, as a New World monkey, also responds with a typical teratogenic pattern to this substance.

We have performed extensive studies with the thalidomide derivative EM12 (N-phthalimidinoglutarimide), a substance with a high teratogenic potency and having the advantage of being more stable to hydrolysis when compared with thalidomide and giving rise to only one major and two minor metabolites (Schmahl et al. 1987). The patterns of malformations observed (Merker et al. 1988) exactly resembled those seen in man (Willert and Henkel 1969). With doses between 100 μg EM12/kg and 10 mg EM12/kg body wt given during the susceptible period (days 58-62 of pregnancy) amelies and phocomelies are induced at a high percentage (Fig. 11); with > 5 mg EM12/kg body wt in 80-100% of the offspring. Surprisingly, the severity of defects induced does not change over the dose-range mentioned; when exposure occurs at the appropriate developmental stage an "all-or-nothing" phenomenon seems to result. There was no indication for a clear-cut embryofeto-lethal effect. Furthermore, the stages susceptible to the action of this substance (very early limb bud formation) have been elucidated in the marmoset (Neubert et al. 1988a).

Using the marmoset as an experimental species we have been able to demonstrate for the first time that the S(-)-form of EM12 (and thereby probably also of thalidomide) is much more active than the R(+)-enantiomer (Heger et al. 1988c; Schmahl et al. 1988).

Possibilities for Assessing Effects of Biotechnology Products

Recently substances produced by biotechnology procedures have gained considerable medical interest. Especially for recombinant human growth factors a therapeutic potential may be expected in the future. Linked to the presumptive use of such substances is the question of the necessity of testing such peptides for possible adverse effects in experimental (preclinical) studies. Such substances pose special problems for a number of reasons:

- they may predominantly be used for "replacement" therapy, and toxic effects of overdosing may be irrelevant,
- toxic manifestations may mostly represent an exaggeration of the physiological effects,
- they may, as human peptides, exhibit a pronounced species specificity, and may exhibit a typical action only in primates (e.g. Table 5),
- they may give rise to (neutralizing) antibody formation, thus making long-term studies difficult or even impossible.

For these reasons it had been suggested that toxicological evaluations of such types of substances have to be routinely performed with non-human primates. However, most regulatory agencies now tend to prefer to decide the necessity for special preclinical toxicological testing of such substances on a "case-by-case" basis, and a generalization of the above mentioned recommendation for exclusively using primates is questionable for several reasons:

- not all human factors exhibit their biological action in primates only,
- antibody formation may also occur in non-human primates,
- many "classical" toxicological evaluations may not be felt to be necessary for some of these agents, especially if only special therapeutic indications exist.

As with other types of substances the rat may be appropriate and the species of first choice to test for embryofeto-toxicity (segment II) in a number of cases of biotechnology products. If the interference by (neutralizing) antibodies is expected, the treatment period may well be shortened and divided. However, such an approach would not be feasible in the case of a segment I or III study.

Nevertheless, in the field of risk assessment the use of non-human primates will be indispensable in the future for the elucidation of some crucial aspects. The marmoset will also prove to be a suitable species in the field of toxicological research, and first studies with a recombinant interleukin in adult marmosets are in progress in our laboratory (Klug et al., in preparation). Longer-term studies with high doses also proved to be difficult or impossible with this species in the case of the rhIL-6 studied, since antibody formation occurred after a few weeks. The fact that the marmoset is small and only small amounts of an expensive substance are needed makes this species especially attractive for the purpose of such testing.

It is noteworthy that, on the other hand, in this area of research much effort has been spent to use *in vitro* methods for evaluating biological effects of such substances and to monitor biological actions. This indicates that for each of the purposes of medical research the appropriate model, from *in vitro* techniques to possibly also studies using non-human primates, has to be selected.

With respect to testing for reproductive and developmental toxicity, in a number of cases such studies were not felt to be necessary. It should be noted that in the case of studies with Old World monkeys most often the number of animals used was too small (4 to 5 per group) to allow any valid conclusion with respect to a possible reproductive hazard. For simple statistical reasons even a 30% (!) malformation rate cannot be excluded when no adverse effect is seen in five pregnant animals with five fetuses (Table 6). It seems fair (also from the point of animal protection) to refrain from such "studies" which use only few individuals and have almost no predictive power.

If a testing for possible teratogenicity is felt to be indispensable, and a primate is the species of choice for good reasons, the marmoset may again be an acceptable alternative: it is quite feasible to perform studies with about 15 to 20 animals per group, which will provide about 35 to 45 fetuses for evaluation.

Examples for Studies with Environmental Substances

The discussion of the applicability and necessity of studies with primates for risk assessment of environmental substances will be confined here to one group of substances which has been studied extensively in our laboratory over the last years: polyhalogenated dibenzo-*p*-dioxins and dibenzofurans.

Assessment of Effects of "Dioxins"

A risk assessment of "dioxins", the popular trivial name for the polyhalogenated dibenzo-*p*-dioxins and dibenzofurans (PHDDs/PHDFs), presents a number of special aspects and difficulties which make the availability of data on non-human primates indispensable. Most of the information on toxicity has been revealed with the most toxic congener of these groups of substances: 2,3,7,8-tetrachlorodibenzo-*p*-dioxin (TCDD). Problematic aspects include:

- Large species differences exist with respect to kinetic variables, the elimination half-life of TCDD for rats being about 3 weeks, that for man apparently about 7 years. This must result in very different rates of accumulation when exposing, e.g., a rat or man to low doses over prolonged periods.

- Large species differences exist with respect to acute toxicity for TCDD.

- The relevance of some teratological data obtained with TCDD in mice (cleft palates), which cannot be reproduced in other experimental species, is obscure for the situation possibly existing in man.

- The relevance of some data obtained with TCDD on male fertility in rats is not clear for the situation possibly existing in man.

- Effects of TCDD on variables of the immune system can, with reasonable doses, apparently only be induced in non-human primates (cf. R. Neubert et al. this book).

With respect to a possible embryofeto-toxicity of TCDD we have performed a number of studies in which marmosets were treated continuously with TCDD. Several pregnancies were evaluated during this period of several years. There was no indication for teratogenicity in these studies, but only 22 fetuses were evaluated. However, the maternal tissue concentrations reached values of 6,000 ppt or more in fat. Some of the animals aborted when the body burden (evaluated in adipose tissue) exceeded 6,000 ppt TCDD (Krowke and Neubert 1988). For this reason it may be concluded that TCDD is certainly not a strong teratogenic substance in primates, but embryomortality may occur at high maternal body burdens; at these doses clear-cut signs of maternal toxicity (e.g. involution of the thymus, and haematological changes) are present.

Table 5. Cross-reactivity of some human growth factors

Human factor		Reactive with tissue of the mouse
IL-1		+
IL-2		+
IL-3		o
IL-4		o
IL-6		+
Tumour necrosis Factor	(TNF-α)	+
Interferon Gamma	(IFN-"τ")	o
Epidermal Growth F	(EGF)	+
Granuloc. Colony stim. F	(G-CSF)	+
Granuloc.-Macroph. Colony stim. F	(GM-CSF)	o

Table 6. Statistical considerations when using non-human primates for reproductive toxicity testing

In most studies on reproductive toxicity Old World monkeys are used, only < *10 animals* per dose-group are used.

If no adverse effect is observed in such a study, it cannot be excluded that such an effect may still occur in:

- *1 / 10* 30% of the cases.

For this reason such "studies" are largely worthless !

They are only capable of revealing extremely strong teratogens!

Additionally, the prenatal transfer and the extent of exposure via mother's milk was studied in the marmoset (Hagenmaier et al. 1990), and some of these data could later be verified by measurements in a human infant (Jödicke et al. 1991; 1992).

In recent years we have performed a variety of further studies with PHDDs/PHDFs using the marmoset as an experimental model. Some of the results have recently been summarized (Neubert 1991; Neubert et al. 1991a; Krüger et al. 1991b). Several of these investigations were part of a joint study program of the departments of pediatrics and toxicology at our university with the aim of developing methods in the marmoset which can later be used as the most sensitive measures to detect possible biological effects of PHDDs/PHDFs and similar substances in children. The goal was to reveal whether any type of effect was detectable in children after breast-feeding when compared to children raised on cow's milk formula, which is known to contain less than 1/10 of the PCDDs/PCDFs measured in breast milk.

No conclusion of a comparison on the basis of doses could be expected. We have always performed experimental studies on biological effects concomitant with comparative kinetic studies (e.g. Neubert et al. 1990a; Hagenmaier et al. 1990), and we have insisted on basing comparative considerations of biological effects on tissue concentrations (Neubert 1991; Neubert et al. 1991a). As most sensitive variables for a biological effect of TCDD the:

- induction of hepatic monooxygenases (cytochrome P450-IA1- and -IA2-dependent), and
- changes in the pattern of lymphocyte subtypes in venous blood

were selected and evaluated. A quantification of both effects could be achieved in the marmoset down to a dose-range at the lower ng/kg-body-wt-level. The induction of hepatic monooxygenases was measured *in vivo* with a non-invasive caffeine- or methacetin-breath-test (Krüger et al. 1990a,b; 1991), and the lowest effective dose was found to lead to TCDD concentrations in the liver of about 30 ppt (Abraham et al. 1989).

A similar susceptibility was found for the marmoset with respect to the induction of changes in the subpopulations of blood lymphocytes (Neubert et al. 1990b;

1991b; 1992). These changes will be discussed in more detail in another presentation of this book. Again, it is most important to mention that similar changes on lymphocytes were not noted in very extensive studies in rats (Korte et al. 1991), thus proving the marmoset to be a suitable and very susceptible species for this type of effect.

Table 7. Lymphocyte subpopulations in *Callithrix jacchus* and man

Lymphocyte subset	% of cells in this subset *Callithrix*	Man	
Lymphocytes	63	35	
Monocytes	3	7	
$CD2^+$	63	85	*all T cells* (E^+ receptor)
$CD2^+HLADR^+$	23	5	*activated T cells*
$CD2^+CD25^+$	6	5	*IL2-receptor*
$CD4^+$	34	48	*helper T*
$CD4^+CDw29^+$	17	39	*helper-inducer T*
$CD4^+CD45RA^+$	12	11	*suppressor-inducer T*
$CD4^+CD45R0^+$	4	8	*helper-memory*
$CD4^+CD56^+$	12	<1	*cytotoxic/killer effector cells*
$CD8^+$	29	32	*suppressor/cytotox T*
$CD8^+CD56^-$	14	20	*suppressor T*
$CD8^+CD56^+$	8	9	*cytotoxic T*
$CD56^+CD2^-$	< 0.1	5	*natural killer cells* (human $CD3^-CD56^+/CD16^+$ equiv.)
$CD20^+$	31	10	*mature and immature B cells*
$CD4^+/CD8^+$	*1.3*	2.3	*ratio: helper/suppressor*

Data of 103 control animals

In the context of this presentation only a comparison shall be given of the pattern of lymphocyte subsets in marmosets and man (Table 7, see above). It can be seen from the data compiled that (using the same selection of monoclonal antibodies) the percentage of some of the subpopulations is practically identical in the two species, while some differences also exist which must be recognized and taken into consideration. Overall, the marmoset has proven to be a suitable and convenient experimental model for studying lymphocyte subsets in venous blood.

In subsequent studies in children, using the above mentioned two biological effects as criteria, no effect of breast-feeding could be demonstrated, but a confounder, maternal smoking, did affect the variables mentioned (Helge et al., personal communication).

Summary and Conclusions

Non-human primates should certainly not be used for "routine" studies. There should always be a good argument why the use of such species cannot be avoided. The main reason will always be that effects to be evaluated cannot be assessed in other species (except in man). Such problems undoubtedly exist in medical research.

Much of the criticism put forth against experimentation with Old World monkeys does not apply for the use of small, New World monkeys. We feel that many studies, even in toxicology, can be performed with this species with minimum distress or harm to these animals.

References

Abraham K, Weberruß U, Wiesmüller T, Hagenmaier H, Krowke R, Neubert D (1989) Comparative studies on absorption and distribution in the liver and adipose tissue of PCDDs and PCDFs in rats and marmoset monkeys. Chemosphere 19: 887-892

Hagenmaier H, Wiesmüller T, Golor G, Krowke R, Helge H, Neubert D (1990) Transfer of various polyhalogenated dibenzo-*p*-dioxins and dibenzofurans (PCDDs and PCDFs) via placenta and through milk in a marmoset monkey. Arch Toxicol 64: 601-615

Heger W, Hoyer G-A, Neubert D (1988a) Identification of the main gestagen metabolite in marmoset (*Callithrix jacchus*) urine by NMR and MS spectroscopy. J Med Primatol 17: 19-29

Heger W, Klug S, Schmahl H-J, Nau H, Merker H-J, Neubert D (1988c) Embryotoxic effects of thalidomide derivatives on the non-human primate *Callithrix jacchus*. 3. Teratogenic potency of the EM12 enantiomers. Arch Toxicol 62: 205-208

Heger W, Merker H-J, Neubert D (1988b) Frequency of prenatal loss in marmoset monkeys (*Callithrix jacchus*). In: Neubert D, Merker H-J, Hendrickx AG, eds, Non-Human Primates - Developmental Biology and Toxicology, Ueberreuter Wissenschaftsverlag, Wien, Berlin, pp 129-140

Heger W, Neubert D (1983) Timing of ovulation and implantation in the common marmoset, *Callithrix jacchus*, by monitoring of estrogens and 6ß-hydroxypregnanolone in urine. Arch Toxicol 54: 41-52

Heger W, Neubert D (1987) Determination of ovulation and pregnancy in the marmoset (*Callithrix jacchus*) by monitoring of urinary hydroxypregnanolone excretion. J Med Primatol 16: 151-164

Hiddleston WA (1976) Large scale production of a small laboratory primate - *Callithrix jacchus*. In: Antikatzides, Erichsen, Spiegel, eds, The Laboratory Animal in the Study of Reproduction, Fischer Verlag, Stuttgart, pp 51-57

Hiddleston WA, Siddall RA (1978) The production and use of *Callithrix jacchus* for safety evaluation studies. In: Neubert D, Merker H-J, Nau H, Langman J, eds, Role of Pharmacokinetics in Prenatal and Perinatal Toxicology, Georg Thieme Publ, Stuttgart, pp 289-297

Jödicke B, Ende M, Helge H (1991) First data on the fecal excretion of polychlorinated dibenzo-p-dioxins and dibenzofurans in a 3-month-old breast-fed infant. Naunyn-Schmiedeberg's Arch Pharmacol 344 (Suppl): R96(252) [Abstract]

Jödicke B, Ende M, Helge H, Neubert D (1992) Fecal excretion of PCDDs/PCDFs in a 3-month-old breast-fed infant. Chemosphere: in press

Korte M, Neubert R, Stahlmann R (1991) No effect of 2,3,7,8-tetrachlorodibenzo-*p*-dioxin (TCDD) on leukocytes in peripheral blood of the Wistar rat. Naunyn-Schmiedeberg's Arch Pharmacol 344 (suppl 2): R115(48) [Abstract]

Krowke R, Neubert D (1988) Embryotoxic potency of 2,3,7,8-tetrachlorodibenzodioxin (TCDD) in marmosets. In: Neubert D, Merker H-J, Hendrickx AG, eds, Non-Human Primates - Developmental Biology and Toxicology, Ueberreuter Wissenschaftsverlag, Wien, Berlin, pp 499-510

Krüger N, Helge H, Neubert D (1991a) CO_2 breath test using ^{14}C-caffeine, ^{14}C-methacetin and ^{14}C-phenacetin for assessing postnatal development of monooxygenase activities in rats and marmosets. Dev Pharmacol Ther 16: 164-175

Krüger N, Helge H, Neubert D (1991b) Bedeutung von PCDDs/PCDFs ("Dioxinen") in der Pädiatrie. Monatsschr Kinderheilkd 139: 434-441

Krüger N, Neubert B, Helge H, Neubert D (1990a) Pre- and postnatal induction of monooxygenases by 2,3,7,8-TCDD in rats and monkeys measured with the ^{14}C-caffeine CO_2-breath-test. Pediatrics 149: 373 [Abstract]

Krüger N, Neubert B, Helge H, Neubert D (1990b) Induction of caffeine-demethylations by 2,3,7,8-TCDD in marmoset monkeys measured with a $^{14}CO_2$ breath-test. Chemosphere 20: 1173-1176

Lehmann H, Niggeschulze A (1971) The teratogenic effects of thalidomide in Himalayan rabbits. Toxicol Appl Pharmacol 18: 208-219

Merker H-J, Heger W, Sames K, Stürje H, Neubert D (1988) Embryotoxic effects of thalidomide derivatives on the non-human primate *Callithrix jacchus*. 1. Effects of 3-(1,3-dihydro-1-oxo-2H-isoindol-2-yl)-2,6-dioxopiperidine (EM12). Arch Toxicol 61: 165-179

Mocarelli P, Marocchi A, Brambilla P, Gerthoux P, Young DS, Mantel N (1986) Clinical laboratory manifestations of exposure to dioxin in children. A six year study of the effects of an environmental disaster near Seveso, Italy. JAMA 21: 2687-2695

Nau H, Neubert D (1978) Development of drug-metabolizing monooxygenase systems in various mammalian species including man. Its significance for transplacental toxicity. In: Neubert D, Merker H-J, Nau H, Langman J, eds, Role of Pharmacokinetics in Prenatal and Perinatal Toxicology, Georg Thieme Publ., Stuttgart, pp 13-44

Neubert D (1991) Peculiarities of the toxicity of polyhalogenated dibenzo-p-dioxins and dibenzofurans in animals and man. Chemosphere 23: 1869-1893

Neubert D, Abraham K, Golor G, Krowke R, Krüger N, Nagao T, Neubert R, Schulz-Schalge T, Stahlmann R, Wiesmüller T, Hagenmaier H (1991a) Comparison of the effects of PCDDs and PCDFs on different species taking kinetic variables into account. In: Banbury Report 35, Biological Basis for Risk Assessment of Dioxins and Related Compounds. Cold Spring Harbor Laboratory Press, Cold Spring Harbor, New York, pp 27-44

Neubert D, Bochert G, Platzek T, Chahoud I, Fischer B, Meister R (1987) Dose-response relationships in prenatal toxicity. Congen Anom 27: 275-302

Neubert D, Heger W, Merker H-J, Sames K, Meister R (1988a) Embryotoxic effects of thalidomide derivatives on the non-human primate *Callithrix jacchus*. 2. Elucidation of the susceptible period and of the variability of embryonic stages. Arch Toxicol 61: 180-191

Neubert D, Merker H-J, Hendrickx AG (eds) (1988b) Non-Human Primates - Developmental Biology and Toxicology. Ueberreuter Wissenschaftsverlag, Wien, Berlin

Neubert D, Siddall RA, Tapken S, Hiddleston WA, Higgins JE (1978) The production and use of *Callithrix jacchus* for safety evaluation studies. In: Neubert D, Merker H-J, Nau H, Langman J, eds, Role of Pharmacokinetics in Prenatal and Perinatal Toxicology, Georg Thieme Publ., Stuttgart, pp 299-309

Neubert D, Tapken S (1988) Comparative studies on the inducibility of benzo(a)pyrene hydroxylases in pregnant marmosets and rats. In: Neubert D, Merker H-J, Hendrickx AG, eds, Non-human Primates - Developmental Biology and Toxicology, Ueberreuter Wissenschaftsverlag, Wien, Berlin, pp 391-411

Neubert D, Wiesmüller T, Abraham K, Krowke R, Hagenmaier H (1990a) Persistence of various polychlorinated dibenzo-*p*-dioxins and dibenzofurans (PCDDs and PCDFs) in hepatic and adipose tissue of marmoset monkeys. Arch Toxicol 64: 431-442

Neubert R, Golor G, Stahlmann R, Helge H, Neubert D (1992) Polyhalogenated dibenzo-p-dioxins and dibenzofurans and the immune system. 4. Effects of multiple-dose treatment with 2,3,7,8-tetrachlorodibenzo-*p*-dioxin (TCDD) on peripheral lymphocyte subpopulations of a non-human primate (*Callithrix jacchus*). Arch Toxicol 66: 250-259

Neubert R, Helge H, Stahlmann R, Neubert D (1991b) Einige Wirkungen von 2,3,7,8-Tetrachlorodibenzo-*p*-dioxin (T4CDD) und von 2,3,4,7,8-Pentachlorodibenzofuran (P5CDF) auf periphere Lymphozyten von Primaten *in-vivo* und *in-vitro*. Allergologie 14: 360-371

Neubert R, Jacob-Müller U, Stahlmann R, Helge H, Neubert D (1990b) Polyhalogenated dibenzo-p-dioxins and dibenzofurans and the immune system. 1. Effects on peripheral lymphocyte subpopulations of a non-human primate (*Callithrix jacchus*) after treatment with 2,3,7,8-tetrachlorodibenzo-*p*-dioxin (TCDD). Arch Toxicol 64: 345-359

Poswillo DE, Hamilton WJ, Sopher D (1972) The marmoset as an animal model for teratological research. Nature (London) 239: 460-462

Schmahl H-J, Nau H, Neubert D (1988) The enantiomers of the teratogenic thalidomide analogue EM12: 1. Chiral inversion and plasma pharmacokinetics in the marmoset monkey. Arch Toxicol 62: 200-204

Schmahl H-J, Winckler K, Klinkmüller K, Heger W, Barrach H-J, Nau H (1987) Pharmacokinetics of the teratogenic and non-teratogenic thalidomide analogue EM12 and supidimide in the rat and marmoset monkey. In: Nau H, Scott WJ, eds, Pharmacokinetics in Teratogenesis, Vol I, Interspecies Differences and Placental Transfer. CRC Press, Boca Raton, Florida, pp 181-192

Schulz T, Neubert D (1988) Comparative studies on the activity of monooxygenases during the perinatal period in the marmoset and rat. In: Neubert D, Merker H-J, Hendrickx AG, eds, Non-human Primates - Developmental Biology and Toxicology, Ueberreuter Wissenschaftsverlag, Wien, Berlin, pp 353-371

Schulz-Schalge T, Heger W, Webb J, Kastner M, Neubert D (1991) Ontogeny of monooxygenase activities in the marmoset monkey (*Callithrix jacchus*). J Med Primatol 20: 325-333

Schulz-Schalge T, Krüger N, Golor G, Wiesmüller T, Hagenmaier H, Neubert D (1990) Induction of ethoxyresorufin O-deethylase by TCDD in liver microsomes of marmoset monkeys (*Callithrix jacchus*) and rats. In: Hutzinger O, Fiedler E (eds) Organo Halogen Compounds, Vol. 1: Dioxin `90 - EPRI-Seminar. Ecoinforma Press, Bayreuth, pp 145-149

Sterz H, Nothdurft H, Lexa P, Ockenfels H (1987) Thalidomide studies on the Himalayan rabbit: new aspects of thalidomide-induced teratogenesis. Arch Toxicol 60: 376-381

Wiedemann HR (1961) Hinweis auf eine derzeitige Häufung hypo- und aplastischer Fehlbildungen der Gliedmaßen. Med Welt 37: 1863-1866

Willert HG, Henkel HL (1969) Klinik und Pathologie der Dysmelie. Springer, Berlin, Heidelberg, New York

Planning and Performance of Segment I, II and III Experiments: Practical Aspects and General Comments

Helmut Sterz

Introduction

In investigations on reproduction toxicology - as is also the case with other toxicity studies - animals are used in place of man. However, if a hazard is to be detected and the risk for man assessed, then certain experimental conditions must be fulfilled.

The First Condition:

There must be the justifiable expectation that the species and strain used are suitable for detecting events that are relevant for humans.

With respect to the selection of a suitable species, this has frequently been carried out in the past with our eyes shut.

One reason for this attitude may be that the guidelines have been too narrowly interpreted and no-one had the courage to take the more time-consuming route and justify, if necessary, an unusual species selection to the health authorities. This is probably not infrequently due to the bitter experience that a particular species choice that has not been previously agreed with the authorities has led to loss of time because they required the usual species as well.

New Guidelines will hopefully improve this research-inhibiting situation since tailor-made studies are required and this must mean that a justified deviation from the usual pattern will be accepted by all authorities.

Up to now, the rat has been used almost exclusively in *segment I studies*. Apart from a few exceptions, we did not even bother to ask whether the rat is a suitable species.

The pro-rat arguments are primarily practicability, the amount of information already available from other toxicity studies including histological findings, historical control data, little seasonal effect on fertility, the fact that small quantities of the test drug are required, etc.

There are, however, several important factors that contradict the choice of the rat, for example sperm analyses intra vitam are virtually impossible, the testicular histology is only of limited informative value for the fertility of the animal, and its libido is difficult to determine. If this is coupled with lack of relevance of the rat from the point of view of kinetics and metabolism, then the hazard estimation may become a matter of chance, of course adequate assessment of reproductive toxicity requires multiple endpoints and approaches. Reliance only on fertility as evidence of reproductive tract damage in the male can be misleading.

It would be better to question the use of the rat as a routine species for segment I studies and to validate alternative species, such as the rabbit, with suitable assessment from the point of view of kinetics and metabolism, or the scientifically questionable treatment of the rabbit with compounds like high-ceiling diuretics, ACE inhibitors or antibiotics, since the exposure of this non-rodent species has to be restricted to very low concentrations of these compounds due to the important species specific sensitivity.

One reason for retaining the usual species in *segment II studies* may be the fact that even the theoretically possible limited use of a study drug in pregnant women would not necessarily permit a better risk estimation than a carefully justified and planned experiment in rats and rabbits - and the monkey is not always a better representative for humans than rodents or lagomorphs. The evolutionary proximity to humans does not necessarily tip the scales in favour of monkeys when it comes to the estimation of teratogenic risk.

In my opinion, the monkey must remain a justifiable exception in teratological screening and should be reserved for those cases in which it shows definite advantages. This would be, for example, considerably closer proximity to humans with respect to kinetics and metabolism.

In our laboratory, we have adopted (not only for the purpose of reproductive toxicology studies, but for the early preclinical development in general) the principle of investigating the *in vitro* metabolism in tissue of human, monkey, dog, rabbit, rat and mouse origin in order to be able to make a preselection of the species early on. Furthermore, in our preliminary experiments on dose selection, we carry out a minimal kinetic study on the species selected for the main experiment. I consider such preliminary investigations to be essential in order to select the most relevant representative for humans. This serves not only to protect humans, but also the animals. Furthermore, it spares us the loss of valuable development time because the licensing authorities demand a repetition of the experiments in other species.

The Second Condition:

The experimental design must be such that the initiation of an adverse reaction is possible.

I consider the justification of the doses used and the length of treatment to be essential elements of the design of reproduction toxicology experiments.

With regard to the rat, it is not necessary in every case to carry out separate preliminary experiments because some of the parallel running subchronic and chronic experiments can provide valuable indicators as will the knowledge about pharmacodynamic effects, mutagenicity, etc. Thus, for example, 80 days' treatment of the males in the segment I experiment is probably redundant if the results of a 13-week study are available. A minimum of 2 weeks pretreatment in males, and also in the females, before mating could be sufficient for the detection of a hazard. Treatment of the male should in this case continue until the corresponding mating partner has been evaluated.

The treatment of 50% of the females in segment I until the end of the lactation period is redundant to segment III and should be deleted. Overlapping treatment lasting a few days between fertility studies of Japanese design and teratogenicity experiments should be sufficient in the initial screening.

Furthermore, I would like to support the movement amongst colleagues to revise the artificial separation of teratogenicity experiments from segment III. It is just not justifiable to begin the treatment in segment III only at the end of the embryonic phase, if the essential target organ in this experiment is to be the CNS. In future, we should combine segments II and III in initial screening and treat from the start of the embryonic phase until the end of the lactation period. Half of the fetuses should be delivered by caesarian section and submitted to the classical examination for anomalies, whilst the other half should be reared and examined as usual in segment III.

If the rat is fundamentally suitable according to experience for this type of study, then I cannot see the reason why in such experiments the number of animals to be used should be greater than for any other suitable species, for example, rabbit or monkey. The reliability of the prediction does not double if we use groups of 24 to 30 rats instead of the usual 12 to 15 animals per group for rabbits.

It appears to me to be completely adequate if 12 litters per dose group are evaluable in the caesarian and rearing groups respectively. The weaned F_1 animals should be examined - unless otherwise intended for further mating - for organ anomalies after weaning. The skeletons of these pups can also be included in the examinations insofar as the caesarian-delivered fetuses raise suspicion. This procedure would have the advantage that fetal skeletal anomalies could be checked for their relevance in adolescent and adult animals without having to start a new experiment.

If one is satisfied with the above-mentioned number of animals per group, then measures such as litter reduction in the postnatal period are no longer justifiable, even for reasons of practicability.

The segment II experiments in rabbits should be carried out as usual. Dosage justification after prior investigations - including pregnant animals - should be a matter of course.

With respect to the question of the maximum doses in segment II, I should like to restrict myself to the following statement: If no relevant toxicity symptoms can be obtained in pregnant animals in the dose range finding experiment, then we should be allowed to accept doses for the main experiment that guarantee an adequate safety factor for humans. ADME-similarity of the species with man should of course be given. I would consider it a retrograde step if it is required in future to increase the dose to the maximum administrable range, although saturation of the pharmacokinetics already exists and completely irrelevant pharmacodynamic and/or toxic effects occur. In principle, this also applies to segments I and III.

The Third Condition:

The methods used for examining the fetuses must be suitable for the detection of anomalies.

You would think that 30 years after the thalidomide catastrophe there would be no doubt at all about the carrying out of fetal inspection. If, however, you read reports in which the tables in the appendix concerning organ examinations are completely free of any anomalies, then the questions are permissible whether the fetuses were looked at at all, whether the examiner had the necessary experience or whether the dissection technique used was inappropriate.

Each of the different proposed and established dissection techniques (Wilson 1965; Barrow and Taylor 1969; Staples 1974; Sterz 1977; Stuckhardt and Poppe 1984) is suitable if carefully carried out to detect minimal anomalies. Each has its particular advantages and disadvantages. None of them can be universely recommended or prescribed as the method to be used, because of the different technical and personnel situations in each laboratory.

The transversal cut technique, better known as "Wilson technique", permit, for example, the transverse sections to be stored as long as necessary after examination. The quality of the sections and their stability, however, is inversely proportional to the thickness of the sections. I do not consider systematic photography of the organ sections to be a satisfactory solution for this problem. One should better sacrifice storage and documentation to the actual aim of the investigation, that means destroy the organs as much as is necessary in dissection to produce a clear diagnosis immediately.

In the transversal cut technique, certain regions seem to be ignored in routine examination or their complexity makes them difficult to examine. If the investigator lacks experience and the ability to think three-dimensionally, then there is a major risk of simply overlooking certain anomalies.

Example 1: Head

There is no reasonable alternative to the "Wilson technique" for the head of the rat. One can, however, modify this technique so that it is more reliable in this region. For example, the pituitary is often poorly detected or even not at all. The question then arises in each case, is it missing or is it simply not exposed on the cut surface?

As there are substances that can initiate pituitary hypoplasia or even aplasia (Roux et al. 1979), we cannot be satisfied with such a doubtful finding. We have therefore introduced the median longitudinal section of the head of the rat fetus, which reliably exposes the pituitary. The usual transversal cuts of the two halves of the head can then be carried out without problems. The longitudinal section simultaneously permits easier assessment of the tubular-shaped structures that run longitudinally to the body axis such as cervical and spinal cord, larynx and trachea.

I would recommend that rabbits be dissected in the fresh state. Fixing of the heads in Bouin's solution with destruction of the skull is not recommendable in view of the small number of rabbit fetuses that are examinable.

Example 2: Neck

As there are few organs in the neck region, I have the impression that it is generally of little interest to the examiners. This is seen in the fact that one progresses with fewer, but therefore thicker segments to the more attractive thorax region. This is, however, not justified since fascinating things take place in this region (Binder 1985) during the early embryonic development. From here, the cephalic neurocrest cells - the origin of many tissues - spread ventrally and any disorder in migration can have severe consequences for the deriving organs which are later found relatively distant from this site, such as derivatives of the pharyngeal pouches like thymus and parathyroid, and the cardiovascular system.

In this respect, the thymus can undertake an important rôle as an indicator for anomalies of the cardiovascular system and warn the investigator to carry out more careful segmentation on his way to the heart. It does not matter whether one uses the "Wilson technique", in which one progresses from the cranial to the caudal direction, or whether an in situ technique is used, in which you progress from ventral to dorsal. You always come first to the thymus, which is for us a sort of "warning sign", since with a certain amount of experience you can already see from this organ whether anomalies are to be expected in the cardiovascular region.

In the case of our investigations in trisomic mice (Bacchus et al. 1987), we have found a very high probability of cardiovascular anomalies if we have previously recognized variations in the form and size of the thymus. Generally, there was a correlation in the severity of the findings in the two organs. These thymus anomalies actually serve to sensitize the technician for deeper lying anomalies.

Example 3: Thorax

In the cardiovascular region, we prefer an in situ technique. We have shown (Sterz and Lehmann 1985) in a comparison with another laboratory that a severe cardiovascular malformation (absence of the ductus arteriosus Botalli) was only found in the laboratory that used a modified Barrow & Taylor technique, while the other laboratory using the Wilson technique was able to find only a displacement of the abdominal aorta because there was simultaneously diaphragmatic hernia and the liver tissue that penetrated into the thorax displaced the aorta from its usual position to the left of the vertebral column.

If the ductus Botalli is not exposed in every fetus by the currently used transversal cut technique, then the risk of detecting an anomaly of the ductus Botalli by chance alone should lead to a switch to one of the microdissection techniques.

This example shows that in routine teratological examinations a validation of the techniques and of the technicians from time to time in an experiment with positive reference substances is advantageous.

The best reference for a valid technique, however, is the historical control data. Anyone who finds regularly in his control populations variations in the position of the carotid arteries and also of the right subclavian artery or, especially in rabbit fetuses, defects in the membranous part of the ventricular septum, does not need to worry about the validity of his techniques for organ examination, no matter which technique he uses to obtain the results.

Example 4: Eye

The examinations of the fetal eye are generally limited to an assessment of the size of the eyeball and of the lens. The latter has become opaque as a result of the use of a fixing medium. On dissecting the eye ball and the lens, the anatomical relationships are frequently destroyed by crushing of the vitreous body, and on the retina one sees fairly regularly only the folds that are known to be mostly artefacts of the fixing process. There still remains an examination of the eye in the segment III experiment shortly before weaning. This consists, however, almost exclusively of an assessment of the pupillary reflex. The majority of the F_1 animals are killed immediately following this examination.

An ophthalmologist would not accept such a type of "eye examination". He would suggest the carrying out with a slit lamp and an ophthalmoscope of an examination of at least a representative sample of the pups. As only some of the animals are killed immediately after weaning and the rest only after completion of behaviour tests or observations on sexual development, these examinations could be carried out in several steps. I agree that this would involve expansion of the experiment and extra effort. The question of the justification of this extra effort, however, will only be raised by someone who has not yet found eye anomalies in his segment III experiments. The reason for this may lie in the absence of methodical possibilities.

One could, perhaps, even things out by omitting some of the redundant and less meaningful observations of physical development during segment III studies.

Summary

The planning and carrying out of reproduction toxicology experiments has become routine. However, critical consideration of the methods used in order to find better ones is surely permissible. If we merely enlarge the size of the study as a result of a falsely understood desire for safety, for example, by gradually increasing the sizes of the groups, by retaining useless parameters and also attempting to satisfy all the health authorities in the world, then the quality of the studies will fall, because even a good team of technicians will not be able to carry out the studies optimally.

Let us take our courage in our hands and after almost 30 years of experimenting throw out the unnecessary ballast in the design of initial screening studies in order to invest the energy released meaningfully - for instance by proving adequate exposure of the targets by carefully justifying species and dose and by performing toxicokinetic investigations regularly.

References

Bacchus C, Sterz H, Buselmaier W, Sahai S, Winking H (1987) Genesis and systematization of cardiovascular anomalies and analysis of skeletal malformations in murine trisomy 16 and 19. Two animal models for human trisomies. Hum Genet 77: 12-22

Barrow MV, Taylor WJ (1969) A rapid method for detecting malformations in rat fetuses. J Morphol 127: 291-306

Binder M (1985) The teratogenic effects of a bis(dichloroacetyl)diamine on hamster embryos: aortic arch anomalies and the pathogenesis of the DiGeorge syndrome. Am J Pathol 118: 179-193

Roux C, Horvath C, Dupuis R (1979) Interpretation of the isolated agenesia of the hypophysis. Teratology 19(1): 39-41

Staples RE (1974) Detection of visceral alterations in mammalian fetuses. Teratology 9: A37-A38

Sterz H (1977) Routine examination of rat and rabbit fetuses for malformation of internal organs: Combination of Barrow's and Wilson's methods. In: D. Neubert, H.-J. Merker, T.E. Kwasigroch (eds), Methods in Prenatal Toxicology, G. Thieme Publishers, Stuttgart, pp 113-122

Sterz H, Lehmann H (1985) A critical comparison of the free-hand razor-blade dissection method according to Wilson with an in-situ sectioning method for rat fetuses. Teratogen Carcinogen Mutagen 5: 347-354

Stuckhard JL, Poppe SM (1984) Fresh visceral examination of rat and rabbit fetuses used in teratogenicity testing. Teratogen Carcinogen Mutagen 4: 181-188

Wilson JG (1965) Methods for administering agents and detect-ing malformations in experimental animals. In: JG Wilson, J Warkany (eds), Teratology: Principles and Techniques, University of Chicago Press, Chicago and London, pp 262-277

Discussion of the Presentation

Neubert: You mentioned "prescreening" metabolic studies with human and animal material. Are you referring to the use of microsomes or the use of isolated cells? Studies with microsomes may give rise to the production of artefacts, having little or no relevance to the *in vivo* situation. The use of isolated cells is often "more physiologic" and more relevant to the situation *in vivo*, but also has many limitations.

Sterz: We actually use hepatocytes.

All sophisticated behavioural tests which are available for postnatal studies have been developed by pharmacologists a long time ago. In fact, mostly for the mouse and with the aim to find desirable effects. I am wondering whether it is meaningful to adopt these tests for the young rat knowing that this model animal does not necessarily behave in the same way as the adult mouse.

Weissinger: If dose response does not indicate toxicity, you suggested that an adequate safety factor relating human to animal dose may be acceptable. Considering that the lack of toxic response may be due to a difference in kinetics, dynamics, or metabolic profile, how can a safety factor based on "dose" be an acceptable alternative?"

Sterz: In fact, I stated in my talk that ADME should be similar in between test species and man. If this is not given and doubts arise based on kinetics, metabolism and dynamics, then one should consider an alternate model or relevance of the model.

Pritchard: You cited a paper claiming microdissection was more effective than Wilson's for cardiovascular abnormalities. The subsequent paper by S. A. Tesh and E. J. Davidson clearly showed the adequacy of Wilson's for identifying cardiovascular abnormalities. In addition, I believe it's inaccurate to say Wilson's is not adequate for routine use. Whilst some laboratories do not do Wilson's adequately, we and other laboratories perform Wilson's adequately *and routinely*. Both techniques depend on the expertise of the operator but *more importantly* depend on the operator examining the foetus *conscientiously*.

Sterz: On the one hand I have said: "...each of the different proposed and established dissection techniques is suitable, if carefully carried out, to detect minimal anomalies..." On the other hand, my examples stress the danger of overlooking anomalies by the transversal sectioning technique. This danger existed in my laboratory many years ago, and probably still exists in others. The danger lies in the fact that quite frequently only a few minutes are accorded for cutting and inspecting the fetuses. If so, the microdissection technique is in my opinion safer than the Wilson technique, since it gives you a natural, three-dimensional impression of the organs *in situ*, which facilitates the inspection and also helps the operator to overcome the aspect of monotony or "routine" of this inspection. Last but not least, the expertise of the operator depends very much on the number of studies he has to inspect per year. If this is only two or three, resulting in longer periods without fetal inspections, I believe that the technician doing the microdissection regains his skill more rapidly since this technique is in essence much easier than the Wilson technique. Everybody should perform the dissection technique in which he has the greatest confidence.

Palmer: As Dr. Pritchard said, conscientiousness in examination is the key. The examiner should know what could be found and make sure that his technique is adequate. We use Wilson technique because we do lots of studies and can detect cardiac abnormalities. However, the Wilson technique must be practised regularly and if someone does not do lots of studies microdissection may be the better way. In short, the examiner should choose the technique best suited for himself and then make sure it can detect anomalies in all areas.

Significance of Postnatal Manifestations of Prenatally-Induced Effects (Behaviour)

Rolf Baß and Beate Ulbrich

Introduction

Guidelines for reproductive toxicity testing in several countries have requested or even required behavioural tests on the offspring for many years. This seems warranted by the assumption that serious problems during human development may arise not only from gross morphological abnormalities but even more so from subtle changes that are not as readily detectable at birth. Detecting the potential for developmental neurotoxicity of a substance in animal experiments, therefore, could prevent widespread exposure of pregnant women, and thereby minimize or eliminate the hazard to the growing and developing child.

Need for Testing

The main uncertainty we have to face with animal studies on offspring behaviour is whether the results will be in any way predictive for what would happen if the same substance acted on the human being. In this respect experience in medicinal products is lacking but data on environmental toxicants (methylmercury, lead) and substances of abuse (ethanol) provide a basis for the assumption that this will be the case. Also, information on what might happen in humans, appears to be very hard to come by for drugs. Epidemiological studies on behavioural parameters in humans are expensive and time consuming. In addition, for many new drugs a confirmation in humans of effects observed in the animal studies will never be achieved, because primary prevention and not verification of such effects in the human has to be the aim of regulatory agencies and their action taken.

According to the indications a new drug may have for use during pregnancy, different testing protocols may be required. For drugs that are contra-indicated during pregnancy but may be given to women of child-bearing potential, inadvertent exposure during *early* pregnancy has to be taken into consideration. Behavioural studies in animals for these compounds, therefore, have to be conducted on offspring exposed during or including the embryonic period. Studies with exposure only during the fetal period and during lactation cannot provide sufficient information for risk identification. For drugs which are likely to be given during pregnancy, exposure should include *embryonic* as well as *fetal* periods and should extend into the postnatal period to take into account the differences in pre- and postnatal development of the commonly used animal species (rat) and humans.

Available Methods

The methods used for behavioural testing vary widely between laboratories. They include everything from "subjective" observation by human beings to "objective" observation employing computer systems. A need for validation, standardisation, or even for the introduction of fixed batteries of test systems and tests has been proposed in order to overcome this situation. So far only few tests out of the available variety have been compared for reproducibility not only within but also between laboratories. The results have been encouraging. Another approach which does not necessarily compare specific endpoints but categories of functions in similar but not identical testing devices has also been tried successfully.

Although no consensus for the recommendation of specific tests has been reached, there seems to be general agreement among regulatory agencies on the functions that should be included in a behavioural testing battery, namely sensory systems, neuromotor development, locomotion/ activity, reactivity/ habituation, learning/ memory, and social/ reproductive behaviour. A list of tests commonly used in behavioural testing of medicinal products is given in Table 1. To integrate behavioural data into the context of developmental toxicity, data on physical development of the offspring have to be available which commonly include data on postnatal weight gain, viability, and physical landmark development. This must not, however, lead to the false impression entertained by some investigators that observations on physical landmark development constitute behavioural testing.

Table 1. Tests that are currently used in behavioural testing for developmental toxicity of medicinal products

Tests for behavioural development	
Righting (surface, air); Cliff avoidance; Grasping; Negative geotaxis; Olfactory discrimination; Homing; Swimming; Visual placing; Rotarod	
Tests for adult behaviour	
Learning/Memory:	Water maze; Avoidance (active, passive)
Locomotion/Activity:	Open field; Hole board; Home cage activity
Social interactions:	Mating; Maternal behaviour

Experiences of an Agency

Experiences in the Federal Health Administration (BGA) with reproduction toxicity studies from applications for marketing authorisation covering the years 1983 to 1991 show that some behavioural tests have been conducted in approximately half of the studies where postnatal parameters (viability, weight gain) were evaluated in the pups (Table 2). The reasons why behavioural parameters were not included in the remainder vary over the years. This group contains a rather large proportion of "old" studies conducted before guidelines began requesting behavioural tests, and newer studies where it was not considered necessary to repeat the testing already performed in the corresponding general reproduction or peri-postnatal studies.

Table 2. Experiences in the Federal Health Administration (BGA). Reproduction studies from applications for marketing authorisation covering the years 1983 to 1991

What is our own experience? I (Dec. 1991)			
Behavioural effects (number of studies)		Effects	
	Not studied +	No	Yes
Segment I *			
Rat	128	110	35
Segment II #			
Rat	67	56	14
Mouse	7	2	2
Segment III			
Rat	109	95	47
Mouse	4	3	1

+ Could have been studied but was not! (reasons unspecified, mostly "old" studies)

* US/European style

"Japanese style"

Behavioural changes have been observed in roughly 25% of the studies where such evaluations were included. The necessity to conduct postnatal behavioural testing also when exposure is limited to the embryonic period is underlined by the 14 out of 70 embryotoxicity studies following the Japanese protocol that showed behavioural changes in the offspring.

Behavioural parameters have been postulated to be extremely sensitive to developmental toxicants. Before widespread testing became "popular" it was presumed that behaviour of the offspring could be affected at lower dosages than those necessary to produce changes in the more traditional parameters of developmental toxicity. Experience so far shows that this cannot be generalized. In some studies, however, behavioural effects in the offspring were noted even in the absence of maternally toxic effects and in the absence of any other adverse effect on postnatal growth and development (Table 3). Relying on detection of non-behavioural effects in these cases would have given the false impression that no developmental toxicity was apparent under the conditions of the study.

Table 3. Effects of drugs on rat offspring behaviour, related to other signs of developmental toxicity and maternal toxicity. Data from applications for marketing authorisation in the years 1983 - 1991

Rat	Segment I	Segment III
No. of studies with behavioural tests	145	142
Behaviour affected	35	47
- Offspring weight, development affected	27	35
- Offspring weight, development normal	8	12
- Maternal toxicity	24	33
- Maternal toxicity but offspring weight, development normal	3	5
- No maternal toxicity and offspring weight, development normal	5	7

As positive findings occurred in most drug classes, albeit at a low frequency, behavioural testing cannot be restricted only to substances of a suspected group. Signs of neurotoxicity in adult animals, however, should motivate investigators to increase the amount of behavioural testing done on a prenatally exposed F1-generation. Behavioural changes were found mainly for three classes of drugs: antibiotics/antimycotics, cardiovascular drugs, and psychoactive drugs (Table 4).

Table 4. Positive behavioural effects in studies with different drug classes

	Segment I studies with Behaviour affected/ Behaviour tested	Segment III studies with Behaviour affected/ Behaviour tested	Substances studied in Segment I and in Segment III	Substance positive in Segment I and in Segment III
Drug class:				
Analgesic drugs	3/6	3/9	4	1
Antiallergic drugs	3/10	2/8	5	0
Antibiotics	9/23	14/34	12	4
Cardiovascular drugs	14/39	11/29	14	2
Anticoagulants	1/4	0/3	2	0
Antiepileptic drugs	0/3	2/2	2	0
Psychoactive drugs	1/13	5/15	7	0
Muscle relaxants	1/2	3/4	2	1
Hyperlipemic drugs	0/3	1/5	1	0
Gastrointestinal drugs	2/10	1/9	5	1
Hormones/Inhibitors	0/3	1/1	0	0
Cytostatic drugs	0/2	1/3	0	0

The meaning of positive (and negative) results for the situation expected to exist in humans has to remain open. We do not know from our current experience with drugs whether the test systems used or behavioural testing in animals as a whole are actually meaningful for the protection of human offspring. However, from the regulators point of view it would be irresponsible to gain certainty on this question by conducting "experiments" in pregnant women for active substances that have shown behavioural effects in animal studies. Experimental approaches in animals to relate behavioural changes in offspring to transient or permanent brain damage may help in the interpretation of effects and in assessing their relevance for humans.

Recommendations

At the present stage of knowledge the following recommendations can be given:

Limited behavioural testing of the prenatally exposed offspring should be part of developmental toxicity testing. The methods used should be standardised within the laboratory and allow assessment of behavioural development as well as of adult Behaviour. There are no objections against using either computerized tests, or human observation, or a combination of both. Collaborative efforts to increase the insight into the usefulness of specific tests are encouraged.

Discussion of the Presentation

Neubert: From which period were your data? Has the percentage of reports with data on behaviour increased recently?

Baß: The reports were evaluated between 1980 and 1991, i.e. they were generated beginning in the early seventies and reaching into the late eighties. During the period evaluated, the percentage of reports including behavioural data increased.

Neubert: What consequences do you see with respect to the studies on "behaviour", if (as it is presently required) the highest dose in a combined segment II and -III test produces signs of maternal toxicity?

Baß: Behaviour changes are expected under such circumstances. If maternal sedation is associated with behavioural effects, this has not been taken too seriously in the past, i.e. it had no consequences for labelling, since maternally toxic doses have to be avoided in humans.

Palmer: People emphasize behaviour tests because of perception of risk. The same as they over-emphasize malformations (teratogenicity) in developmental toxicity. Dr. Baß' number of positives look high but this will not be resolved until detailed assessment is made. Mostly analysis of behaviour/development tests is poor. Values often show marked kurtosis so variance analysis is useless. Categorical analysis should be used. Also for landmarks of early events it is necessary to express values in postcoital time, not post-partum time.

Scallet: A rather low point in the history of psychology took place 20 to 25 years ago when the RNA molecule was proposed as the medium of memory storage and "RNA transfer" experts thought to confer memories from one animal to another were debunked. However, many experiments of this time showed interference of antibiotics with performance of learning and memory tasks. Even if the "RNA synthesis interference" explanations of the effects proposed at that time are wrong, the empirical observations remain. Might this not help explain the pattern of developmental effects of antibiotics you reported?

Baß: Not restricting ourselves to RNA, yes.

Frohberg: The degree of variation in behavioural studies is especially large. For this reason, it would be of interest to know how many compounds submitted in connection with behavioural studies include the corresponding laboratory control values.

Baß: Usually no historical data are submitted. Sometimes "ranges" are given.

Peters: At this moment I prefer an accurate multigeneration test. We cannot use, evaluate or assess data from behavioural studies, the same holds for *in vitro* results. As I indicated before, I want to have a motivation from the experimenter why such tests are done.

Baß: Toxicity studies on "unknown" substances are mostly done for prospective safety reasons. If the apical multigeneration test is performed, it seems wise to include behavioural endpoints.

Significance of Prenatally or Early Postnatally-Induced Organ Dysfunctions - Other than CNS Defects

Ralf Stahlmann, Ibrahim Chahoud, Hans-Joachim Merker and Diether Neubert

Introduction

Within the framework of registration of medicinal products extensive studies for reproductive toxicity are required to identify possible hazards. However, the routinely conducted studies on fertility, embryo/fetotoxicity - including teratogenicity - and postnatal manifestations of prenatally induced lesions are not always sufficient to detect the toxic potential of chemical substances on reproductive and developmental functions. Furthermore, the induction of adverse effects is not confined to the prenatal period, and the newborn and infant may be more vulnerable to a given chemical than the adult organism. Consequently, such hazards are not recognized and therefore potential risks cannot be fully assessed. In this paper we will discuss three examples to elucidate that, e.g., "functional anomalies" are not detected with the administratively required and routinely performed tests.

This review compiles data from studies with three antimicrobial agents which are widely used to treat bacterial or viral infections. All of them have been proven to induce functional defects in different organ systems (other than the CNS) if the developing organism is exposed to them prenatally or early postnatally. A more detailed description of the methods used and a discussion of the results is published in the original papers as cited throughout the text.

I. Alterations of Kidney Function in Rats after Prenatal Exposure to Gentamicin

Gentamicin is widely used as a parenterally applied antibiotic to treat bacterial infections. The use of this antibiotic during pregnancy is usually considered contraindicated since it has been known for a long time that aminoglycosides can be ototoxic to the fetus if they are applied prenatally. Evidence for such effects exist mainly for streptomycin (Conway and Birt 1965), but caution is indicated for the whole group of closely related drugs. Despite the "contraindication" in some publications treatment with gentamicin for pregnant women is considered justified in the case of serious infections (Wise 1987).

Beside being ototoxic aminoglycosides are well-known nephrotoxic agents. However, little is known on the possibility of prenatally induced kidney changes with these drugs. With gentamicin we induced functional and ultrastructural kidney alterations in offspring of rats treated with daily s.c.-injections of rather high doses of 100 or 110 mg gentamicin/kg body wt on days 10 through 15 or 15 through 20 of gestation (Stahlmann and Chahoud 1985). An example of such symptoms of functional alterations is given in Figure 1. The figure shows the serum urea concentrations in 60-day-old rat offspring exposed to gentamicin *in utero* (100 mg/kg body wt from day 10 through day 15 of pregnancy) in comparison to controls of the same age.

A detailed electron microscopic evaluation was performed on the kidney tissue of these animals. The alterations were detectable in the tubular as well as the glomerular system. All changes were more pronounced in female than in male offspring. Sex-related differences were also recognizable in a further functional test: one year after exposure to 110 mg/kg body wt on days 10 to 15 or days 15 to 20 of gestation a statistically significant *elevation of the arterial blood pressure* (Figs. 2 and 3) could only be seen in the female offspring (Chahoud et al. 1988b).

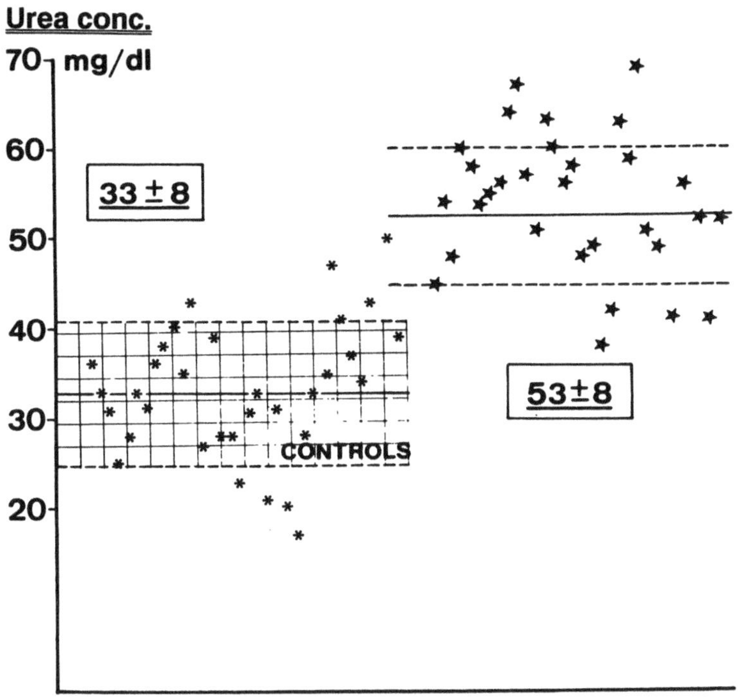

Fig. 1. Concentrations of urea in plasma of 60-day-old rat offspring (individual values and mean ± SD). Pregnant rats were treated from day 10 through day 15 of gestation with daily s.c.-injections of 100 mg gentamicin/kg body wt. Control rats remained untreated. Two months after birth the urea concentration was measured in plasma of the offspring

During the 6-day treatment period we collected multiple blood samples from a tail vein of each pregnant animal and measured the drug levels as well as the urea concentrations in plasma. The data varied considerably. For example, 6 hours after the first injection of gentamicin on day 15 of gestation ("group II") the concentrations ranged from 0.9 to 6.8 mg gentamicin/l plasma (mean ± SD: 3.5 ± 1.8 mg/l), and under corresponding conditions on the 6th day of treatment the concentrations ranged from 17 to > 80 mg gentamicin/l plasma (Stahlmann et al. 1988a).

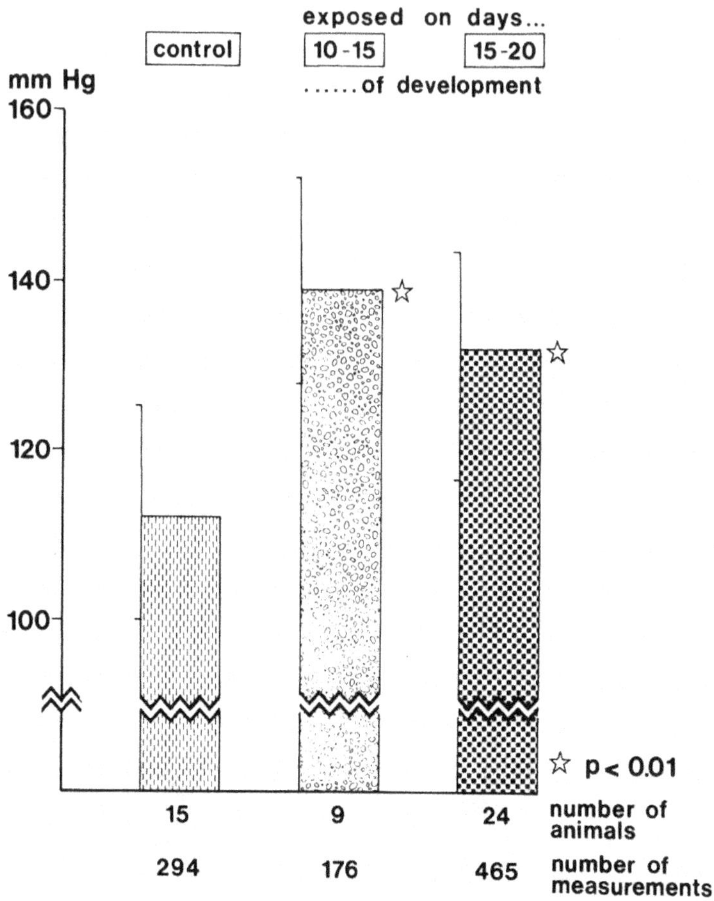

Fig. 2. Median values (Q_1/Q_3) of blood pressure measurements in *female offspring* of rats treated from day 10 through 15 or from day 15 through 20 of gestation with daily s.c.-injections of 110 mg gentamicin/kg body wt. The systolic arterial pressure was measured several times in each animal. The differences between the treated animals and the controls are statistically significant (t-test; $p < 0.05$)

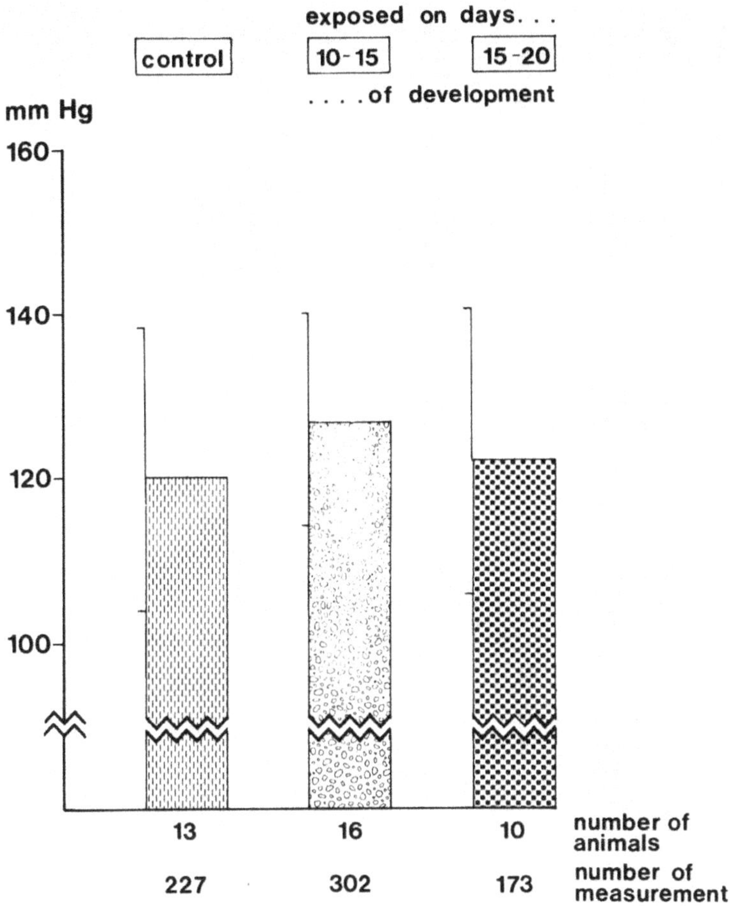

Fig. 3. Median values (Q_1/Q_3) of blood pressure measurements in *male offspring* of rats treated from day 10 through 15 or from day 15 through 20 of gestation with daily s.c.-injections of 110 mg gentamicin/kg body wt. The systolic arterial pressure was measured several times in each animal

Table 1. Correlation between gentamicin and urea plasma concentrations in pregnant rats and functional anomalies in one-year-old offspring. Rats were treated from day 15 through day 20 of gestation with daily s.c.-injections of 110 mg gentamicin/kg body wt; maternal plasma concentrations of gentamicin and urea were determined in blood samples taken from a tail vein 6 hours after the last injection; systolic arterial pressure (SAP) and plasma urea concentrations in the offspring were determined at the age of 8 to 12 months. (Data are presented as mean ± SD)

	Pregnant rats		Male offspring			Female offspring		
	Genta (mg/l)	Urea (mg/dl)	n=	SAP (mm Hg)	Urea (mg/dl)	n=	SAP (mm Hg)	Urea (mg/dl)
Control	--	33 ± 9	13	118 ± 21	33 ± 3	15	112 ± 9	32 ± 6
Gentamicin	38 ± 14	64 ± 36	10	122 ± 14	37 ± 5	24	132** ± 17	39* ± 10

* $p < 0.05$
** $p < 0.01$ (t-test)

During this study we did not investigate a possible dose-response relationship. But we could classify the outcome of the postnatal study according to the gentamicin plasma concentrations measured during the treatment period:

(i) None of the offspring from the dams with the highest gentamicin concentrations (> 48 mg gentamicin/l) could be evaluated due to resorptions and early postnatal mortality,

(ii) in the remaining offspring the gentamicin concentrations showed a good correlation with the postnatal results, e.g., an increase in blood urea or an increase in systolic arterial pressure which was determined one year postnatally (Chahoud et al. 1988b).

Obviously, the degree of renal *impairment depends on the sex of the animal.* The overall increase in arterial blood pressure was only statistically significant in females [132 ± 17 mm Hg (treated) vs. 112 ± 9 mm Hg (controls)]. A similar result was demonstrable for the urea concentrations: when the results from all exposed animals are compared with the corresponding controls, the difference is significant for the females only, although in five male rats the urea concentrations were significantly higher than in the controls (41 ± 3 vs. 33 ± 3 mg/dl; $p < 0.01$; t-test).

For the adult organism sex-related differences in the toxic reaction of kidney tissue to aminoglycosides have also been described for rats and man (Kourilsky et al. 1982).

Surprisingly, in our experiments we found no clear-cut differences with regard to the treatment period. This might be explained on the one hand by the typical way in which the kidney develops pre- and postnatally (new nephrons are formed over a period of several weeks) and, on the other hand, by the pharmacokinetic characteristics of the drug which is only slowly released from the fetal compartment. On day 21 of gestation we examined the kinetic behaviour of gentamicin in the fetus (Fig. 4). At several time intervals after a single s.c.-injection of the drug the animals were sacrificed and the concentrations were determined in maternal and fetal blood plasma. We found a rapid transfer of gentamicin to the fetal blood: already 45 minutes after treatment we measured a mean concentration of 5.9 ± 1.9 mg/l (maternal blood: 171 ± 36 mg/l). However, eleven hours after injection we measured mean concentrations (± SD) of 4.4 ± 0.6 mg/l in the fetal and 1.2 ± 0.6 mg/l in the maternal blood indicating much slower elimination of gentamicin from the fetal compartment than from the maternal organism.

In summary, we showed that persistent functional and ultrastructural alterations could be induced in the kidney of rats by prenatal exposure to gentamicin. No gross-structural defects were observed under these experimental conditions. Further studies should elucidate whether similar changes can also be induced with lower doses, which might be expected due to the extraordinary characteristics of the gentamicin transfer into the fetal compartment.

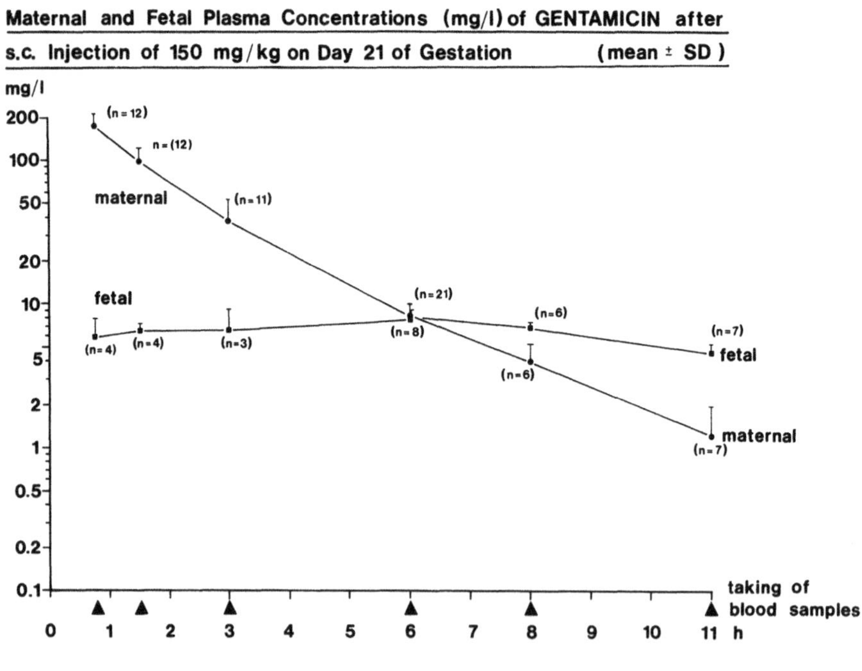

Fig. 4. Plasma concentrations of gentamicin in pregnant rats and fetuses after a single s.c.-injection of 150 mg gentamicin/kg body wt on day 21 of gestation. Blood was taken from a tail vein 45, 90, 180 and 360 min after injection in 12 rats. At corresponding times and, in addition, 8 and 11 hours after injection, groups of rats were sacrificed

II. Alterations of the Immune System in Rats after Prenatal Exposure to Aciclovir[1]

A number of chemicals have been reported to alter functions of the immune system in laboratory animals and man. Very little is known, however, on the possibility to induce immunological deficiencies by *prenatal* exposure to xenobiotics. Aciclovir is an example of a drug which induces morphological thymus alterations when given to rats prenatally (Neubert et al. 1989).

This drug is widely used for the therapy of herpes virus infections. First indications of the teratogenic potential of this drug were found with *in vitro* methods ("whole-embryo culture") and some evidence exists that its prenatal toxicity is more pronounced than that of several related drugs (Klug et al. 1985; 1991). The effect was not seen in routinely performed segment II studies since aciclovir treatment with doses > 25 mg/kg body wt causes cristalluria and reversible nephropathia making the interpretation of results of studies after administration of multiple high doses over a *longer time period* difficult. However, after the application of high doses for a *short time period* (e.g. 100 mg/kg body wt only on day 10 of gestation), the teratogenic effect was demonstrable in rats under *in vivo* conditions (Chahoud et al. 1988a; Stahlmann et al. 1988b).

Gross-structural abnormalities of the skeletal system and multiple internal organs were induced by aciclovir. Beside other defects *thymus alterations* were seen after three injections of 100 mg aciclovir/kg body wt on day 10 of gestation; a striking finding was the weight reduction of this organ. In 21-day-old fetuses the mean thymus weight was only 3.7 ± 2.4 mg (n = 56), corresponding to a reduction of almost 50% compared to controls (6.9 ± 2.3 mg; n = 140), and in 16 out of 56 fetuses (= 29%) this organ was macroscopically not detectable at all. These alterations persisted postnatally: e.g. in 15-week-old female rat offspring we found a mean thymus weight of 239 ± 34 mg after prenatal

1 Beside the above listed authors the following colleagues contributed to the experiments described in this part: *Maria Korte, Renate Thiel, Michael Förster, Henk van Loveren**, *Joseph G. Vos** *[*National Institute of Public Health and Environmental Protection, Bilthoven, Netherlands]*

exposure to 1 x 100 mg aciclovir/kg body wt on day 10 of gestation. This means a 25% weight reduction in comparison to controls (307 ± 54 mg). In Table 2 another example of organ weight alterations from a different experimental series is given: the relative weights of spleen and thymus from 12-week-old male rats after prenatal exposure to one or three injections of aciclovir on day 10 of gestation are significantly increased or decreased, respectively. A detailed description of the morphological findings of the thymus in these experiments has been given elsewhere (Stahlmann et al. 1990b).

Table 2. Relative organ weights of 12-week-old male rats after prenatal exposure to aciclovir (% of body weight; mean values ± SD)

		Spleen	Thymus
I.	Control (n=10)	0.224 ± 0.014	0.179 ± 0.037
II.	Aciclovir (1 x 100 mg/kg body wt) (n=12)	0.230 ± 0.017	0.158 ± 0.073
III.	Aciclovir (3 x 100 mg/kg body wt) (n = 11)	0.253* ± 0.039	0.141* ± 0.016

* $p < 0.05$ (t-test)

To check a specific *function of the immune system* of these rats a host-resistance experiment was conducted (for a more detailed description cf. Stahlmann et al. 1992). Six-week-old male rats, exposed prenatally to aciclovir (two groups; one or three injections of 100 mg/kg body wt on day 10 of gestation), were orally infected with the parasite *Trichinella spiralis*. *Before* and several times *after* this infection blood samples were taken from a tail vein and the immunoglobulin IgA, IgE, IgM and IgG titers in the plasma of these animals were measured. In addition, the number of *Trichinella* larvae were counted in muscle preparations at the end of the experiment six weeks after infection. We found a significant increase in the number of muscle larvae in the group exposed prenatally to 3 x 100 mg aciclovir/kg body wt to 2.3 ± 0.7 muscle larvae/mm^2 in comparison to the control group (1.5 ± 0.8 muscle larvae/mm^2) (p = 0.016; t-test).

There were several further indications of altered functions of the immune system in the prenatally aciclovir-exposed rats in comparison to the controls. For example, the mean IgE titer *six weeks* after the infection was reduced from 4.8 ± 1.9 (control) to 2.9 ± 0.9 in the 3 x 100 mg/kg body wt group (p = 0.016; t-test). A similar situation was found for the IgG titers: whereas no significant difference was observed three weeks after infection, we found a significant decrease six weeks after infection (Fig. 5).

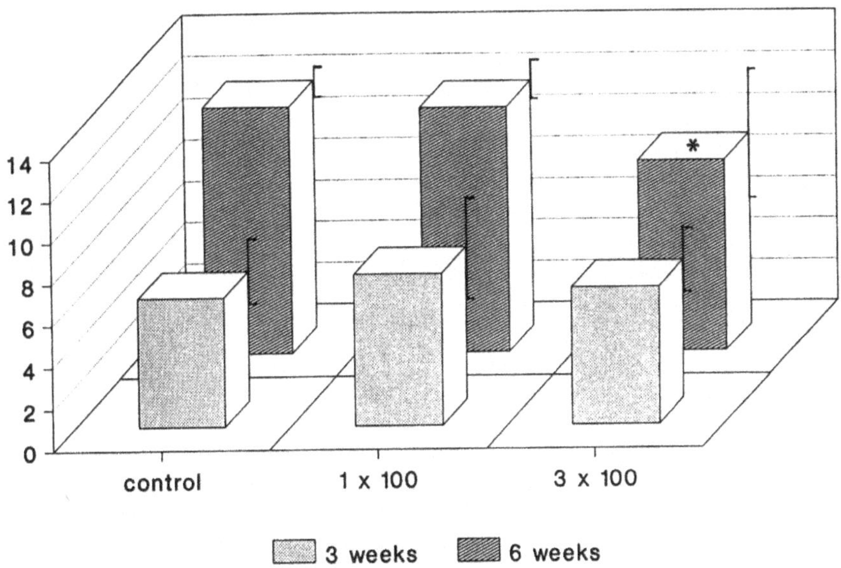

Fig. 5. *Trichinella spiralis* antibody titers (**Immunoglobulin G**) in three groups of rats (n= 10 to 12) three and six weeks after infection with *T. spiralis* muscle larvae. Randomly selected male offspring from dams treated on day 10 of gestation with 1 or 3 injections of 100 mg aciclovir/kg body wt (1 x 100 and 3 x 100) in comparison to untreated controls. Blood samples were taken from a tail vein. [* = $p < 0.05$; t-test]

Figure 6 shows that for the IgM titers a different situation exists: for this immunoglobulin significant differences were found *two weeks* after infection: the mean values (± SD) of the titers were 1.3 ± 1.0 (control), 2.7 ± 1.8 (1 x 100 mg aciclovir/kg body wt) and 5.4 ± 2.8 (3 x 100 mg aciclovir/kg body wt).

Human data on the possible influence of aciclovir on prenatal development are scarce. Since the drug is only poorly absorbed after oral administration (approx. 20%), concentrations achieved under these conditions are probably too low to represent a specific hazard. The situation might be different if (a) the drug is given intravenously, (b) higher doses are recommended and/or (3) derivatives with a better bioavailability are developed.

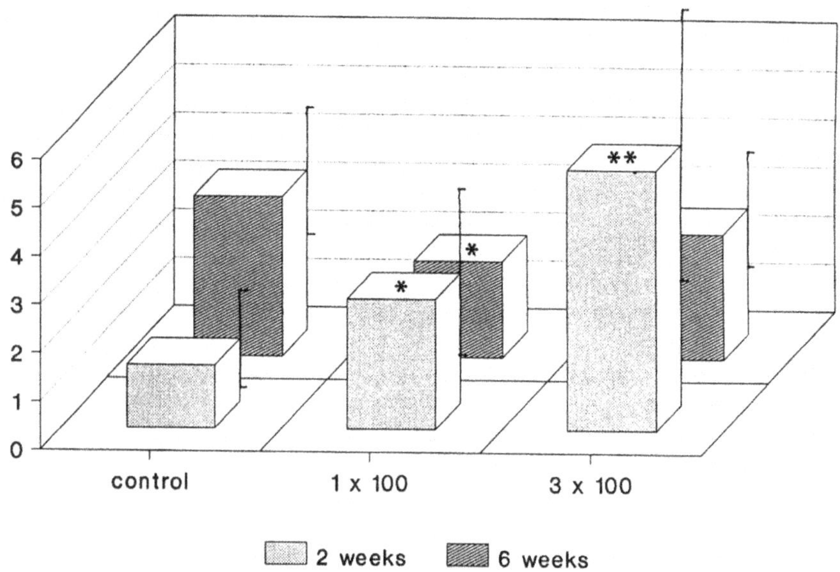

Fig. 6. *Trichinella spiralis* antibody titers **(Immunoglobulin M)** in three groups of rats (n= 10 to 12) two and six weeks after infection with *T. spiralis* muscle larvae. Randomly selected male offspring from dams treated on day 10 of gestation with 1 or 3 injections of 100 mg aciclovir/kg body wt (1 x 100 and 3 x 100) in comparison to untreated controls. Blood samples were taken from a tail vein [* = $p < 0.05$; t-test]

In summary, our findings indicate that weight alterations of thymus and spleen are - beside other malformations - typical and persistent findings in rats after prenatal exposure to aciclovir. The function of the immune system of these animals is also impaired as the results of the challenge experiment with *T. spiralis* show.

III. Arthropathia in Juvenile Animals after Treatment with Ofloxacin[2]

Arthropathia induced by quinolones in *juvenile* animals is an extraordinary toxic effect which is not known in this form from other drugs or chemicals. In 1977 - i.e. more than 10 years after these drugs had been introduced for antibacterial therapy - *gait anomalies* were detected in juvenile dogs after treatment with nalidixic acid (Ingham et al. 1977). At autopsy the authors noted severe damage of the cartilage in weight bearing joints correlating to this *unusual functional defect*. Up till now, nothing is known on the mechanism of this toxic effect.

All quinolones studied so far induce this damage in several animal species, therefore these findings have led to an important restriction in the use of these antibacterial agents: they are contraindicated for children and adolescents during the growing phase and for women during pregnancy and lactation. On the other hand, it is obvious from retrospective studies that the risk for man at therapeutic doses is rather small. Ciprofloxacin has been used in altogether more than 1000 children - these included mainly patients with cystic fibrosis or patients with severe infections due to multiresistent bacteria. In approximately 1 to 2% of the patients reversible arthropathia was diagnosed, however, the causal relationship remains unclear in most cases (Chysky et al. 1991). A study from France indicates that differences might exist between the arthropathogenic potential of different quinolones: arthralgia was diagnosed in 14% of young patients with cystic fibrosis treated with pefloxacin but not in patients treated with ofloxacin (Pertuiset et al. 1989).

---

2 Beside the above listed authors the following colleagues contributed to the experiments described in this part: *Christian Förster, Norbert Hinz, Jessie Webb, Gerd Bochert*

Since there are little experimental data published on this phenomenon which would allow a better risk assessment we performed some experiments in juvenile rats and marmosets to investigate the quinolone-induced chondrotoxicity more closely. We used ofloxacin since among all quinolones available so far, ofloxacin possesses a relatively long elimination half-life. It is only metabolized to a small extent and therefore best qualified for a study with oral administration.

Electron microscopically we found alterations in the cartilage of knee joints in 4- to 6-week-old rats after a 5-day treatment period with 2 x 600 mg/kg body wt daily. The alterations affected the cellular components as well as the intercellular matrix. Typical findings were a destroyed cartilage surface, matrix-free areas around the chondrocytes and swelling of mitochondria and endoplasmic reticulum of the chondrocytes. Very similar effects were detected in juvenile marmosets (9- to 18-week-old) after a 5-day treatment period with 2 x 200 mg ofloxacin/kg body wt per day. Additionally, we used immunomorphological methods in this non-human primate to detect the ofloxacin-induced alterations in the matrix composition. In the area of vesicle formation (detachment of the upper cartilage layer) a fibronectin enrichment was recognizable. On the other hand, there was a loss of proteoglycanes in the matrix (Stahlmann et al. 1988c; Stahlmann et al. 1990c).

In contrast to the findings in dogs described in the literature we never observed gait alterations or other signs of functional defects in the two species studied. A comparison of the data from animal studies with the human situation is given in Table 3. Up till now, the risk for arthropathia in children and adolescents due to quinolone therapy remains unclear. Some case reports on this topic are difficult to interpret with respect to the causal relationship (Jüngst and Mohr 1987). The sensitive periods differ over a wide range and are not clearly defined in man. There is also a wide range of daily dosages applied for therapy or in toxicological studies (factor 200 in rats). However, plasma concentrations do not differ over such a wide range (less than factor 10 in marmosets).

Since it is still unclear whether the primary target of this toxic effect is the chondrocyte or the cartilage matrix, we tried to identify those matrix components to which the quinolones bind and to detect differences between the members of this group of antibacterial agents. A 24-hour-incubation of Processus xiphoides from sterna of rats of different ages in solutions of two fluoroquinolones showed that ciprofloxacin is bound to a greater extent to this tissue than ofloxacin under identical conditions (Fig. 7). The result of this simple test does of course not mean that ciprofloxacin is more toxic than ofloxacin, but it represents a first step to compare the degree of the cartilage binding of different quinolones without complication by pharmacokinetics and metabolism.

Further studies are essential to study the chondrotoxicity of quinolones and to provide a better basis for a rational risk assessment. These considerations will become more and more important since the use of these drugs is continuously increasing and children and adolescents will be inadvertently treated with quinolones or will have to be treated for a vital indication.

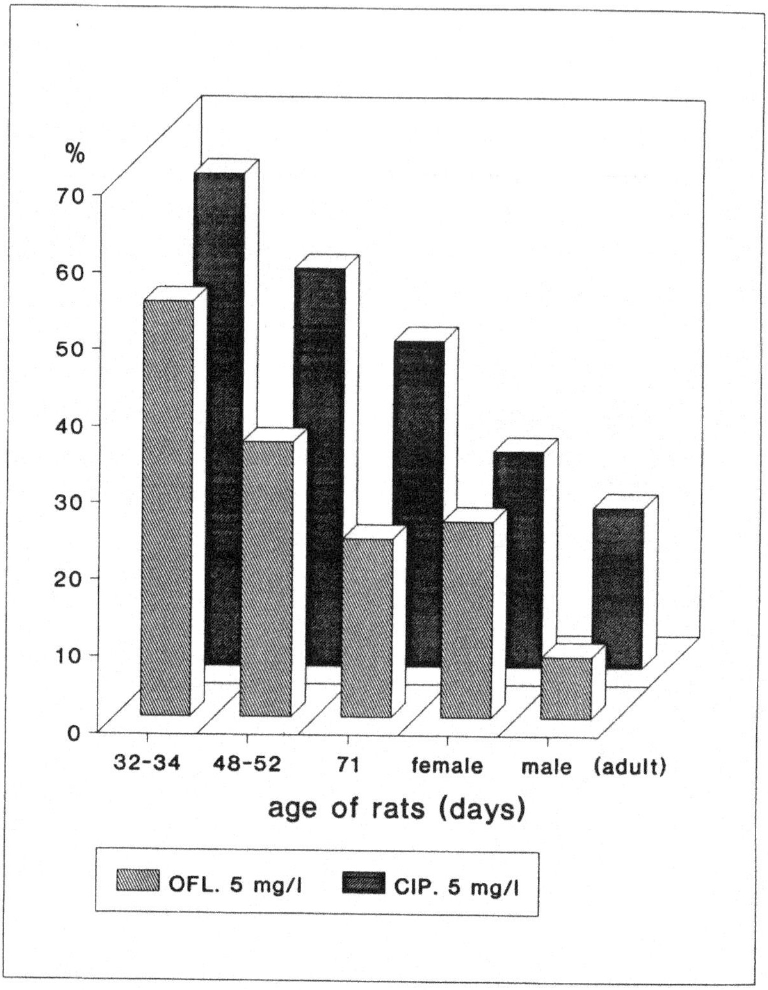

Fig. 7. Binding of ciprofloxacin and ofloxacin to cartilage (proc. xiphoideus) from rats of different ages. Cartilage tissue was incubated for 24 hours with one of the quinolones and the concentration was determined with a spectrofluorometer before and after incubation. [These data are part of the doctoral thesis of Christian Förster to be submitted to the Fachbereich Pharmazie, Freie Universität Berlin]

Table 3. Chondrotoxicity of ofloxacin in different species

Species	Sensitive period (months pn)	Daily dose^x (mg/kg body wt)	Peak plasma concentration (mg/l)	Functional defect	Morphology
Man[1)]	? ? ?	2 x 6	3.5 ± 0.7	? ? ?	? ? ?
Dog[2)]	3 - 6	20	13	lameness, gait alterations	blisters, erosions
Rat[3)]	1 - 2	2 x 600	34 ± 10*	not observed	altered cartilage surface, loss of matrix
Marmoset[3)]	2 - 4 (?)	2 x 200	43 ± 17**	not observed	blisters, loss of proteoglycanes

x = The therapeutic dose of ofloxacin is 2 x 100 to 2 x 400 mg daily (= 2 x 1.4 mg/kg body wt to 5.7 mg/kg body wt in patients with 70 kg body wt; the dose used in dogs is the lowest dose to induce arthropathia; the LOELs were not determined for the marmosets and rats
* = 45 min after a single dose given orally to juvenile rats (n = 12)
** = One hour after last dose (5th day of treatment; n = 6)

References: 1) Jüngst & Mohr 1987 / Lode et al. 1987; 2) Okazaki et al. 1984 / Mayer 1987; 3) Stahlmann et al. 1990c

Summary and Conclusion

Three examples of frequently used antimicrobial drugs which induce *functional anomalies in animals* when given during pre- or postnatal development have been reviewed:

(1) *Gentamicin:* permanent alteration of renal function and of blood pressure was induced by treatment when given during the *second half of gestation* in rats.

(2) *Aciclovir:* abnormal development of thymus and spleen in rats and also alteration of functions of the immune system was induced *during organogenesis.*

(3) *Quinolones:* chondrotoxicity and gait alterations (in dogs) - representing an extraordinary functional defect - was induced during the *early postnatal period* in several animal species.

It is quite feasible that a closer analysis will also reveal effects on other organ systems with other substances. In order to allow a risk assessment for the alterations mentioned with respect to the situation possibly existing in man under therapeutical conditions, we have combined all these experiments with pharmacokinetic studies.

In no case have these findings been clearly shown to represent a hazard to man under the present therapeutic conditions, but the arthropathy induced by quinolones in animals is the background for a clear-cut restriction of the use of these drugs.

In all cases where functional defects were observed, morphological alterations were also found (at least at an ultrastructural level), but on the other hand, all morphological as well as functional anomalies were not detected by routinely applied methods used for todays safety examination of drugs. It can be concluded that although quite extensive studies are performed with newly developed drugs to identify all possible hazards to reproduction, many possibilities still might exist for the occurrence of unexpected toxic effects which cannot be included in the overall "risk assessment" since they remain unrecognized.

References

Chahoud I, Stahlmann R, Bochert G, Neubert D (1988a) Gross-structural defects in rats after aciclovir application on day 10 of gestation. Arch Toxicol 62: 8-14

Chahoud I, Stahlmann R, Merker H-J, Neubert D (1988b) Hypertension and nephrotoxic lesions in rats one year after prenatal exposure to gentamicin. Arch Toxicol 62: 274-284

Chysky V, Kapila K, Hullmann R, Arcieri G, Schacht P, Echols R (1991) Safety of ciprofloxacin in children: world-wide clinical experience based on compassionate use. Emphasis on joint evaluation. Infection 19: 289-296

Convay N, Birt B D (1965) Streptomycin in pregnancy: Effect on the foetal ear. Brit Med J 2:260-263

Ingham B, Brentnall DW, Dale EA, McFadzean JA (1977) Arthropathy induced by antibacterial fused N-alkyl-4-pyridone-3-carboxylic acids. Toxicol Lett 1: 21-26

Jüngst G and Mohr R (1987) Side effects of ofloxacin in clinical trials and in postmarketing surveillance. Drugs 34: 144-149

Klug S, Lewandowski C, Blankenburg G, Merker H-J, Neubert D (1985) Effect of acyclovir on mammalian embryonic development in culture. Arch Toxicol 58: 89-96

Klug S, Lewandowski C, Merker H-J, Stahlmann R, Wildi L, Neubert D (1991) *In vitro* and *in vivo* studies on the prenatal toxicity of five virustatic nucleoside analogues in comparison to aciclovir. Arch Toxicol 65: 283-291

Kourilsky O, Solez K, Morel-Maroger L, Whelton A, Duhoux P, Sraer J-D (1982) The pathology of acute renal failure due to interstitial nephritis in man with comments on the role of interstitial inflammation and sex in gentamicin nephrotoxicity. Medicine 61:258-268

Lode H, Höffken G, Olschewski P, Sievers B, Kirch A, Borner K, Koeppe P (1987) Pharmacokinetics of ofloxacin after parenteral and oral administration. Antimicrob Ag Chemother 31: 1338-1342

Mayer DG (1987) Overview of toxicological studies (Ofloxacin). Drugs 34: 177-178

Neubert D, Neubert R, Stahlmann R, Helge H (1989) Immuno-toxicology and -pharmacology. Brazilian J Med Biol Res 22: 1457-1473

Okazaki O, Kurata T, Hashimoto K, Sudo K, Tsumura M, Tachizawa H (1984) Metabolic disposition of DL-8280. The second report: absorption, distribution and excretion of 14C-DL-8280 in various animal species. Chemotherapy 32: 1185-1202

Pertuiset E, Lenoir G, Jehanne M, Douchain F, Guillot M, Menkes CJ (1989) Tolérance articulaire de la péfloxacine et de l'ofloxacine chez les enfants et adolescents atteints de mucoviscidose. Revue du Rhumatisme 56: 735-740

Stahlmann R, Chahoud I (1985) Prenatally induced kidney impairment by treatment with gentamicin and pharmacokinetics of the antibiotic in pregnant rats. Naunyn-Schmied Arch Pharmacol 329: R32 (Abstract 125)

Stahlmann R, Chahoud I, Meister R, Düerkop C, Neubert D (1988a) Gentamicin plasma concentrations in pregnant and non-pregnant rats and fetuses after single and multiple injections. Arch Toxicol 62: 232-235

Stahlmann R, Klug S, Lewandowski C, Bochert G, Chahoud I, Rahm U, Merker H-J, Neubert D (1988b) Prenatal toxicity of acyclovir in rats. Arch Toxicol 61: 468-479

Stahlmann R, Blankenburg G, Chahoud I, Webb J, Merker H-J, Hinz N, Neubert, D (1988c) Effects of quinolones on joint cartilage in juvenile rats and marmosets. In: Neubert D, Merker H-J, Hendrickx A (eds), Non-human Primates - Developmental Biology and Toxicology, Überreuter Wissenschaftsverlag, Berlin, pp 547-565

Stahlmann R (1990a) Safety profile of the quinolones. J Antimicrob Chemother 26 (Suppl D): 31-44

Stahlmann R, Chahoud I, Korte M, Thiel R, van Loveren H, Vos JG, Förster M, Merker H-J, Neubert D (1990b) Structural anomalies of thymus and other organs and impaired resistance to *Trichinella spiralis* infection in rats after prenatal exposure to aciclovir. In: Kendall MD and Ritter MA, eds. Thymus Update, Vol 4, Harwood, Chur, London, Paris, New York, Melbourne, pp 129-155

Stahlmann R, Merker H-J, Hinz N, Chahoud I, Webb J, Heger W, Neubert D (1990c) Ofloxacin in juvenile non-human primates and rats. Arthropathia and drug plasma concentrations. Arch Toxicol 64: 193-204

Stahlmann R, Korte M, van Loveren H, Vos JG, Thiel R, Neubert D (1992) Abnormal thymus development and impaired function of the immune system in rats after prenatal exposure to aciclovir. Arch Toxicol (in press)

Wise, R (1987) Prescribing in pregnancy. Antibiotics. Brit Med J 294:42-46

Discussion of the Presentation

Frohberg: At the age of 1 year female rats, prenatally exposed to gentamicin, showed still elevated BUN-plasma levels and an increase of systolic blood pressure. Male rats, prenatally treated the same way showed normal values. What was the result of the histological examination of the kidneys? What might be the reason for the observed sex difference?

Stahlmann: Kidneys of these offspring were investigated by light and electron microscopy. We observed hyaline casts in the tubules and glomerular lesions which were characterized by a varying degree of sclerosis. Electron microscopically a very early indication of an alteration was the merging of the podocyte processes on the basement membrane. The more severely affected glomeruli were characterized by a considerable increase in matrix and cells in the mesangia. In the final stage of sclerosis, the original glomerulus consisted of a continuous matrix mass into which fibrocyte-like or mesangial cells were incorporated. The surface of these matrix masses was covered by the former podocytes, which had, however, lost their morphological characteristics and showed an extremely wide lumen. All histological findings were found in the female kidneys only and showed a considerable degree of variability from animal to animal, which might be a reflection of the considerable individual differences in plasma concentrations measured during the treatment. A detailed description of the morphological alterations was made in the publication by Chahoud et al. (*Arch. Toxicol. 62:274-284, 1988*).

With respect to the second part of your question I would like to indicate that sex-dependent differences in the reaction of the kidney tissue to toxic concentrations of aminoglycosides were first described approximately ten years ago for the *adult* organism. This was shown for laboratory animals as well as in patients treated with these drugs. The kidney of adult male rats reacts in the form of severe acute tubular necrosis, whereas the kidneys of female rats - beside tubular necrosis - also show signs of an interstitial nephritis. To my knowledge the exact reason for this different reaction is not known and - of course - it is not *proven* that the sex-dependency of the reaction of the adult as well as that of the fetal kidney can be explained by the same mechanism.

Kavlock: What is the consequence of your data? We cannot evaluate everything, especially not all possible postnatal manifestations of prenatally-induced lesions.

Neubert: To me this seems to be one of the crucial questions in all of our attempts to evaluate a possible risk in the area of reproductive and developmental toxicity. For a considerable period of time, gross-structural abnormalities were the predominant outcome to be considered in risk assessment. Subsequently, postnatal manifestations were discussed, but only "behavioural" abnormalities were considered for testing, and this was (in our opinion) prematurely introduced in many guidelines. The problem seems to be whether possible effects on the hormonal status, and especially on the immune system are not of much more importance. We will have to develop techniques to also monitor for such possible endpoints, especially in the case of substances to which many people are exposed, possibly also at rather high doses. Such experimental data will be important for a risk assessment, since it is unlikely that appropriate data in man will become available in these areas in the near future.

Stahlmann: I agree with this statement. Although "we cannot evaluate everything" - as Dr. Kavlock said - we should first provide experimental data to find out which functional abnormalities can principally occur in mammals and try to find a feasible way to screen for those which seem to be the most important ones.

If we consider the two large human disasters in this area, before the thalidomide tragedy it was known that xenobiotics could induce malformations, but nobody asked for systematic testing. Long before the DES-induced carcinoma in young girls, transplacental carcinogenesis was well established in experimental animals, but apparently few people foresaw that this outcome could also be relevant for man. We do not have to wait for the next human tragedy to occur, if the technology for primary prevention is available.

Quantitative Morphological Tests for Evaluation of Testicular Toxicity

Gabriele M. Rune, Jutta Hartmann and Philippe De Souza

Introduction

Numerous recent publications attest to the substantial increase in public and scientific concern about the potential of industrial and environmental chemicals to interfere with male reproductive functions. While morphological investigations cannot reveal much about the mechanism of action of a compound, they can provide very useful information for direct subsequent studies specifically on affected cells or regions. This is certainly true for the testis. A good general tool for detecting testicular toxicity is undoubtedly a good histopathological evaluation. On the other hand, for defining the toxicological potency of a compound, quantification of toxic effects is necessary, which morphological studies do not automatically include, but which is able to provide valuable information for risk assessment of male reproductive toxicity.

It has been demonstrated in our laboratory and by other authors that 2,3,7,8-tetrachlorodibenzo-*p*-dioxin (TCDD), which is generated as an undesired industrial by-product, is able to impair spermatogenesis and male fertility (Norback and Allen 1973; Kociba et al. 1976; Chahoud et al. 1989) and in particular to induce endocrine changes (Moore et al. 1985; Mebus et al. 1987; Astroff and Safe 1988; Umbreit et al. 1988; Rune et al. 1991a,b). In addition to our qualitative investigation of TCDD-influenced testicular morphology (Rune et al. 1991a,b), in the present investigation we tried to quantify these morphological results with the following tests: 1) As a first and rather simple step we weighed the testes of TCDD-treated animals, since it has been shown that if there is severe toxicity for a period of time sufficient to induce germ cell loss, the weight of the testis drops significantly until no more germ cells are present (Chapin 1988); 2) We determined the presence of certain tubular stages in the testes, since it could recently be demonstrated by Hess et al. (1990) that the different ratios of tubular stages can indicate at which stage spermatogenesis is inhibited; 3) We counted spermatids and sperm in testes, cauda epididymidis and ductus deferens 4) and determined the ratio of morphologically deformed sperm in the ductus deferens.

Material and Methods

Animal maintenance. Testes of sexually mature marmosets (*Callithrix jacchus*) and rats (Wistar) were used. The marmosets were bred in our colony at the Institute of Toxicology. The animals were kept at a constant day/night cycle (light from 6 a.m. to 6 p.m.) at 22 ± 1°C, and marmosets, additionally at 55 ± 5% relative humidity. They were fed a standardized Altromin^R rat or marmoset diet and water ad libitum (for marmoset breeding see also Heger and Neubert 1983).

Treatment. TCDD was given subcutaneously (into the back) to assure absorption and to minimize contamination of the cages and exposure of the personnel. For injection the substance was dissolved in DMSO and diluted with castor oil to obtain a mixture of 1 (DMSO) plus 3 (castor oil). From this solution a volume of 1 ml/kg body weight was applied using a Hamilton microsyringe.

^{14}C-TCDD was used for all studies in order to check possible contamination of the surroundings. The substance was purchased from Cambridge Isotope Lab. (Mass., USA). It had an indicated activity of 33 mCi (1.22 + 10^9 Bq)/mmol. It was dissolved in DMSO and kept in the dark. ^{14}C-TCDD was diluted with cold TCDD to a specific activity of 3.9 mCi (1.44 + 10^8 Bq)/mmol (100 dpm = 3.73 ng TCDD).

Monkeys and rats were treated with TCDD at doses of either 300 ng/kg body wt, 1000 ng/kg, 3 μg/kg and 10 μg/kg. Two untreated animals and two, injected with the vehicle, were used as controls.

All animals were observed for changes in appearance at least twice a day and sacrificed after seven days by an overdose of Evipan[R]. Testes were weighed. One organ was cut into small pieces and specimens were immediately fixed according to Karnovsky (1967) for 24 h. For the morphological examination sperm was collected by perfusion of the corresponding ductus deferens. For counting of sperm and spermatids the second testis and cauda epididymidis were homogenized (see below).

Morphology. The fixed specimens were then thoroughly washed in cacodylate buffer (0.1 M, pH 7.4) and postfixed for 1 h in cacodylate-buffered (0.1 M) 1% OsO_4 at pH 7.4. After dehydration in acetone they were embedded in Epon 218. Sections were cut on an ultra microtome (Reichert-Jung, OmU 3, Vienna, Austria).

Contrast of ultrathin sections was enhanced with uranyl acetate followed by lead citrate (Reynolds 1963). The sections were then examined on Zeiss electron microscopes (EM 109, EM 10, Oberkochen, FRG). "Thick" sections (1 μm) were mounted on slides and stained with methylene blue/azure II according to Richardson et al. (1960).

Morphometry. "Thick" sections (1 μm) were used for the determination of tubular stages. The stage was determined according to the developmental steps of spermiogenesis after Leblond and Clermont (1952) for the rat and after Holt and Moore (1984) for the marmoset. In each testis 200 cross-sectioned tubules were estimated, as recommended by Hess et al. (1990), and the ratio of tubular stages was given as percent.

Counting of spermatids and sperm. Testis - after removal of the tunica albuginea - and cauda epididymidis were homogenized in 0.9% NaCl containing 0.05% Triton X-100. Spermatids in the testis and sperm in the cauda epididymidis were counted using a hemocytometer.

Results

Organ weight. None of the animals died during the week after administration, not even at a dose of 10 μg/kg body wt TCDD. With rats the average body weight (320 - 350 g) of the controls, vehicle-treated, and experimental groups increased up to 106% of the initial value with no significant difference between the three groups. The average body weight of marmosets ranged between 350 - 380 g. Differences were not found between the three groups after one week of treatment. No weight loss of the testes under TCDD influence could be found, neither with rats nor with marmosets. The absolute weights (rats: 1.7 g; marmosets: 1.8 g) were not lower in treated animals compared with controls or vehicle-injected animals (Table 1).

Table 1. Testis weight after TCDD treatment. Absolute weights (mean ± standard deviation) of testes given in gram after different doses of TCDD

Dose (μg/kg)	Rat	Marmoset
0	1.73 ± 0.2 (n=4)	1.85 ± 0.3(n=4)
0.5	1.68 ± 0.4 (n=4)	1.83 (n=2)
1	1.72 ± 0.2 (n=4)	1.90 (n=2)
3	1.78 ± 0.4 (n=4)	1.87 (n=2)
5	1.62 ± 0.3 (n=4)	1.85 (n=2)
10	1.70 ± 0.3 (n=4)	1.84 (n=2)

Presence of tubular stages. Determination of the tubular stages in semi-thin sections, as judged by the presence of developmental steps of spermiogenesis (for the rat according to Clermont and Leblond (1952) and for the marmoset according to Holt and Moore (1984)) and quantitative evaluation of the presence of "tubular stages" according to Hess et al. (1990) showed no differences between the untreated and treated rats but clear-cut differences between treated and untreated monkeys. By calculating the percentage of tubular stages within a group of 200 tubules (except for a small proportion that could no longer be categorized due to the degree of affection by TCDD, particularly at high doses), it became evident that stages I to III and VII to IX were underrepresented and stages V and VI appeared more frequently in the treated monkeys. This trend, already seen at a dose of 1000 ng TCDD/kg body wt, became more obvious at 3 μg TCDD/kg body wt and was highly distinct at 10 μg TCDD/kg body wt. A comparable change in the presence of tubular stages was not found with rats. Although differences were found, these differences were not significant at any stage or dose (Fig. 1 A,B).

Morphological lesions of sperm. Anomalies of the spermatozoa collected from the ductus deferentes were found in all groups with rats as well as with marmosets. Anomalies of the head included double heads, invaginated acrosomes, double acrosomes, malformed acrosomes and invaginated nuclei. Disorganized middle pieces with an irregular number of dense fibres or microtubules were occasionally found. Cytoplasmic droplets were common findings although spermatozoa with droplets did not exhibit signs of immaturity. According to these criteria we found less than 1% of the examined spermatozoa to be affected in marmosets and an average of 2% to be affected in rats. No differences were seen between the controls, vehicle-treated or TCDD-treated animals (Table 2).

Counting of sperm and spermatids. A decrease of spermatids in the testis and of sperm in the cauda epididymidis was seen at 1 μg TCDD/kg body wt in marmosets as well as rats. In both animals the reduction became significant in the testis at a dose of 3 μg TCDD/kg body wt. A reduction in the number of sperm in the vas deferens was not found (Table 3).

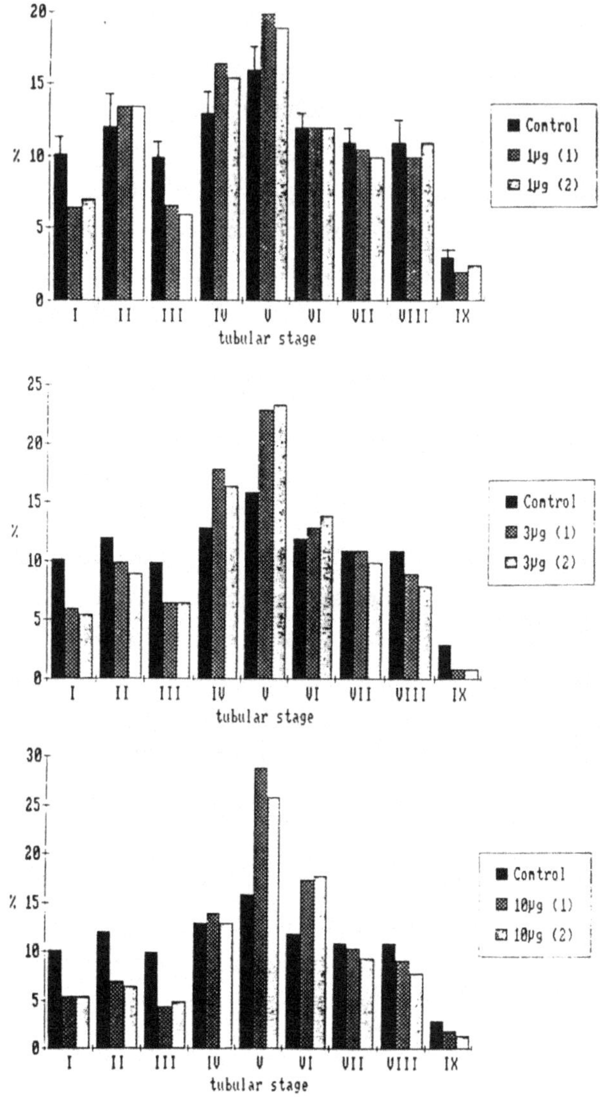

Fig. 1 A. The ratio of present tubular stages in marmoset testes after TCDD treatment at various doses (2 animals of each group) in comparison with untreated animals (2 animals). The ratio of tubular stages IV and V increased dose-dependently after TCDD treatment whereas the subsequent stages tended to become underrepresented. It is most obvious at 10 μg/kg body wt TCDD where the ratio of stage V is nearly two-fold compared with the control

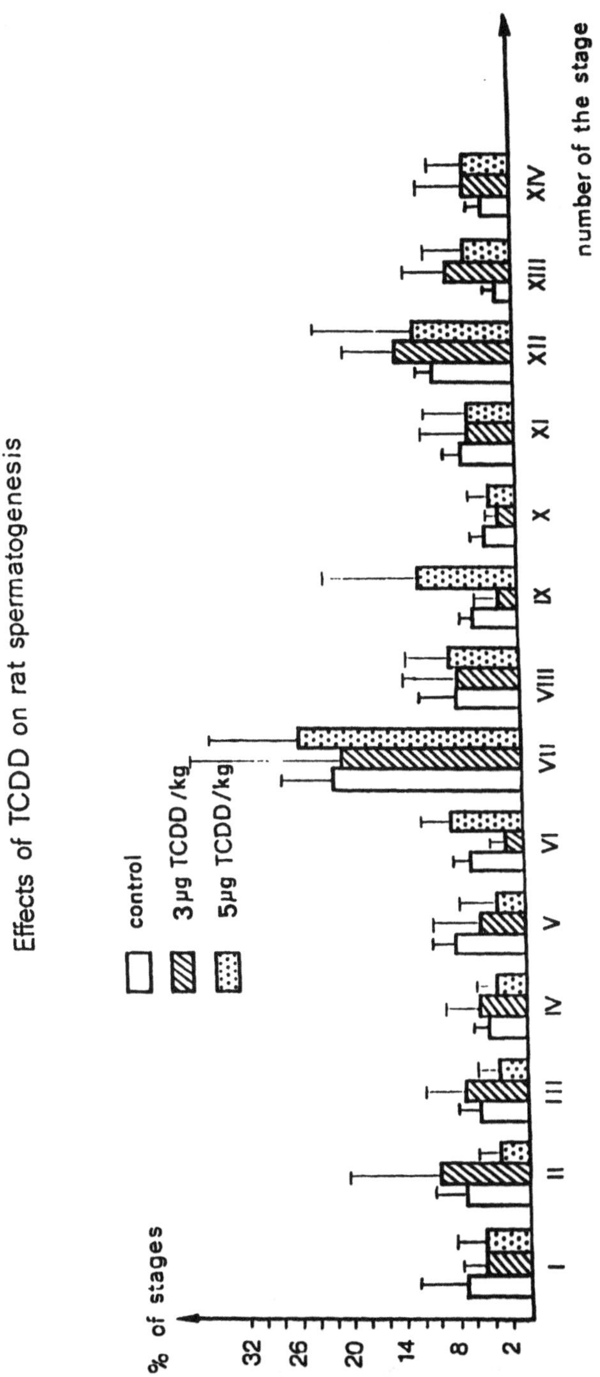

Fig. 1 B. The ratio of present tubular stages in rat testes after TCDD treatment (4 animals of each group) in comparison to untreated animals (4 animals). Although differences are seen, these differences are not significant and no specific pattern is seen

Table 2. Morphological alterations of spermatozoa in the vas deferens after TCDD treatment. Percentage of morphological alterations (mean ± standard deviation) of rat and marmoset spermatozoa in the ductus deferens after different doses of TCDD

Dose (μg/kg)	Rat	Marmoset
0	2.3 ± 1.0 (n=4)	0.9 ± 0.3(n=4)
0.5	2.0 ± 0.8 (n=4)	0.7 (n=2)
1	2.1 ± 1.8 (n=4)	1.0 (n=2)
3	3.0 ± 1.5 (n=4)	0.8 (n=2)
5	1.8 ± 0.9 (n=4)	1.85 (n=2)
10		0.7 (n=2)

Table 3. Number of spermatids and sperm (mio) in TCDD-treated rats after one week. Number of spermatids and sperm in rat testes, cauda epididymidis and vas deferens (Mio, mean ± standard deviation). Numbers of spermatids or spermatozoa are reduced at a dose of 3 μg TCDD/kg body wt in the testis and cauda epididymidis

Dose (μg/kg)	Testis	Cauda epididymidis	Ductus deferens
0	129 ± 30	50.9 ± 21.3	6.3 ± 1.3
0.5	113 ± 23	36.3 ± 11.8	4.1 ± 0.8
1	105 ± 22	33.9 ± 9.9	4.2 ± 1.4
3	67 ± 10*	31.0 ± 10.1	5.8 ± 2.0
5	66 ± 19*	30.5 ± 19.2	5.5 ± 2.9

* $p < 0.05$

Discussion

One almost confounding feature of the testis is that degeneration and subsequent necrosis of germ cells reliably correlates with a loss of testis weight. This has frequently been shown (see: Chapin 1988) and as a first step in evaluating a new compound this may be sufficient. Our findings, however, indicate that toxic effects, in the case of TCDD, are not necessarily accompanied by a decrease in testis weight or that these toxic effects can be revealed by a histopathological investigation before the loss of testis weight due to germ cell loss becomes significant. In our previous investigation (Rune et al. 1991a,b) we found clearly degenerative germ cells in the rat and in the marmoset after one week of TCDD exposure. In the rat almost all steps of germ cell development were affected, whereas in the marmoset degenerative signs were only perceptible during early spermiogenesis. Thus, our findings confirm those of Chapin et al. (1984) who investigated the effects of dimethyl methylphosphonate on testicular morphology. These authors found severe tissue damage with no change in testicular weight. Using organ weight as a parameter for testicular toxicity may be useful for defining an end-stage lesion, but one runs the risk of ignoring initial effects and target cells.

There are several examples of stage-specific lesions described in the literature (for review see: Chapin 1988) because the testis is composed of many different cell types and different compounds have been shown to produce different selective lesions in just one or a few of these cells, in particular germ cells. One possible methodological approach to detect stage-dependency is to determine the presence of tubular stages in semi-thin cross-sections, as recently shown by Hess et al. (1990). In our investigation the determination of the ratio of present tubular stages, according to the presence and morphology of spermatids, revealed that the ratio of stages V and VI increased dose-dependently under TCDD influence in marmosets. Both stages contain early round spermatids which will normally continue to develop up to elongated spermatids throughout the following stages VII to IX and I to III. The decreased ratio of tubular stages V and VI confirms most of the degenerative signs found in the course of spermiogenesis in our qualitative electron microscopic investigation (Rune et al. 1991a). They were perceptible in round spermatids of early developmental steps, like they occur during stages V and VI. Thus, these degenerative phenomena appeared to inhibit further development. When determining the tubular stages according to the developmental steps of spermiogenesis, this inhibition will lead to a relatively high proportion of stages V and VI and to a lower ratio of the following stages, as our morphometric determinations revealed. A comparable shifting of present tubular stages was not found in rats. Although we found differences between untreated and TCDD-treated animals, these differences were not significant and no relation to the applied dose was found. On the one hand, this result corresponds to our electron microscopic investigation (Rune et al. 1991b) where we could not find any specific pattern of germ cell degeneration. On the other hand, it indicates that TCDD effects are obviously species-specific.

In the context of germ cell degeneration it also has to be mentioned that under physiological conditions germ cells in most developmental stages (except for spermatogonia and preleptone spermatocytes) are located in a protected area within the germinal epithelium due to the blood-testis barrier (for details see: Waites and Gladwell 1982). This barrier is formed by tight junctions between adjacent Sertoli cells and separates the epithelium into a basic compartment that communicates with the intertubular space and a adluminal compartment which is closed for substances originating from the intertubular space. Thus, normally blood-borne substances have no access to most of the developing germ cells. In our electron microscopic investigation we found wide intercellular spaces between neighbouring Sertoli cells in rats as well as in marmosets (Rune et al. 1991a,b) and concluded from this that the blood-testis-barrier may be affected by TCDD. This could explain germ cell degeneration and the stage-specific lesions. Provided this is true, the effects on Sertoli cells must precede germ cell degeneration which we, however, are not able to decide at present. It points to the need for further investigations in which animals are dosed for as short a period of time as practical and then serially sacrificed at known intervals.

The decrease of spermatids in the testis and the decrease of sperm in the cauda epididymidis, which became significant at a dose of 3 μg TCDD/kg body wt corresponds to our qualitative morphological investigation (Rune et al. 1991a,b). Clear-cut morphological changes were not seen with doses lower than 3 μg TCDD/kg body wt. In addition, the decrease of sperm could be an explanation for the reduced male fertility (Chahoud et al. 1989). On the other hand, from the morphological point of view our results may indicate that those spermatozoa that escape from the fate of degeneration are obviously functional, since the ratio of morphological anomalies was not increased in our investigation. From this finding it becomes explicable why male fertility is only reduced to a certain degree. In this context, however, it has to be mentioned that an exposure time of one week, as in our study, is not sufficient to state this definitely, since the development of e.g. a TCDD-affected pachytene spermatocyte up to a mature spermatozoa present in the ductus deferens takes longer than one week. At the same time, the exposure time may also account for the fact that we did not find any reduction in sperm number in the vas deferens.

Acknowledgements: This investigation was supported by the Deutsche Forschungsgemeinschaft (Sfb 174). The authors are indebted to Mrs. A. Steuer, H. Wohlfeil, and S. Baar for technical assistance and to B. Steyn for linguistic help.

References

Astroff B, Safe S (1988) Comparative antiestrogenic activities of 2,3,7,8-tetrachlorodibenzo-*p*-dioxin and 6 methyl-1,3,8-trichlorodibenzofuran in the female rat. Toxicol Appl Pharmacol 95: 435-443

Chahoud I, Krowke R, Schimmel A, Merker H-J, Neubert D (1989) Reproductive toxicity and pharmacokinetics of 2,3,7,8-tetrachlorodibenzo-*p*-dioxin. 1. Effects of high doses on the fertility of male rats. Arch Toxicol 63: 432-439

Chapin RE, Dutton SL, Ross MD, Sumrell BM, Lamb JC (1984) Development of reproductive tract lesions in male F344 rats after treatment with dimethyl methylphosphonate. Exp Mol Pathol 41: 126-140

Chapin RE (1988) Morphologic evaluation of seminiferous epithelium of the testis. In: JC Lamb, PM Foster (eds) Physiology and Toxicology of Male Reproduction, Academic Press, New York, pp 155-177

Heger W, Neubert D (1983) Timing of ovulation and implantation in the common marmoset, *Callithrix jacchus*, by monitoring of estrogens and 6ß-hydroxypregnanolone in urine. Arch Toxicol 54: 41-52

Hess RA, Schaeffer DJ, Eroschenko VP, Keen JE (1990) Frequency of stages in the cycle of the seminiferous epithelium in the rat. Biol Reprod 43: 517-524

Holt WV, Moore HDM (1984) Ultrastructural aspects of spermatogenesis in the common marmoset (*Callithrix jacchus*). J Anat 138: 175-188

Karnovsky MJ (1967) A formaldehyde-glutaraldehyde fixative of high osmolarity for use in electron microscopy. J Cell Biol 35:213

Kociba RJ, Keeler PA, Park CN, Gehring PJ (1976) 2,3,7,8-Tetrachlorodibenzo-*p*-dioxin (TCDD): Results of a 13-week oral toxicity study in rats. Toxicol Appl Pharmacol 35: 553-574

Leblond CP, Clermont J (1952) Definition of the stages of the cycle of the seminiferous epithelium in the rat. Ann NY Acad Sci 55: 548-557

Mebus CA, Reddy VR, Piper WN (1987) Depression of rat testicular 17-hydroxylase and 17,20 lyase after administration of 2,3,7,8-tetrachlorodibenzo-*p*-dioxin (TCDD). Biochem Pharmacol 36: 727-731

Moore RW, Potter CL, Theobald HM, Robinson TA, Peterson RE (1985) Androgenic deficiency in male rats treated with 2,3,7,8-tetrachlorodibenzo-*p*-dioxin. Toxicol Appl Pharmacology 79: 99-111

Norback DH, Allen HR (1973) Biological response of the non-human primate, chicken and rat to chlorinated dibenzo-*p*-dioxin ingestion. Environ Health Perspect 5: 233-240

Reynolds SE (1963) The use of lead citrate as an electron opaque stain in electron microscopy. J Cell Biol 17: 208-212

Richardson UE, Jarrett I, Finke EH (1960) Embedding in epoxy resins for ultrathin sectioning in electron microscopy. Stain Technol 35: 313-323

Rune GM, De Souza PH, Krowke R, Merker H-J, Neubert D (1991a) Morphological and histochemical effects of 2,3,7,8-tetrachlorodibenzo-*p*-dioxin on marmoset testes. Arch Androl 26: 143-154

Rune GM, De Souza PH, Krowke R, Merker H-J, Neubert D (1991b) Morphological and histochemical pattern of response in rat testes after administration of 2,3,7,8-tetrachlorodibenzo-*p*-dioxin (TCDD). Histol Histopathol 6 (in press)

Umbreit TH, Hesse EJ, MacDonald GJ, Gallo MA (1988) Effects of TCDD-embedded interactions in three strains of mice. Toxicol Lett 40: 1-9

Waites GMH, Gladwell RT (1982) Physiological significance of fluid secretion in the testis and blood testis barrier. Physiol Rev 62: 624

Discussion of the Presentation

Wickramaratne: The target cell and mechanism of TCDD effect can be evaluated using Sertoli cell cultures. As the pictures and data presented indicated that the Sertoli cell was the target, how soon after treatment was the lesion detectable?

Rune: The lesions described in our study were documented one week after TCDD treatment. However, the determination of the exact time point of the occurrence of morphological effects would necessitate investigations after consecutive short time intervals. TCDD is known to be resorbed slowly after subcutaneous administration, which makes this approach rather difficult. To approach this problem, we have already started to study the effects of TCDD on cultured staged seminiferous tubules. Preliminary results show that the same effects on Sertoli cells are evident already after 24 h at a concentration of 1 ng/ml medium TCDD. Moreover, in pure rat and marmoset Sertoli cell cultures in a bicameral chamber system, the formation of tight junctions was obviously inhibited by TCDD (1 ng/ml medium). In contrast to the control cultures, TCDD prevented the formation of an impermeable barrier (comparable with the blood testis barrier *in vivo*) between the inner and the outer chamber.

Schulz-Schalge: Have you measured the TCDD concentration in the testis? It might be interesting to correlate the TCDD concentration to the biological effect.

Rune: We did not measure the concentration of TCDD, but others have. TCDD was not detected in testicular tissue one week after treatment when injected subcutaneously. This raises the question whether the effects are indirect. Our *in vitro* findings, however, suggest a direct mechanism of action. If this were so, the effective concentrations must be very small *in vivo*.

Yamashita: You clearly showed that the intercellular spaces between Sertoli cells open after administration of TCDD. In addition to this finding, it seemed to me from your electron micrographs that the width of tight junctional complexes sealing the luminal surface from the basolateral part of the Sertoli cells became thin in the TCDD group. Were there any changes in the permeability or in the tightness of the junctional complexes, for example, in the sense of the number of junctional strands under the influence of TCDD?

Rune: We have to admit that we did not determine the number of junctional strands, which may be an interesting aspect. From our sections we would tend to say that the existing tight junctions in TCDD treated testes were morphologically unchanged. However, they were rarely found. For judging the tightness of existing junctions and testing whether the blood testis barrier is intact after TCDD treatment, we intend to perfuse TCDD influenced testes with a fixation solution containing a tracer such as Lanthan. This work is now in progress in our laboratory.

Scallet: I wonder if it is possible that you would obtain the same effects by directly injecting TCDD into one testis, saving the other for a control? It has been suggested recently (SOT 91 meeting) that TCDD may affect the hypothalamus as shown by its decreased content of the neuropeptide ß-endorphin. Effects of TCDD on the hypothalamus may mediate testicular effects or the hypothalamic effect may result from changed function of the testis.

Rune: I think your proposal is an interesting methodological approach. It would be helpful to distinguish between direct effects of TCDD on testicular morphology and those possibly mediated by the hypothalamus. At present we cannot exclude this possibility. Nevertheless, there are two good reasons why we think that the described effects are direct ones: (1) because we found similar results *in vivo* as *in vitro*; (2) because *Russell et al., (Tissue and Cell 13: 3369-3380, 1981)* described a uniform change in the morphology of the testis after interruption or affection of the hypothalamo-hypophyseal-testis axis, whose pattern clearly differs from the pattern of degeneration after TCDD treatment.

Morphological Changes in Cultures of Hippocampus Following In Vitro Irradiation

Jorge Cervós-Navarro, Gundula Hamdorf, Angela Becker and Andreas Scheffler

Introduction

In our previous study (Hamdorf et al. 1990) we showed that irradiation of the rat caused a number of morphological changes in neurons and glial cells in the developing rat brain.

It has been argued that the effect of radiation on developing postnatal hippocampus could be mediated by break down of the blood-brain-barrier (Griffin et al. 1977), by ischemia produced by decreased blood flow (Suzuki et al. 1983; Kirino et al. 1985) or by radiation-induced decreases in brain blood flow and blood pressure (Chapman and Young 1968; Cockerham et al. 1986; Hornsey et al. 1990). The present study was initiated to determine the *direct* effects of radiation on developing neuronal tissue. Hippocampus was chosen because of its particular sensitivity to ionizing radiation (Tolliver and Pellmar 1987). In addition, the hippocampal slice provides an excellent section since its laminar organization makes observation easy.

Material and Methods

Culture Preparation and *Irradiation Procedure*

Hippocampal cultures were prepared after the technique described by Gähwiler (1984). Hippocampi of neonate rats (7 days old) were removed, cut parasagitally into 350 - 400 μm thick slices, placed in a plasma clot, and cultured in a roller drum at 36.5°C dry incubator.

The cultures were fed weekly with 1 ml of medium consisting of horse serum (25%), basal medium Eagle (50%) and Hank's or Earles basal salt solution (25%) supplemented with 2 mM glutamine and glucose to a final concentration of 6 mg/ml. No antibiotics were added to the medium.

Prior to the addition of culture medium, slices were exposed to irradiation from a ^{60}Co-unit (yielding about 0.6 Gy/min) at doses of 1, 2 and 4 Gy.

Light and Electron Microscopy

After 14 days *in vitro*, cultures were processed for Transmission Electron Microscopy (TEM). Semi-thin sections were stained with toluidine blue for light microscope observation and morphometric evaluation. Observations were performed mainly in the area of the gyrus dentatus.

Cultures were fixed in glutaraldehyde, postfixed in 2% osmium tetroxide, dehydrated in graded alcohols, and embedded in Araldite while still on the plastic cover slip. Ultra-thin sections were obtained with a diamond knife in an Reichert-Jung-ultramicrotome (Cambridge Instruments GmbH). The sections were stained with uranyl acetate in 70% ethanol and subsequently with lead citrate. Observations were made with a Zeiss-10-electron microscope at 60 kV.

Morphometric and Quantitative Evaluation

Evaluation of the size of the perikarya of pyramidal cells and their nuclei was carried out with the aid of the BIOQUANT IV-computer program.

Approximately 100 squares of 10 x 10 μm (total area sampled: 10 000 μm^2) of the molecular layer in electron micrographs (magnification x 2100) were taken randomly. Counts were made of synaptic contacts, identified by synaptic vesicles and by the electron dense membranes limiting the synaptic cleft.

Statistics

The data were statistically evaluated by analysis of variance (ANOVA) and subsequently by Mann-Whitney-U-Test. The $p < 0.05$ level was selected for rejection of the null hypothesis.

Results

Light Microscopy

In semi-thin sections of the cultures the main neuronal cell layers as *in vivo* could be recognized. Both pyramidal and granule cell types survived in cultures and revealed a normal appearance of cell bodies and neuropil.

In the irradiated cultures a variable amount of vacuolization and a large number of small, dense granules could be seen in the cytoplasma of pyramidal and granule cells. In the pyramidal band as well as in the hilus, necrotic neurons were present.

Morphometric analysis of the size of the perikaryon of pyramidal cells revealed a significant decrease after irradiation in comparison to controls (Table 1). A dose-dependence was not apparent. The reduction of nuclear size was also significant. The ratio size of perikaryon to size of nucleus remained constant except for a slight decrease after irradiation with 4 Gy.

Table 1. Size of perikarya and nuclei of pyramidal neurons in irradiated hippocampus and controls

	Size of Perikarya (μm^2)		Size of Nuclei of Pyramidal Cells (μm^2)		Ratio Cell/Nucleus
Controls	288 ± 99		107 ± 29		2.69
irradiated 1 Gy	212 ± 88	$p < 0.0001$	79 ± 24	$p < 0.01$	2.68
irradiated 2 Gy	224 ± 92	$p < 0.05$	83 ± 27	$p < 0.05$	2.70
irradiated 4 Gy	200 ± 94	$p < 0.001$	81 ± 18	$p < 0.01$	2.47

Electron Microscopy

Control Cultures:

Pyramidal neurons, like pyramidal cells *in vivo*, preserved their lamellar order. The perikarya exhibited all cytoplasmic organelles characteristic of hippocampal neurons (Fig. 1).

The granular endoplasmic reticulum was arranged concentrically in the narrow rim of cytoplasm surrounding the apical part of the nucleus. Mitochondria and Golgi apparatus were situated mostly in the area close to the dendrite.

Cultures exhibited mature *synapses*, often surrounded by astrocytic processes, and morphologically equivalent to those observed in the molecular layer *in vivo*. Some synapses contained few dense core vesicles. Several axons were myelinated and the *myelin* sheath was composed of many lamellae.

Irradiated Cultures:

A great majority of *pyramidal cells* showed enlargement of cisternae and vacuoles of the Golgi-complex. In some neurons Golgi apparatus consisted only of masses of vacuoles.

Swellings of the mitochondria were most pronounced in cultures exposed to doses of 4 Gy. Neurons also contained isolated multivesicular bodies, numerous lysosomes and lipofuscine granules in the perikaryon (Fig. 2). Alterations in mitochondria included swellings and disappearance of cristae and were most prominent in neurons showing advanced necrotic changes.

In contrast to the orderly arrays of endoplasmic reticulum in the pyramidal cells, in irradiated cells the granular reticulum sometimes seemed to be diminished. Sometimes it consisted only of isolated cisternae or small clumps in disorderly arrays and situated towards the periphery.

The majority of nuclei of neurons in irradiated cultures appeared condensed and showed an accumulation of hetero-chromatine under the nuclear membrane.

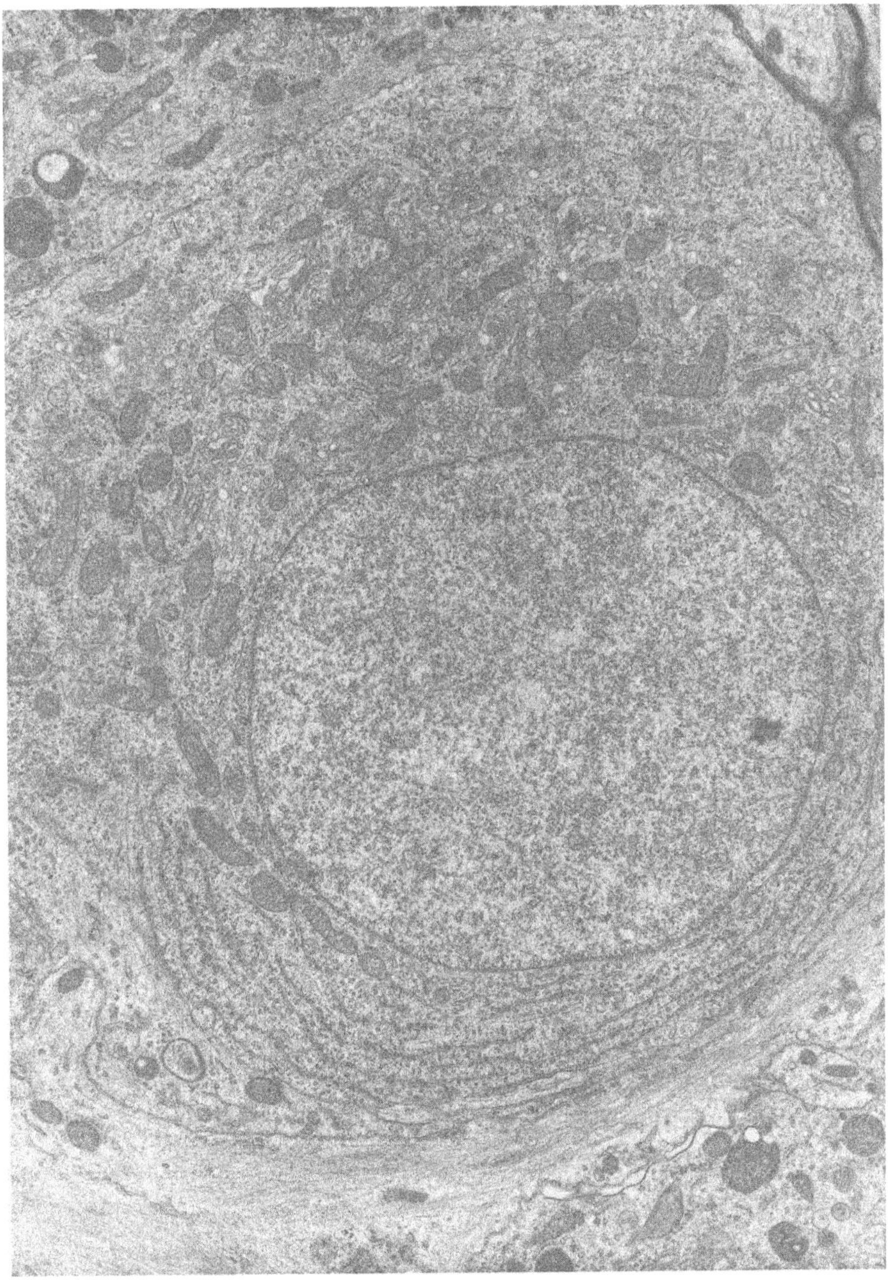

Fig. 1. Micrograph of a hippocampal neuron in control culture, 14 days *in vitro*. Magnification x 5000

Some of the *granule cells* appeared affected and contained vacuoles and electron dense granules.

Astrocytes sometimes appeared vacuolated and filled with lysosomes, the amount of gliofilaments seemed increased in comparison to controls. *Several oligodendrocytes* displayed in their cytoplasm multilamellar bodies and lipid drops (Fig. 3).

Synapses appeared not to be affected by irradiation in their morphology. However, a decrease in number occurred in all groups of irradiated cultures. At a dose of 4 Gy the reduction of intact synaptic contacts was highly significant (Table 2), whereas there were no statistically reliable differences between 1 and 2 Gy irradiated slice cultures and controls.

Table 2. Number of synaptic contacts per 100 μm^2

Controls	*6.08* ± 3.90	
Irradiated cultures		
1 Gy	*4.59* ± 3.31	n.s.
2 Gy	*5.41* ± 3.45	n.s.
4 Gy	*2.88* ± 2.10	$p < 0.001$

Though a quantitative study of number and thickness of myelin sheaths has not been performed, qualitatively the formation of *myelin* in cultures seemed to be affected by irradiation. The amount of myelinated axons in irradiated cultures appeared reduced in the majority of slices and the myelin sheaths appeared thinner. Multilamellar bodies were very conspicuous compared with the control.

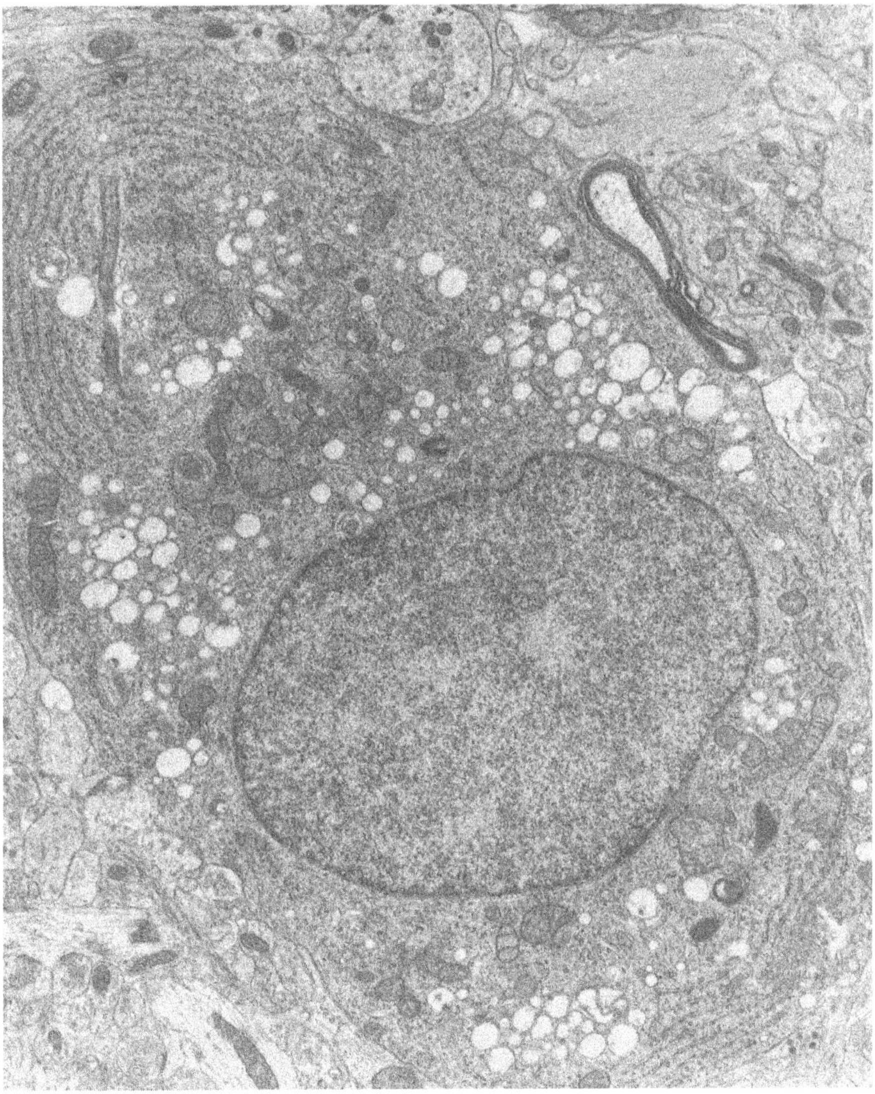

Fig. 2. Micrograph of an affected neuron in *in vitro* irradiated hippocampus slice culture, (4 Gy), 14 days *in vitro*. Magnification x 5000

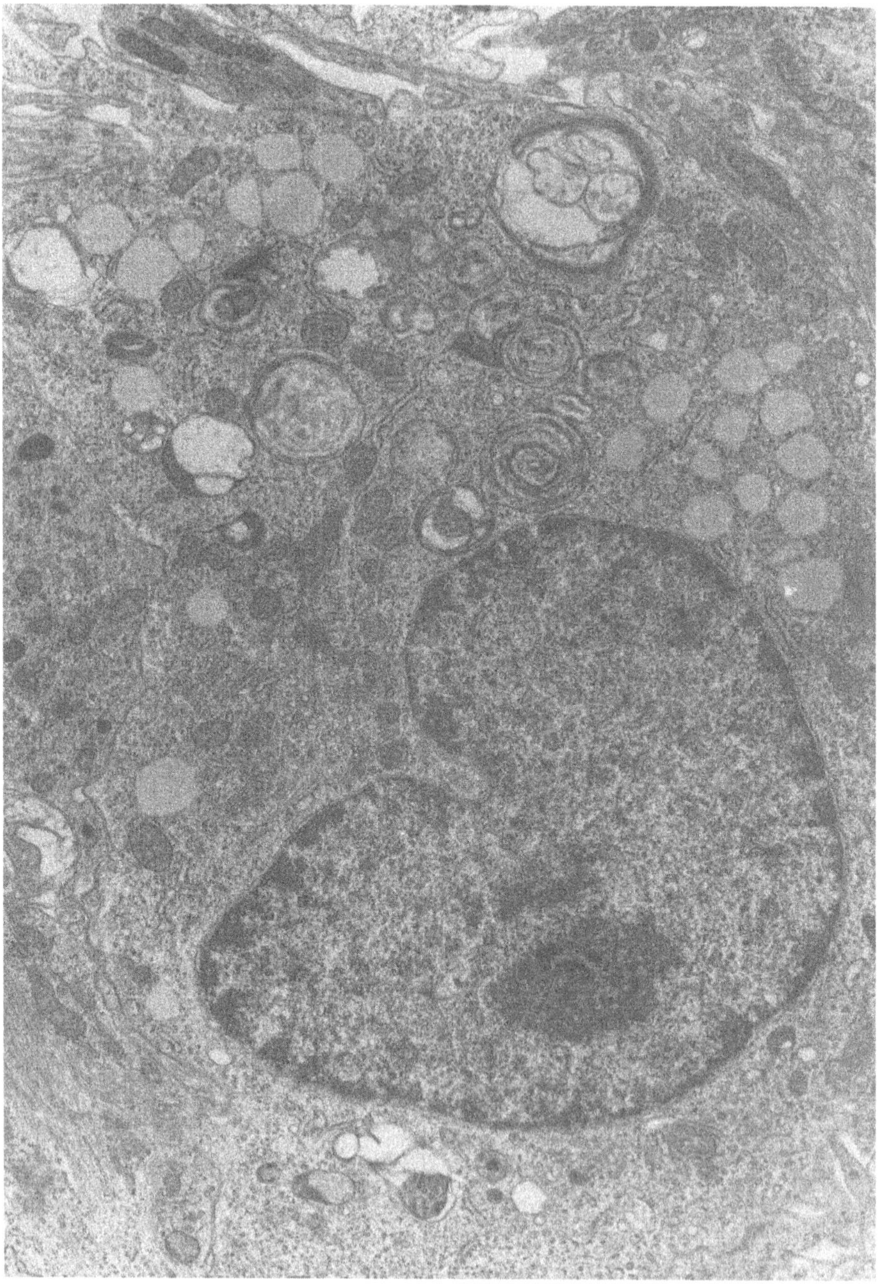

Fig. 3. Micrograph of an oligodendrocyte in an *in vitro* irradiated hippocampus slice culture (2 Gy), 14 days *in vitro*. Magnification x 5000

Discussion

Damage to the developing nervous system due to irradiation was previously reported (for a review see Kriegel et al. 1986). The extent of damage was related to the time of gestation and irradiation dose. According to Hicks and D'Amato (1963) damage to the brain was more pronounced when irradiation insult happened during neuroblast formation than in further stages of differentiation.

Our results demonstrate that damage to the central nervous system and especially to neurons does not only occur in prenatal stages of development. *In vitro*, even at low doses of only 1 Gy there is damage to the neurons and myelin formation following irradiation in the early postnatal period.

Pyramidal neurons revealed a similar amount and type of destruction as reported in the cultures obtained from prenatally irradiated newborn.

The mechanism of the reduction of pyramidal cell size and the size of nuclei following irradiation is yet unknown. However, it could indicate an impaired metabolism.

Swellings of the mitochondria in irradiated neurons was the most conspicuous change. The disappearance of dense granules seen in the mitochondria of controls as well as cristae likely led to the functional impairment of all the respiratory electron transport mechanisms, and many dehydrogenases linked to the inner membrane of the mitochondria.

The reduction of the amount of granular endoplasmic reticulum and number of ribosomes indicates a reduction of cytoplasmic protein synthesis, as most of the RNA exists in the form of ribosomes which are associated with cytoplasmic protein synthesis. Golgi apparatus plays a central role in the interchanges between the granular endoplasmic reticulum and the plasmalemma, exchanging membranous elements with both of them. It also completes the assembly of glycoproteins and has a central role in the formation of lysosomes. Furthermore the Golgi apparatus may have a part in supplying or repairing the critical permeability mechanisms-receptors, ion channels, and structural proteins.

Enlargement of the Golgi apparatus and the increased number and complexity of lysosomes are to be regarded as reconstructive changes, not as a degenerative phenomenon; a restorative process aimed at reconstituting the neuron to its original state (Peters et al. 1991).

Multivesicular bodies are believed to be involved in the degradation of membranes of synaptic vesicles. Although neurons are known to contain lysosomes, neurons in control slices did not. Perhaps they were too young. Lysosome formation in irradiated neurons was evident. They contain acid phosphatase and other hydrolytic enzymes and are products of the Golgi apparatus (Peters et al. 1991). As the animal ages, or if the nervous system is subjected to disease, trauma or toxic substance (Borges et al. 1986), many of the lysosomes develop into lipofuscine granules. Initially the accumulation appears to be without detriment to the cell, but when a certain point is reached there may be a progressive reduction in the amount of cytoplasmic ribonucleic acid.

All morphological signs indicate that neuronal metabolism is disturbed after postnatal irradiation and there are attempts of reconstruction which may be successful occasionally.

Although a clear-cut distinction between neuronal and non-neuronal cells during perinatal gliogenesis and maturation is difficult, it became evident that glial cells were included in morphological changes induced by irradiation.

Similar morphological changes such as vacuolization, swellings of mitochondria and Golgi apparatus, have been reported due to several agents trimethyllead and triethyllead (Nolan and Brown 1989), hypoxia, hypoglycemia, ischemia, status epilepticus (Siesjö et al. 1980).

Changes in the structure of *chromatin* within the neuronal nuclei may be secondary to a different or reduced metabolism after cell injury, although not all of the evaluated cells obviously presented signs of an injury. Our finding of altered chromatin is consistent with the results of Jaberaboansari et al. (1988), who reported that neuronal chromatin structure had been changed after exposure of rats' heads to ionizing radiation.

The *granule cells* exhibited some features of damage following *in vitro* irradiation, contrasting to the results in experiments with prenatal *in vivo* irradiation (Hamdorf et al. 1990) in which they were apparently not affected. Among all hippocampal neurons granule cells of the dentate gyrus are the latest to differentiate. More than half of them arise postnatally and their development into mature neurons continues until the twentieth postnatal day (Angevine 1965). During prenatal irradiation (day 13 *in utero*) in our former study the majority of granule cells had not been formed yet. In our present investigation, radiation insult affected them in the later stage of multiplication and differentiation, while influences of blood-brain-barrier, hormones and immune system were excluded.

In all cultures, axon terminals formed *synaptic contacts* with cell bodies, dendritic shafts and dendritic spines. The synaptic density in our control cultures was at a slightly lower level in comparison to values reported by McWilliams and Lynch (1984) in hippocampus *in vivo*, due to the fact that synapses in cultures lack afferent stimuli. It has been reported that afferent fibers exert an inductory effect on the formation and maintenance of postsynaptic elements (Hamori 1973; Frotscher et al. 1977). Their morphology, however, did not appear altered in spite of a loss of afferent innervation in slice cultures.

There were differences in the amount of synapses in controls and irradiated cultures. The reduction in irradiated cultures was more pronounced and significant after irradiation with 4 Gy than with lower doses. This agrees with the results of Bassant and Court (1978) who found that 4 Gy irradiation altered neuronal firing patterns, whereas Gaidamakin and Ushakov (1989) reported severe destructive changes in the majority of synapses in adult rats only after 200 Gy.

Irradiation at a later stage of development affected synapse formation since synaptogenesis takes place mainly postnatally (Crain et al. 1973).

It is suggested that low dose irradiation affected formation of synapses only slightly. Synapses that had already been formed, could not be destroyed by irradiation of the low doses used. In 4 Gy irradiated cultures, there was a reduced amount of synapses. Synapse function of hippocampal slices was affected by radiation only at higher doses: Tolliver and Pellmar (1987) reported that synaptic deficits occurred only at 50 Gy which contributed to damage at 150 Gy.

We could rely only on the morphological aspect of synapses, however, it cannot be excluded that synapses have been incorrectly situated thus causing functional irregularities.

Myelinogenesis was obviously affected by *in vitro* irradiation. Our results confirm the results of Jacobs et al. (1986), who described *in vivo* altered cellular differentiation leading to "a specific disruption in the synthetic machinery required for normal myelination" following postnatal exposure of the central nervous system to ionizing radiation. A reduction in the quantity of myelin was seen after irradiation of the rat in early postnatal life *in vivo* (Gilmore 1963), whereas high doses of postnatal irradiation inhibited myelinogenesis completely as a result of a radiation induced marked reduction of the normal glial population (Sims and Gilmore 1989; Sims et al. 1987). Impairment in the production of oligodendrocytes is regarded as a late effect of radiation (Reyners et al. 1982). Lower doses caused a delay in the onset of myelin formation with quantities of myelin which eventually reached normal levels (Gilmore 1966). Our finding that less myelin is found after irradiation even *in vitro* may be a confirmation that irradiation injured oligodendrocytes.

It has been suggested that the number of neurons present at the end of neuronal proliferation determines the extent of the subsequent glial cell proliferation which in turn determines the amount of myelin produced. The net effect of a loss of neurons is the production of a brain which has the same concentrations of cells and myelin as control brains and is normal in shape but is irreparably deficient in size and function (Wanner and Edwards 1983). In humans randomly exposed to radiation during pregnancy, microencephalopathy is the most commonly detected abnormality (Miller and Mulvihill 1976).

A variety of conditions are associated with neuronal lesions of relatively similar appearance and localization. A common mechanism seems probable. There is evidence of involvement of oxidative events in the development of cell damage, the possibility of free radical damage to neurons as a common mechanism of damage has been discussed (Siesjö et al. 1980; Siesjö 1981). The generating of free radicals is an integrated part of metabolism after hypoxia and ischemia, hypoglycemia and status epilepticus (Siesjö 1981), and intoxication with hydrogen peroxide (Tolliver and Pellmar 1987). Free radicals are also known to be effective in damage by irradiation. If this is the factor inducing injury, the use of free radical scavengers should have favourable effects not only after radiation insults but also in hypoxia and ischemia.

Lipid peroxidation, a consequence of ionizing radiation, that modifies membrane properties might contribute to neuronal dysfunction. It is readily produced at lower doses (less than 1 Gy/min) (Raleigh et al. 1977) and is an unlikely mechanism for deficits at the synapse. It might underlie, however, the damage at the postsynaptic site.

On the other hand, it has been shown that ionizing irradiation at a dose of 4 Gy (Tolliver and Pellmar 1987) and less (Peimer et al. 1986) altered the neuronal firing pattern, thus indicating an influence on neuronal excitability which in its turn can result in an uncontrolled synaptic release of excitatory amino acid neurotransmitters. As a result, specific membrane receptors are persistently overstimulated, triggering a cascade of biochemical events responsible for cell death.

We could not find a clear dose-dependence in the amount of injury except in formation of synapses. Maybe at these low doses it cannot easily be detected. Otherwise this could give evidence to the hypothesis of Mole (1986) that it is only a question of probability of damage, which could rise with higher doses.

Conclusions

Using hippocampal explants we were able confirm the deleterious effect of low dose *in vitro* irradiation on the developing nervous system, especially on neurons, synapse and myelin formation. Irradiation produces similar effects *in vivo* and *in vitro*, thus giving evidence of a direct effect on the developing brain.

Our model in tissue culture of the nervous system provides a good model for further experiments to test radioprotective substances.

Acknowledgements: This investigation was supported by grants from the Deutsche Forschungsgemeinschaft to the Sonderforschungsbereich 174.

References

Angevine JB (1965) Time of origin in the hippocampal region. Exptl Neurol Suppl 2: 1-70

Bassant MH, Court L (1978) Effects of whole body-gamma-irradiation on the activity of rabbit hippocampal neurons. Radiat Res 75: 593-606

Borges MM, Paula-Barbosa MM, Volk B (1986) Chronic alcohol consumption induces lipofuscin deposition in the rat hippocampus. Neurobiol Aging 7: 347-355

Chapman PH, Young RJ (1968) Effect of cobalt 60 gamma irradiation on blood pressure and cerebral blood flow in the Macaca mulatta. Radiat Res 35:78-85

Cockerham LG, Cervany TJ, Hampton JD (1986) Postradiational regional cerebral blood flow in primates. Aviat Space Environ Med 57: 578-582

Crain B, Cotman C, Taylor D, Lynch G (1973) A quantitative electron microscopic study of synaptogenesis in the dentate gyrus of the rat. Brain Res 63: 195-204

Frotscher M, Hamori J, Wenzel J (1977) Transneural effects of enthorhinal lesions in the early postnatal period of synaptogenesis in the hippocampus of the rat. Exptl Brain Res 30: 549-560

Gähwiler BH (1984) Slice cultures of cerebellar, hippocampal and hypothalamic tissue. Experientia 40: 236-243

Gaidamakin NA, Ushakov IB (1989) The state of synapses of the cortex of cerebral hemispheres on gamma-irradiation. Neurosci Behav Physiol 19: 483-488

Gilmore SA (1963) The effects of X-irradiation on the spinal cords of neonatal rats. II. Histological observations. J Neuropathol Exptl Neurol 22: 294-301

Gilmore SA (1966) Delayed myelination of neonatal rat spinal cord induced by X-irradiation. Neurology 16: 749-753

Griffin TW, Rasey JS, Bleyer WA (1977) The effect of photon irradiation on blood brain barrier permeability to methotrexate in mice. Cancer 40: 1109-1111

Hamdorf G, Shahar A, Cervós-Navarro J, Scheffler A, Sparenberg A, Skoberla A (1990) Morphological changes in cultures of hippocampus following prenatal irradiation in the rat. J Neurosci Res 26: 327-333

Hamori J (1973) The inductive role of presynaptic axons in the development of postsynaptic spines. Brain Res 62: 337-344

Hicks SP, D'Amato CJ (1963) Low dose radiation of the developing brain. Science 141: 903-905

Hornsey S, Myers R, Jenkins T (1990) The reduction of radiation damage to the spinal cord by post-irradiation administration of vasoactive drugs. Int J Radiat Oncol Biol Phys 18: 1437-1442

Jaberaboansari A, Nelson GB, Roti JL, Wheeler KT (1988) Postirradiation alterations of neuronal chromatin structure. Radiat Res 114: 94-104

Jacobs AJ, Maniscalco WM, Parkhurst AB, Finkelstein JN (1986) *In vivo* and *in vitro* demonstration of reduced myelin synthesis following early postnatal exposure to ionizing radiation in the rat. Radiat Res 105: 97-104

Kirino T, Tamura A, Sano K (1985) Selective vulnerability of the hippocampus to ischemia - reversible and irreversible types of ischemic cell damage. Prog Brain Res 63: 39-58

Kriegel H, Schmahl W, Gerber GB, Stieve FE (eds) (1986) Radiation Risks to the Developing Nervous System. Gustav Fischer, Stuttgart, New York

McWilliams JR, Lynch G (1984) Synaptic density and axonal sprouting in rat hippocampus: stability in adulthood and decline in late adulthood. Brain Res 294: 152-156

Miller RW, Mulvihill JJ (1976) Small head size after atomic irradiation. Teratology 14: 335-358

Mole RH (1986) Problems related to prenatal exposure of the nervous system. History and perspective. In: Kriegel H, Schmahl W, Gerber GB, Stieve FE (eds) Radiation Risks to the Developing Nervous System, Gustav Fischer, Stuttgart, New York

Nolan CC, Brown AW (1989) Reversible neuronal damage in hippocampal pyramidal cells with triethyllead. The role of astrocytes. Neuropathol Appl Neurobiol 15: 441-457

Peimer SI, Dudkin HO, Swerdlow AG (1986) Response of hippocampal pacemaker-like neurons to low doses of ionizing radiation. Int J Radiat Biol 49: 597-600

Peters A, Plalay SL, Webster HF (1991) The fine structure of the nervous system. Neurons and their supporting cells. Oxford University Press, New York

Raleigh JA, Kremers W, Gabourg B (1977) Dose-rate and oxygen effects in models of lipid membranes: linoleic acid. Int J Radiat Biol 31: 203-213

Reyners H, Gianfelici-de-Reyners E, Malsin JR (1982) The beta astrocyte: a newly recognized radiosensitive glial cell type in the cerebral cortex. J. Neurocytol 11: 967-983

Schmidt SL, Lent R (1987) Effects of prenatal irradiation on the development of cerebral cortex and corpus callosum of the mouse. J Comp Neurol 264: 193-204

Siesjö BK, Rehncrona S, Smith, D (1980) Neuronal cell damage in the brain: possible involvement of oxidative mechanisms. Acta Physiol Scand Suppl 492: 121-128

Siesjö BK (1981) Cell damage in the brain: A speculative synthesis. J Cereb Blood Flow Metab 1: 155-185

Sims TJ, Gilmore SA (1989) Interactions between Schwann cells and CNS axons following a delay in the normal formation of central myelin. Exp Brain Res 75: 513-522

Sims TJ, Gilmore SA, Waxman SG (1987) Temporary adhesions between axons and myelin-forming processes. Develop Brain Res 40: 223-232

Suzuki R, Yamaguchi T, Li CL, Klatzo I (1983) The effects of 5-minute ischemia in Mongolian gerbils. II. Changes of spontaneous neuronal activity in cerebral cortex and CA1 sector of hippocampus. Acta Neuropathol (Berlin) 60: 217-222

Tolliver JM, Pellmar TC (1987) Ionizing radiation alters neuronal excitability in hippocampal slice of the guinea pig. Radiat Res 112: 555-563

Wanner RA, Edwards MJ (1983) Comparison of the effects of radiation and hyperthermia on prenatal retardation of brain growth of guinea-pigs. Brit J Radiol 56: 33-39

Discussion of the Presentation

Neubert: Do the changes you observe only develop when you culture the brain explants *in vitro*? Or have you also seen comparable changes directly *in vivo*?

Cervós-Navarro: We have examined brains of neonate rats that have been exposed to irradiation of the same dose *in utero*. Indeed, there were similar changes *in vivo*. In addition, we could detect conspicuous changes in the vessels which did not disappear during the first four weeks of postnatal life, on the contrary, seemed to increase. Thus, we suggest that the deleterious effect of radiation *in vivo* could be multiplied by additional vascular impairment.

Neubert: In your very interesting studies, from the first data you presented, 1 Gy was apparently not a no-observed-effect-level (NOEL). Do you have data with even lower doses and is it possible to assess the smallest effective dose? In my opinion this would be a very important aspect, and you have the technology to do such studies.

Cervós-Navarro: Assuming that the deleterious effect of radiation could be a question of probability and the number of cells injured are dose-dependent, the electron microscopy is not the right method to face the problem of the no-observed-effect-level. More subtle methods that allow conclusions on functional aspects should be used to answer this interesting question, as we will do in future experiments.

Scallet: Have you measured the number of cells by counting nucleoli in the cultures in order to relate this measure to the number of synapses per unit area in the hippocampal cultures?

Cervós-Navarro: So far, we have not correlated the number of synapses and the number of cells. This could give interesting results, we will follow your advice, thank you very much.

Possible Contribution of *In Vitro* Methods to Risk Assessment in Reproductive and Developmental Toxicology

Stephan Klug and Diether Neubert

Introduction

Risk assessment of reproductive and developmental toxicology is presently almost exclusively performed using whole-animal studies. Furthermore, in this area of toxicological risk assessment the largest number of animals are used (more than 50% of all animals used for routine toxicological evaluation, if exposed or possibly affected offspring are also considered).

More recently the use of *in vitro* techniques has also been propagated for this area of toxicology, but beside an impressive number of scientific publications on a variety of *in vitro* methods using developing systems, no convincing data or strategies have been presented to provide the basis for replacing the presently applied *in vivo* methods with *in vitro* techniques.

In this presentation we shall discuss the reasons for the major difficulties in this area of research, and we will draw conclusions from our own, extended experience in this field. However, it should be stressed that *in vitro* methods may have clear-cut benefits within the framework of risk assessment, although not in primary-stage testing.

Definition of the Problem

Since routine toxicological testing is predominantly an applied science, it largely depends on the political stipulation. The society decides what margin of safety is acceptable, and this sets the frame for the necessary toxicological studies. So far, in our country this demand for safety has been extremely high for medicinal substances as well as for many environmental chemicals, also when compared with the willingness to accept other risks in our daily lives.

If, for example, the society should decide that animal studies are no longer acceptable, then many more adverse health effects would have to be accepted; although in our dishonest society it is more likely that people would just expect that the necessary experiments to be performed elsewhere.

Special Difficulties when Attempting to Assess Reproductive and Developmental Toxicity with *In Vitro* Techniques

In contrast to all other fields of toxicology (mostly assessment of single and defined effects), the possibilities of adverse outcomes (morphological defects, functional defects, disturbed fertility) are numerous in the field of developmental toxicology. In fact, an adverse outcome may manifest itself during the entire lifespan, from formation of the gonads up to life expectancy. For this reason, the desire for replacement of intact animals with *in vitro* systems in this field can hardly be fulfilled completely, maybe it is even impossible to reach this goal. Today there is not one single *in vitro* system, not even the most complex one - the whole-embryo culture -, which can simulate the whole period of embryonic and fetal development and which permits the assessment of every kind of abnormal development.

There is an immense number of *in vitro* systems available which allow simulation of a variety of developmental processes. However, all the systems simulate a very restricted period of development (in most of the cases less than three days in the case of mammalian gestation) and always in an isolated form, neglecting the variety of interfering factors. Therefore, directly exposing, e.g., sperm to a compound in order to assess its influence on fertility is completely useless since such an attempt would ignore important aspects of development (different susceptibility of various developmental stages, stages of development, hormonal influences, blood-testes-barrier, the essential interaction with Sertoli cells, etc.). This arbitrary example may vicariously demonstrate one of the main difficulties when *in vitro* systems should be used to assess a possible hazard to complex developmental processes.

An apparently logical, but false, conclusion could be to establish a huge battery of different *in vitro* systems. This strategy is bound to fail from the beginning since a meaningful interpretation of the results obtained from such a mass of data would be prevented by a variety of "false negatives" and "false positives", and functional abnormalities which manifest themselves late postnatally could still not be excluded and revealed.

The situation is somewhat easier when the problem is confined to the evaluation of "teratogenic" effects, i.e. the induction of abnormal prenatal development. This means restriction to the assessment of morphological development and to a limited period of time.

Nevertheless, an important problem remains: according to our present knowledge there are two types of interfering effects: they may be called *uniform* and *specific* (Neubert et al. 1980). It will be rather easy to detect uniform effects (general cytotoxicity) with *any in vitro* system (unless bioactivation is necessary). In contrast, the situation will be much more complicated if the effect which should be assessed is specific and well defined. This type of hazard is confined to one or a few organ anlagen, in some cases even to a certain tissue. Detection of such effects using *in vitro* systems will only be successful if an appropriate *in vitro* system allowing the evaluation of the appropriate endpoint is selected, or has been used by chance.

"False negative" results will always be expected when the abnormal development induced by the agent cannot be evaluated using the endpoints measurable in the system selected. As an example: abnormal limb development cannot be assessed in the whole-embryo culture, because the limbs do not develop at least up to the cartilage stage.

Aside from these facts there are a number of *a priori* difficulties which may complicate the work with every *in vitro* system:

1. Metabolic activation may be a main problem if the risk of an unknown chemical is attempted to be assessed solely on the basis of *in vitro* data. It will be impossible to be satisfied with a "negative outcome" since a possible risk due to an active metabolite in man cannot be excluded.
2. In many cases the solubility in the culture medium and not the toxic potential will be the limiting factor during *in vitro* studies. This is highly relevant since many biologically active substances show a high lipophilicity and are therefore poorly water-soluble.
3. Almost any substance can interfere via cytotoxicity with normal *in vitro* development, if the applied concentration is high enough.

Most of these restricting problems are not specific for the use of *in vitro* systems in reproductive toxicology, but may equally arise when assessing other toxic effects in culture. Furthermore, tests using intact animals may also have limitations, and an intelligent use of *in vivo* and *in vitro* methods may show many advantages. This implies that several *in vitro* systems must be available in the laboratory in order to select the most appropriate one.

Assessments of Developmental Toxicity for which the Use of *In Vitro* Techniques is Feasible

In the area of developmental toxicity the use of *in vitro systems* is mainly confined, and probably most effective, in the field of prenatal toxicity. *In vitro* systems will *not serve to replace* intact animals in this field (i.e. replacement of segment or multigeneration studies) but they can play an *important role* in *supplementing in vivo studies*. In this respect, two basically different approaches can be distinguished.

"Primary-stage testing". This term refers to testing without having any data from previous studies available. *In vitro* methods are primarily unsuited for obtaining extensive and final information for the purpose mentioned, and there is not a

single case in which a possible prenatally-toxic risk was assessed exclusively with *in vitro* systems. As mentioned before, this is especially the case when no abnormal development is found *in vitro*. However, appropriate *in vitro* methods may successfully be used for a "screening" of a large series of chemicals at an early stage of development, thus allowing the selection of compounds with no embryo- or cytotoxic hazard among structural analogues. Testing in this way can only be successful if the original or mother compound has already demonstrated its potential to induce dismorphogenic effects *in vivo* or (and) in the chosen *in vitro* system. Nevertheless, this procedure may contribute to reducing the number of larger scale animal tests and thus lead to a reduction of the number of experimental animals used. In general, the substances selected in such a "primary-stage" testing will have to be additionally evaluated in whole-animal studies, since it has to be excluded that the substance selected with this "screen" induces no further abnormal development of other organ systems.

"Secondary-stage testing". In this case data from *in vivo* experiments are available, and subsequent *in vitro* studies may be based on the results of the *in vivo* studies. Often routinely performed *in vivo* tests (segment or multigeneration studies) reveal results which cannot be clearly interpreted. The question whether an observed teratogenic effect is due to maternal or due to a direct influence to the developing organism comes up rather often. Elucidation of this problem is essential if a risk assessment should be performed. Subsequent testing of the compound with complex or selected *in vitro* systems will provide a chance to answer this question. Another problem often concerns the question whether the original compound or a metabolite is responsible for the induction of an observed adverse effect. Solving this problem may be important when considerable species differences exist with respect to drug metabolism of the substance to be evaluated. Subsequent testing with an appropriate *in vitro* system (separate testing of the original compound and possible metabolites) will help to distinguish between the effects of the different derivatives.

In general, *in vitro* systems are *well suited* (and often more so than *in vivo* setups) for studying *expected or defined effects* in the field of prenatal toxicology, and to evaluate a possible mode of action, provided a suitable test system has been chosen.

In contrast, *in vitro* systems are *not well suited* for the investigation of an unknown effect on the reproductive system. And, furthermore, they are *not suited* for overall studies on postnatal development or possible toxic effects on fertility or other complex dysfunctions.

Choice of Techniques to be Used for Assessing Risk for Abnormal Development

An *in vitro* system should satisfy a number of criteria in order to be useful for a risk assessment in reproductive toxicology. In the ideal case the system should have a high predictive value with respect to the reproductive system of the species used. This does not necessarily mean a predictive value for human prenatal development. Furthermore, an *in vitro* system should be sensitive, specific and should be able to react quantitatively (i.e. the response should increase concentration-dependently).

A priori, the applicability of an *in vitro* system is limited. For this reason and especially for use in reproductive toxicity assessment, the most suitable system has to be selected for solving a defined question; this automatically implies that several systems with different endpoints and biological complexity have to be available. It makes no sense to set-up only a single *in vitro* system and then attempt to solve all occurring problems with such a test. If meaningful and successful, the strategy must run the other way around: an occurring and specific scientific problem must be tackled with the most appropriate *in vitro* test system. The basic rule when planning such *in vitro* studies in reproductive toxicology should be: the simpler the hypothesis to be confirmed or dismissed, the simpler the test system which should be chosen, and therefore there is a good chance that the data obtained can be interpreted and are useful.

This strategy which we have favoured for the last decade and which has been proven to be useful may be outlined with two examples:

- If the problem were to assess the "teratogenic potential" or potency of a given compound, and e.g., the mouse limb bud culture was chosen as the test system for answering this question, such an approach would be worthless in the case of a "negative" outcome. No single *in vitro* system is capable of solving this problem!

- If the problem were to assess whether a limb abnormality observed in *in vivo* studies must be considered as the result of a direct embryotoxic action, or whether it might rather be caused by "maternal toxicity", evaluating such a substance with the limb bud culture may well provide evidence for helping to answer this question.

Summarizing these general comments, the following can be stated:

1. It is presently not feasible to assess *all possible manifestations* of reproductive and developmental toxicity solely with *in vitro* techniques. This situation is unlikely to change in the near future, and this goal may never be achievable.

2. If the problem is confined to all manifestations of *prenatal toxicity*, the situation is somewhat more favourable. However, prenatally-induced dysfunctions (e.g. of the immune or the hormonal system) cannot be assessed with *in vitro* methods. These dysfunctions may be of increasing concern in the next years.

3. If "*primary*-stage testing", in the strict sense (no information from previous *in vivo* or *in vitro* studies is available), of a "*teratogenic*" potential or potency is the goal then this again *cannot* be achieved with single *in vitro* methods, because the biological "complexity" of none of these systems is sufficient for such an evaluation. This especially holds true for the case of a "negative" outcome of such tests.
   A special situation exists when specific data (from *in vivo or in vitro* tests) on a substance exist, and a large series of derivatives are to be examined. Such a "screening" is quite feasible if an appropriate *in vitro* test can be chosen, allowing an assessment of a defined biological endpoint. However, the substances selected from such a "screening" will certainly have to be subsequently, additionally tested *in vivo* with conventional methods.

4. The strength and advantage of *in vitro* methods in the area of reproductive and developmental toxicology clearly lies *not* in attempts to *replace*, but to *supplement in vivo* studies. This is as such quite important, and it may contribute to reducing the number of *in vivo* tests and the number of animals used, and especially to improving the scientific quality of the information required. Often the outcome of initial *in vivo* studies is insufficient for a risk assessment and further studies are necessary. "Secondary-stage testing" (information from previous *in vivo* studies is available) with *in vitro* techniques may be considered superior to additional testing with conventional *in vivo* methods if a specific and well-defined suitable system is available.

If, with respect to developmental toxicity, an institution wants to use *in vitro* techniques, it should be an undebatable prerequisite that *several* such systems are available with different biological endpoints and complexities.

Examples of Results obtained with *In Vitro* Techniques

In the following we will present some examples of results obtained with *in vitro* techniques at our institute over the last ten years. These examples shall show some of the benefits of such *in vitro* systems in the field of developmental toxicology, and furthermore demonstrate what type of problems may be tackled.

Comparative Evaluation of Abnormal Development In Vitro

As mentioned above it is very difficult, maybe impossible, to use *in vitro* systems for "primary-stage testing" (no information from previous *in vivo* or *in vitro* studies is available) with the purpose of detecting unknown embryotoxic potentials. The situation is different when the "specific" embryotoxic potential and potency (and the specifically affected biological endpoint) of a given test compound is known. In this case it may be of great help to "screen" a large number of related compounds in order to select substances "negative" in this test for further *in vivo* testing. This may greatly reduce the need for *in vivo* testing.

In vitro systems for this purpose should be sufficiently complex (whole-embryo culture) or should mimic the abnormal development observed *in vivo* (eg. palate culture, limb bud culture, etc.).

It should be stressed that *in vitro* studies do not abandon the need for additional studies to elucidate the "pharmacokinetic situation" in the culture system (e.g. the substance concentration within the explant, protein binding at various concentrations, etc.). The need for such additional data has been demonstrated before (Neubert 1988).

A study performed in our laboratory with six *ß-blockers* may serve as an example. Tested in a rat whole-embryo culture system these substances induced a very similar pattern of abnormal development *in vitro*. However, these compounds exhibited EC_{50} values (medium-concentration inducing a 50% abnormality rate) with differences of three orders of magnitude (*see:* Fig. 1). Measurements of the substance levels within the cultured embryos at corresponding EC_{50} concentrations showed quite different values (*see:* Table 1).

Table 1. EC_{50} values, substance concentrations and values for lipophilicity of six ß-blockers

Substance	Concentration in the medium EC_{50}* (µM)	Concentration in the embryo (µM, M ± SD)	Relative concentration in the embryo (% medium conc.)	Lipophilicity**
Propanolol	30	5.20 ± 1.80	17.3	5.0
Alprenolol	30	8.42 ± 2.32	28.1	5.0
Metoprolol	100	4.53 ± 1.47	4.5	0.2
Pindolol	150	8.98 ± 0.33	6.0	0.2
Acebutolol	450	12.52 ± 2.34	2.8	0.2
Atenolol	5000	77.03 ± 1.97	1.5	0.003

* Concentration producing a 50% abnormality rate
** Octanol/buffer distribution (data from the literature)

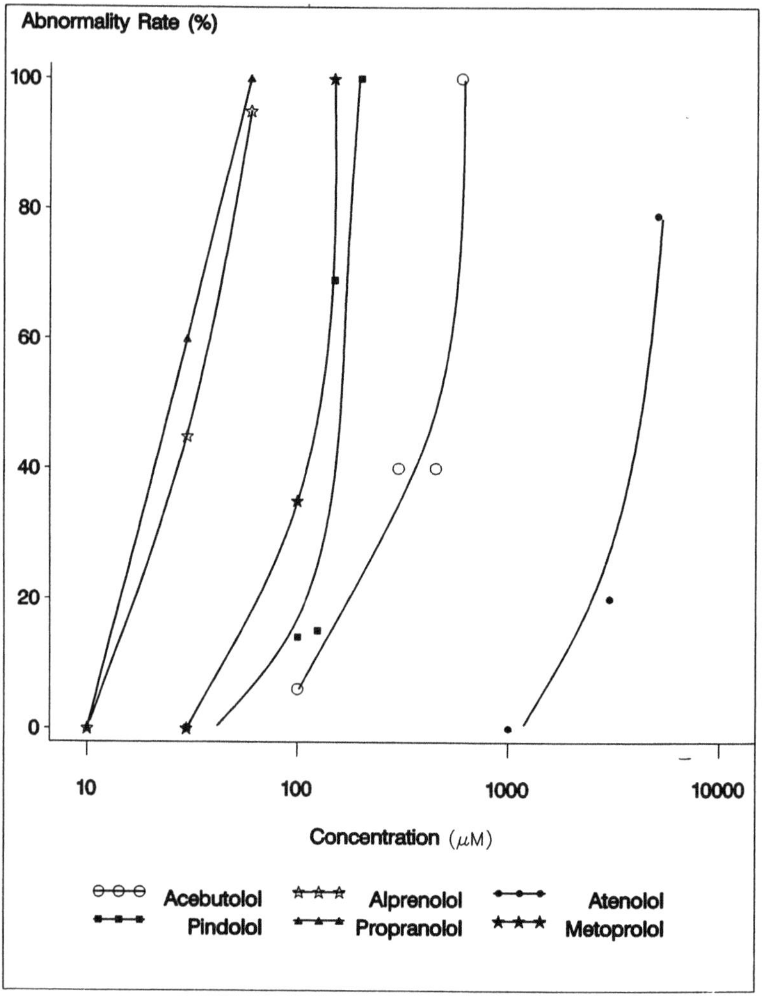

Fig. 1. Concentration-effect relationships found in the whole-embryo culture (rat) for six ß-blockers. 9.5-day-old rat embryos were cultured for 48 hours in bovine serum in the presence of the ß-receptor blockers. The concentrations indicated corresponded to the amounts added to the culture medium.

According to the EC_{50} values and the degree of substance transfer to the embryo (relative substance concentration in the embryo; *see:* Table 1) the following statement can be made: propranolol and metoprolol revealed a much higher intrinsic embryotoxic potency than atenolol. Apparently the lipophilicity of the substance (which seems to correlate with the transfer of the ß-blocker to the embryo) seems to be a rate limiting factor in this case.

Similar studies performed in our laboratory with six calcium channel blockers (Stein et al. 1990) also revealed characteristic changes induced in the whole-embryo culture which were identical for all of the calcium channel blockers tested, despite the fact that they belonged to different classes of chemicals. Apparently, the abnormal development was induced in connection with the extensive calcium channel blockage, the pharmacologic action of these substances. In this case the abnormal development predominantly concerned the yolk sac circulation and morphology. If this were the primary defect induced by these calcium channel blockers, this could only be revealed with difficulties with *in vitro* studies. In the discussion of the paper by Stein et al. (1990) some of the difficulties when attempting to extrapolate data from *in vitro* studies to the situation possibly existing *in vivo* are outlined. Since the explants are continuously exposed to the substance *in vitro* and the pharmacokinetics proceeds quite differently *in vivo*, the entire testing strategy would have to be changed when a direct comparison of *in vitro/in vivo* is the goal.

The two examples clearly indicate that the culture method used was capable of revealing "group characteristics", since the pattern of abnormalities were quite different in the case of ß-blockers or calcium channel blockers.

Combination of In Vivo and In Vitro Studies

Routinely performed *in vivo* studies (e.g. according to "Guidelines") may not always offer satisfying conditions to detect an embryotoxic potential of a substance to be tested. Therefore, we suggested a modification (Klug et al. 1991; Neubert et al. 1986; *see:* Fig. 2) of the test strategy: the substance is given on one day of pregnancy only (day 10 of pregnancy in the rat). This additional testing offers at least three advantages:

- Maternal toxicity, possibly restricting the dosing over an extended period, is reduced and may no longer complicate the experimental set-up.
- This strategy may also be used for biotechnology products (e.g. human recombinant proteins) which form (neutralizing) antibodies.
- This dosing period during a main period of organogenesis can be exactly mimicked *in vitro* with the whole-embryo culture, thus allowing a combined *in vivo* /*in vitro* approach.

As an example studies with the virustatic agent *aciclovir* shall be mentioned. In this case *in vivo* studies were complicated by a nephrotoxic action seen in the dam at higher doses of aciclovir due to the poor water solubility and crystallization of the substance in the maternal kidneys.

Aciclovir is, to our knowledge, the only substance for which evidence for a possible embryotoxicity/teratogenicity was first obtained in *in vitro* studies, using the whole-embryo culture (Klug et al. 1985). With the background of the results of these previous *in vitro* experiments it was reasonable to suggest a teratogenic potency of aciclovir also *in vivo*, and the data from the whole-embryo culture suggested that day 10 of gestation may be within the susceptible period. Therefore, higher doses of aciclovir were given *in vivo* over a restricted period during pregnancy (Stahlmann et al. 1988). With different dose regimens on day 10 of gestation exactly the same dismorphogenetic effects were inducible in experiments using intact animals when the evaluation was performed on day 11.5 (*see:* Table 2 and Fig. 3). At the end of pregnancy or even postnatally, with this dosing regimen numerous abnormalities could be observed.

This is a good example of a mutual confirmation of the data from *in vitro* and *in vivo* experiments, and it indicates that under certain conditions it may be possible to detect an embryotoxic potential of a drug first by the use of an *in vitro* system, with a subsequent confirmation *in vivo*.

However, it should be stressed that it was, also in this case, almost impossible to predict the final outcome at the end of pregnancy from the *in vitro* studies.

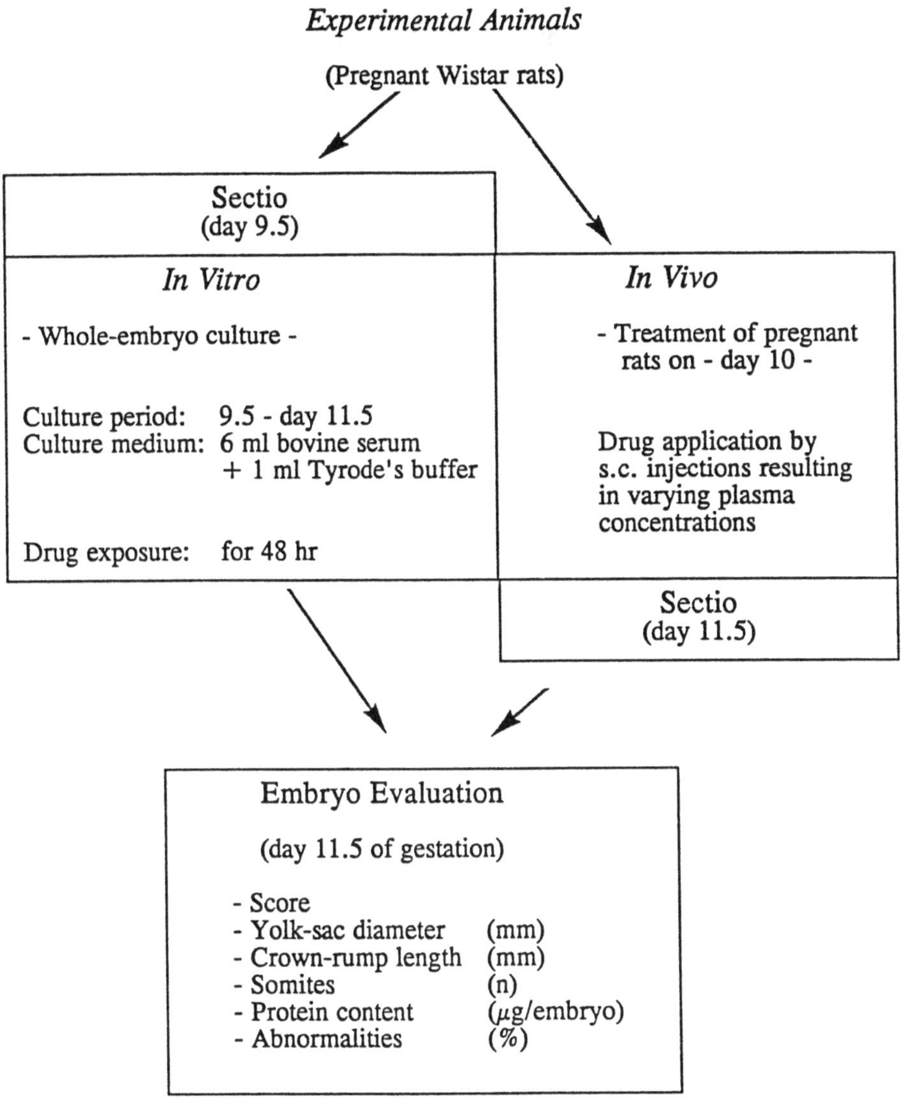

Fig. 2. Proposal for a combined *in vivo/in vitro* approach to assess possible "teratogenic" effects. This strategy is suggested when the usual long-term treatment (10 days) is not feasible, and when a direct comparison of effects obtained *in vivo* and *in vitro* is the goal

We chose to test aciclovir *in vitro* because it may initiate a chain termination during DNA replication, a mechanism highly suspect of interfering with normal development. In this respect, it is interesting that another virustatic agent with a presumed very similar mode of action, azidothymidin (used for HIV infections) had a significantly lesser potency to induce abnormal development in the same *in vitro* system (difference of a factor of 60 in comparison to aciclovir). This was later also confirmed in corresponding *in vivo* studies.

These examples show that under certain prerequisites effects on development may be well monitored *in vitro*, and that comparative studies with various derivatives are possible.

Table 2. *In vivo/in vitro* comparison of the influence of aciclovir on the development of rat embryos on day 11.5 of gestation

		n	CRL (mm)	Somites	Protein (μg/emb)	Score	Abn (%)
			3.60	*27*	*284*	*38*	
In vitro	Control	44	**3.36**	**26**	**235**	**37**	**0**
			3.18	*25*	*176*	*35*	
			3.36	*27*	*175*	*33*	
In vitro	100 μM	19	**3.00**	**26**	**138**	**32**	**95**
			2.82	*25*	*122*	*31*	
			3.18	*25*	*166*	*35*	
In vivo	3x100 mg/kg*	31	**3.12**	**24**	**140**	**34**	**100**
			2.88	*20*	*86*	*32*	

Data as median values (**bold**) and the first and third quartile (*italic*)
CRL: Crown-rump length
n: Number of embryos
Abn: Abnormality rate
*: Administration on day 10 of gestation (s.c. injections)

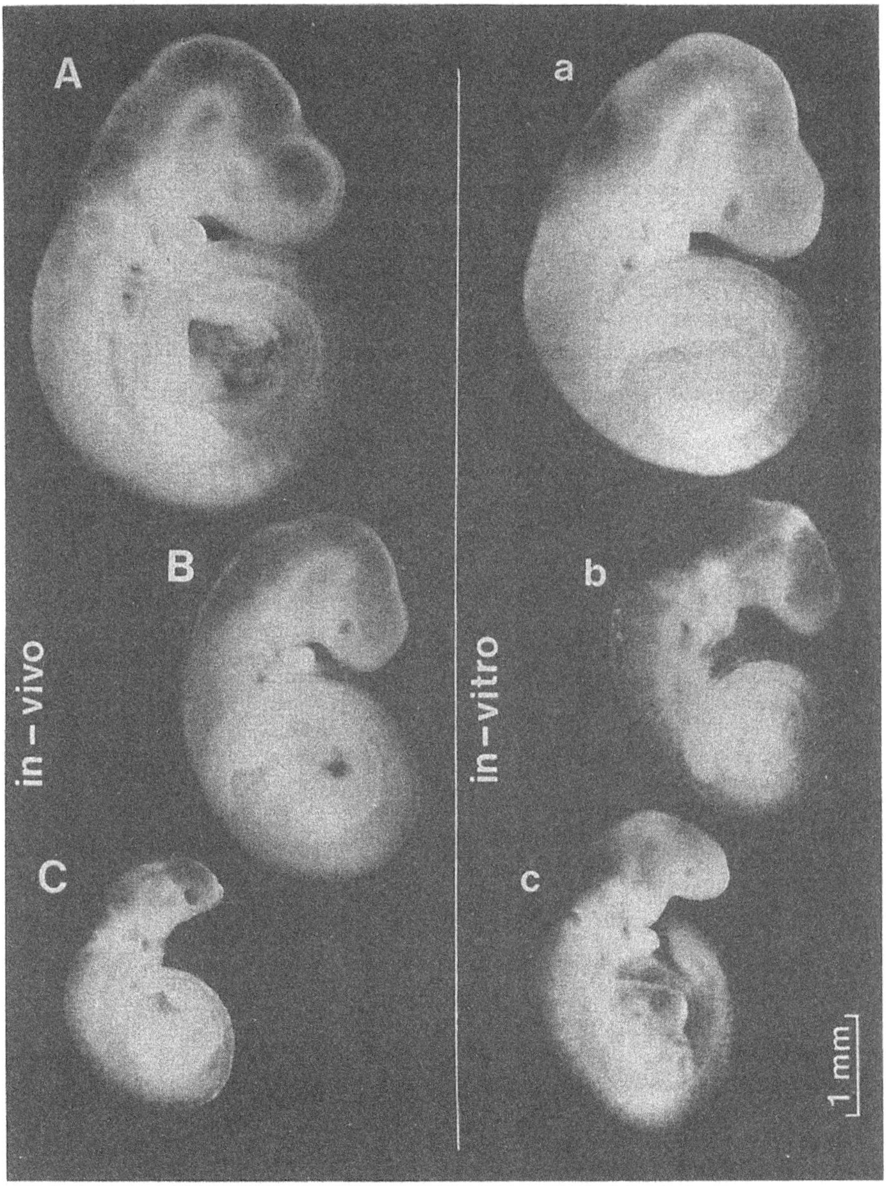

Fig. 3. Comparison of aciclovir-induced dismorphogenetic effects *in vivo* and *in vitro* in the rat. Experimental design as outlined in Figure 2. The embryos shown are from parallel experiments performed *in vitro* and *in vivo*. In this special case the data from the whole-embryo culture were available first, and the *in vivo* studies were performed subsequently to confirm the findings. There is an excellent agreement in the types of defects induced both *in vitro* and *in vivo*

In Vitro Studies, "In Vitro Pharmacokinetics" and Drug Development

Valproic acid, an anticonvulsant drug, was suggested to be a (weak?) human teratogen, and it was proven to be capable of inducing "teratogenic" effects in animals (Kao et al. 1981; Nau et al. 1984). Therefore, it was of considerable interest to develop a drug with the same anticonvulsive activity but with a significantly lower embryotoxic potency.

Valproic acid (VPA) and eight analogues or metabolites were tested in the whole-embryo culture system in our institute (Lewandowski et al. 1986; Klug et al. 1990). The most significant result was a different embryotoxic potential of VPA and 2-en VPA (both substances differing in their structure by only one double-bond): already at concentrations of 0.3 mM in the culture medium VPA induced a pronounced abnormal development of the cultured embryos, whereas 2-en-VPA, even at concentrations of 1.8 mM, did not lead to morphogenetic defects (Table 3). Extensive pharmacokinetic studies (Nau et al. 1988) of the conditions *in vitro* revealed that 2-en VPA reached the embryo at the no-observed-adverse-effect-level (NOEL) in an almost 20-fold higher concentration than VPA (Table 4). Thus, pharmacokinetic reasons were not responsible for this difference in biological activity.

These results clearly demonstrate that 2-en VPA shows a significantly lower intrinsic embryotoxic potency while exhibiting a similar anticonvulsive activity as VPA. These *in vitro* and further *in vivo* studies provided the basis for presently performed clinical trials with 2-en VPA in patients with epilepsy.

In Vitro Systems and Elucidation of Mechanisms of Teratogenic Actions

One of the most important applications of *in vitro* methods in the field of reproductive and developmental toxicology may be the elucidation of mechanisms of action, and the possibility to obtain more indepth information on a substance.

Table 3. Effect of VPA and 2-en-VPA on the *in vitro* development of 9.5-day-old rat embryos

Concentrations (mM)	n	CRL (mm)	Somites	Protein (μg/embryo)	Score	Abn (%)
Control	18	*3.60* **3.40** *3.10*	*26* **25** *24*	*249* **203** *146*	*37* **36** *35*	**0**
0.3 mM VPA	15	*3.60* **3.20** *2.80*	*26* **24** *23*	*166* **141**** *114*	*36* **35*** *33*	**0**
0.6 mM VPA	19	*3.30* **3.10*** *2.70*	*26* **21*** *18*	*167* **132**** *77*	*35* **33**** *25*	**32**
1.2 mM VPA	17	*2.70* **1.80**** *1.50*	*21* **19**** *16*	*110* **80**** *49*	*23* **15**** *12*	**100**
1.8 mM 2-en-VPA	15	*3.30* **3.20** *3.10*	*27* **26** *23*	*173* **136*** *122*	*37* **36** *35*	**0**

Median values (**bold**) and the 1. and 3. quartiles (*italic*) are given
CRL: Crown-rump length
n: Number of embryos
Abn: Abnormality rate
*: Significantly different to control values ($0.05 < p < 0.01$)
**: Significantly different to control values ($p < 0.01$)

Table 4. Substance concentration of VPA and 2-en-VPA in different compartments of the culture systems

Substance		Concentration			
		In the medium			
	(mM)	Total (μg/ml)	Free (μg/ml)	In the embryo (μg/g)	Ratio (E/Mf)
VPA	0.3	**44.6** ± 6.3	**5.5** ± 1.1	**12.9** ± 3.1	**2.3**
VPA	0.6	**87.0** ± 7.7	**21.8** ± 1.6	**37.6** ± 7.4	**1.7**
2-en-VPA	0.6	**93.0** ± 12.0	**13.4** ± 1.7	**21.8** ± 9.8	**1.6**
2-en-VPA	1.8	**292.0** ± 7.1	**133.0** ± 16.0	**193.0** ± 34.0	**1.5**

E/Mf: Concentration ratio: embryo/culture medium (free concentration)
Mean values ± standard deviation (n = 5)

The main reason for this special suitability of *in vitro* methods is the high degree of *reproducibility* achievable *in vitro*, due to a strict standardization of the experimental conditions (e.g. the developmental stages at the beginning of the culture). This is in sharp contrast to the huge variability usually observed in *in vivo* studies when assessing reproductive toxicity. A high degree of standardization and of reproducibility is the prerequisite for performing further biochemical studies or electron microscopic investigations.

Furthermore, concentration-response relationships can be evaluated much better and easier *in vitro*. Additionally, it is possible to expose the explants to the agent for defined periods at defined concentrations only, and to continue culturing in a normal culture medium.

As an example we present some of our recent results obtained with *2,3,7,8-TCDD*. The induction of cleft palates in mice by TCDD (2,3,7,8-tetrachloro-dibenzo-*p*-dioxin) is the only convincing teratogenic effect of this compound observed *in vivo*, except for dilatation of the renal pelvis (which is not to be considered a typical prenatal toxicity). Up till now, little is known about the mode of action of this induction of palatal clefts. Results of Abbott and Birnbaum (1990), Abbott and Buckalew (1992) and first results in our institute suggested that TCDD may interfere with the palate closure after or shortly before attachment of the shelves.

Since we had established a system which allows to simulate closure of the secondary palate *in vitro* (Shiota et al. 1990), we consequently studied the palate closure *in vitro* and the possible effects of TCDD in this system. In addition to gross-morphological evaluations we used histological *immunofluorescence* with polyclonal antibodies to a variety of tissue and extracellular matrix components as a tool. Antibodies against a variety of collagens, to basement membrane components, and to molecules such as laminin and fibronectin were used. Staining of the cultured palate anlagen at the end of culture revealed obvious differences in the localization of the antigens as well as differences in the morphology during the process of closure. Here only a few of the results obtained can be mentioned.

Exposure to TCDD (625 pg/ml culture medium) was found to induce an increased expression of collagen type I in the cultured shelves. Staining with polyclonal antibodies against, e.g., laminin allowed us to very closely follow the boundaries of the basement membranes of the two attaching shelves. In the TCDD-treated explants these boundaries were found to be, in contrast to the controls, not parallel and were interrupted with bleb-like structures (*see:* Fig. 4A,a). Another finding (*see:* Fig. 4B,b) was the less densely packed fibronectin in the area of the fusion of the shelves in the presence of TCDD when compared to the controls. These examples indicate that *in vitro* methods in the field of reproductive toxicology show numerous advantages and allow elucidations which are not that easily, or even not at all, possible with *in vivo* techniques.

With the modern methods of molecular biology, and especially since *in situ* hybridization has now become available, such techniques may also be used for elucidating possible modes of action during abnormal development. Although it has presently not been used, this approach may gain considerable importance in the near future. Such studies are best performed with *in vitro* techniques because of the high reproducibility of the development of embryonic tissue in culture.

Summary and Conclusions

It was our intention to outline that *in vitro* techniques in the field of reproductive and developmental toxicology have *a priori* serious *limitations* when compared with *in vivo* tests (and this will remain so in the future), but at the same time they exhibit numerous *advantages* which guarantee them an established place within the framework of risk assessment.

It may be expected that these limitations as well as the advantages may become even more pronounced in the future:

- It is to be foreseen that the main emphasis of risk assessment in developmental toxicology will shift from the presently predominating morphological defects and focus more on congenital *functional* alterations (e.g. of the immune and the hormonal system). Since such dysfunctions are bound to manifest themselves late postnatally, the chance to detect and to predict them with *in vitro* methods will be *negligible*.

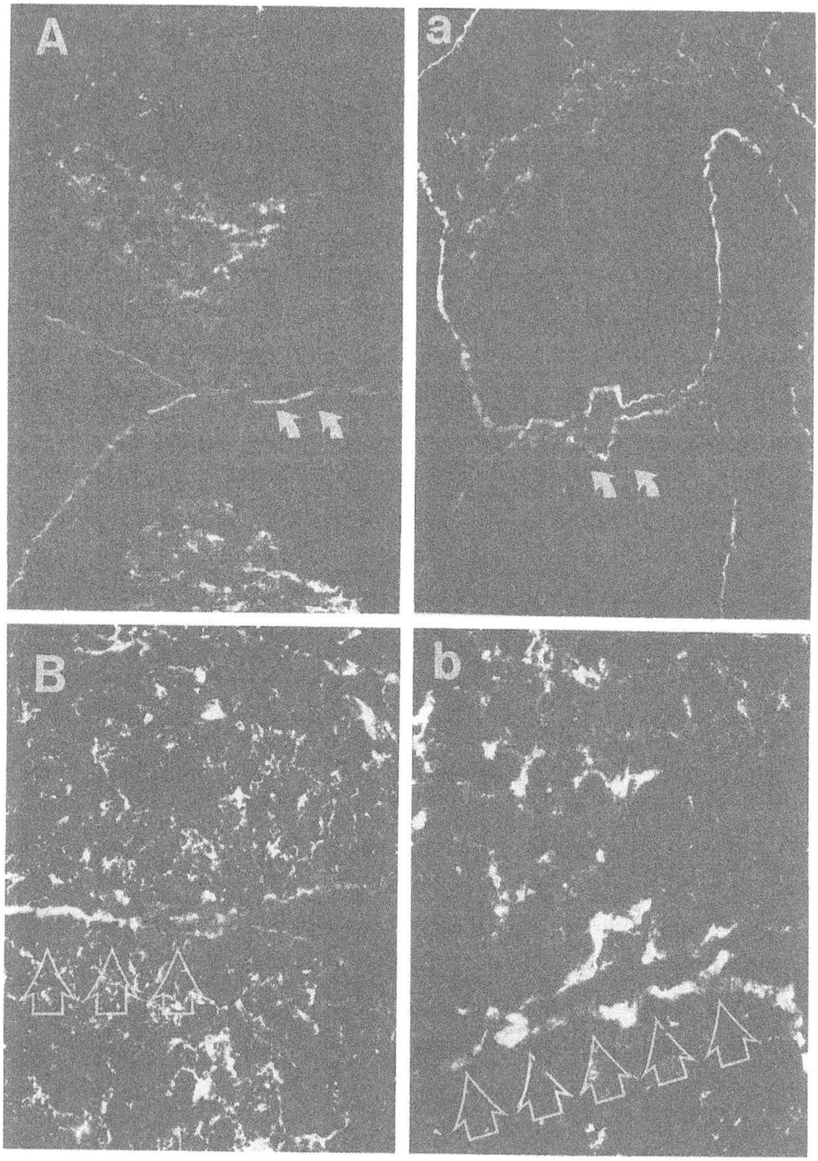

Fig. 4. Immunofluorescence analysis with polyclonal antibodies against laminin and fibronectin in histological sections of cultured palate anlagen treated with TCDD *in vitro*. *A + a: Laminin; B + b: Fibronectin. A + B: Controls*. Palate anlagen of mouse embryos (day 12 of gestation) were cultured in a chemically-defined medium in the presence or absence of 625 pg TCDD/ml. The surfaces of the two adhering shelves are normally in close contact (*see:* arrows), but a quite uneven contact area is seen in the presence of TCDD (for more details: *see* text). Magnification: X 40

- Alternatively, with increasing knowledge of basic processes connected with normal prenatal development, the possibilities to include mechanisms of actions of agents disturbing such processes into the framework of the risk assessment will become attractive and even essential. Information for such *mechanistic* considerations may be expected to predominantly come from results of *in vitro* studies. However, such *in vitro* approaches cannot be "dictated".

We strongly feel that it would be deleterious to confine the efforts of developing and using *in vitro* methods in the field of reproductive and developmental toxicology to the aim of *replacing* animal studies ("primary-stage testing"). Investing much effort in that direction, as predicted according to our present knowledge, would be a waste.

In vitro methods for assessing developmental toxicity should by no means be confined to the purpose stated above. Their strength and superiority to *in vivo* tests is in the field of "secondary-stage testing" and in the possibility of elucidating mechanisms leading to abnormal development. These latter applications may, in the long run, be more fruitful than a "standard testing procedure". However, such an approach to risk assessment will increase the scientific demands to be put on the investigator, because no longer fixed test recipes and "guidelines" are demanded but intelligence and mental flexibility combined with solid information of the toxicologist in basic morphological and functional developmental biology and medicine. Are our young scientists prepared for such a task?

It is to be hoped that the future scientific development of *in vitro* techniques will no longer be dominated by political restraints and by what is considered "modern", including the one-sided approach to "animal protection". As outlined, there is an intrinsic and much more important scientific aspect for the development, expansion, and also the use of *in vitro* techniques, which are by no means limited to the presently favoured culture techniques, in the field of developmental toxicology. Without doubt risk assessment will benefit from such developments.

References

Abbott BD, Birnbaum LS (1990) Rat embryonic palatal shelves respond to TCDD in organ culture. Toxicol Appl Pharmacol 99: 276-286

Abbott BD, Buckalew AR (1992) Embryonic palatal responses to teratogens in serum-free organ culture. Teratology 45: 369-382

Kao J, Brown NA, Schmid B, Goulding EH, Fabro S (1981) Teratogenicity of valproic acid: *In vivo* and *in vitro* investigations. Teratogen Carcinogen Mutagen 1: 367-382

Klug S, Lewandowski C, Blankenburg G, Merker H-J, Neubert D (1985) Effect of acyclovir on mammalian embryonic development in culture. Arch Toxicol 58: 89-96

Klug S, Lewandowski C, Zappel F, Merker H-J, Nau H, Neubert D (1990) Effects of valproic acid, some metabolites and analogues on prenatal development of rats *in vitro* and comparison with effects *in vivo*. Arch Toxicol 64: 545-553

Klug S, Lewandowski C, Merker H-J, Stahlmann R, Wildi L, Neubert D (1991) *In vitro* and *in vivo* studies on the prenatal toxicity of five virustatic nucleoside analogues in comparison to aciclovir. Arch Toxicol 65: 283-291

Lewandowski C, Klug S, Nau H, Neubert D (1986) Pharmacokinetic aspects of drug effects *in vitro*: effects of serum binding on concentration and teratogenicity of valproic acid and 2-en-valproic acid in whole-embryos in culture. Arch Toxicol 58: 239-242

Nau H, Lewandowski C, Klug S, Neubert D (1988) Pharmacokinetic aspects of drug effects *in vitro* (II): Placental transfer to the embryo and activity of some carboxylic acids structurally related to valproic acid in whole-embryos in culture. Toxicol In Vitro 2: 169-174

Nau H, Löscher W, Schäfer H (1984) Anticonvulsant activity and embryotoxicity of valproic acid. Neurology 34: 400-401

Neubert D, Blankenburg G, Chahoud I, Franz G, Herken R, Kastner M, Klug S, Kröger J, Krowke R, Lewandowski C, Merker H-J, Schulz T, Stahlmann R (1986) Results of *in vivo* and *in vitro* studies for assessing prenatal toxicity. Environ Health Perspect 70: 89-103

Neubert D (1988) Significance of pharmacokinetic variables in reproductive and developmental toxicity. Xenobiotica 18,1: 45-58

Neubert D, Barrach H-J, Merker H-J (1980) Drug-induced damage to the embryo or fetus (Molecular and multilateral approach to prenatal toxicology). In: E Grundmann (ed) Current Topics in Pathology. Vol 69, Springer-Verlag, Berlin, Heidelberg, New York, pp 241-331

Shiota K, Kosazuma T, Klug S, Neubert D (1990) Development of fetal mouse palate in suspension organ culture. Acta Anat 137: 59-64

Stahlmann R, Klug S, Lewandowski C, Bochert G, Chahoud I, Rahm U, Merker H-J, Neubert D (1988) Prenatal toxicity of acyclovir in rats. Arch Toxicol 64: 468-479

Stein G, Srivastava MK, Merker H-J, Neubert D (1990) Effects of calcium channel blockers on the development of early rat postimplantation embryos in culture. Arch Toxicol 64: 623-638

Discussion of the Presentation

Scott: Your institute has had a long-standing effort utilizing the chick embryo to evaluate teratogenic activity. Could you tell us your opinion of this species?

Neubert: That's a very relevant question. Our *in vitro* department has extensively studied the chick embryo *in ovo* for the past seven years. Dr. Barbara Heinrich-Hirsch, who is now with the Max-von-Pettenkofer Institute (BGA) did this work for us. I would like to give Dr. Heinrich-Hirsch the opportunity to answer this for you.

Heinrich-Hirsch: For several decades the chick embryo has represented one of the classical models for studying *developmental biology*, therefore it is also an obvious choice to consider for studying *abnormal* development, including the investigation of the embryofeto-toxic and especially teratogenic potential of xenobiotics. An approach like this might appear all the more reasonable and attractive since this species constitutes a model of development independent from the maternal organism.

In regard to our own experience with this system (we have evaluated more than ten thousand chick embryos over the past seven years), we started with the initial hope of a wide suitability of this system. Otherwise, we would not have put that much time and effort into these studies. We have now abandoned these lines of research, being rather disappointed in the potential of this system. With respect to the chick embryo's special features in revealing or analysing developmental toxicity of chemicals, we are now convinced that the chick embryo *in ovo* has only a very limited applicability as a model for this purpose. The situation is similar to the applicability of mammals in *in vitro* systems, and in many respects even less favourable. Thus, in our opinion, all claims of the suitability of the chick embryo *in ovo* for "primary-stage" testing are completely unjustified and unproven.

There are a number of important factors which cannot be controlled or which hamper any attempt to extrapolate results obtained to the situation possibly existing in mammals and in man. These include:

- While the effects of some xenobiotics seen in mammals was found to be *reproducible* in the chick embryo, others were not.
- Of the various possible methods of drug *application* only direct exposure of the embryo is suitable.
- It is almost impossible to control the *concentration* of the substance reaching the embryo, depending on the different methods of application (yolk, egg white, direct exposure, etc.). This problem has not been considered so far, and it has not been sufficiently investigated (if that were even possible).
- It is hardly possible to separately evaluate all of the *various* embryotoxic effects. Thus, dose-related effects can only very poorly be studied with this model.

The chick embryo *in ovo* is possibly useful for a *subsequent* investigation and confirmation of the teratogenic effect of a substance known to induce developmental toxicity ("secondary-stage testing"). This could be: confirmative studies or experiments on possible mechanisms of action, or even the "screening" of derivatives of a substance known to induce specific effects in this system. For example, with this model we demonstrated (*Arch Toxicol 65: 402-408, 1991*) that the teratogenic effect of aciclovir is not confined to mammalian species, and that abnormal development results from a direct action on the embryo. Thus, the model may be useful for providing evidence that effects seen in mammalian *in vivo* tests are not the result of maternal toxicity.

Another important aspect is the problem of metabolic conversion of xenobiotics in the chick embryo *in ovo*. It has been shown that some substances requiring metabolic activation (e.g. cyclophosphamide) are capable of inducing abnormal development in this model. We have found (*Arch Toxicol. 64: 128-134, 1990*) that drug metabolizing capacities (i.e. certain monooxygenase activities) are present within the chick embryo *in ovo* (e.g. in the yolk sac) already at early stages. However, we are far from having a complete picture of the metabolic capabilities of the chick embryo. Extensive kinetic investigations would have to be included in all studies, which has so far not been done by any investigator. For this reason all of the evidence available indicates that the chick embryo *in ovo* offers no advantages but several disadvantages, when compared with mammalian *in vitro* test systems. Thus, the use of this system could hardly be scientifically justified, but only politically (i.e. use of mammals vs. chick). However, in such a case an *in vitro* culture of chick embryos (still to be developed and standardized) will be the better choice.

Kavlock: In the United States several non-mammalian tests have been proposed to screen potential developmental toxicants, among them Xenopus, Drosophila and Hydra, would you comment on whether these tests offer any advantage over the criticisms you presented for *in vitro* screens using culture systems.

Neubert: For several years now we have explored the possibilities, in large scale studies, of using simpler vertebrates (fish, amphibia) and invertebrates (hydra, hydractinia, ciona) to obtain information on abnormal development induced by chemicals. Hydra and hydractinia have turned out, as expected, to be largely worthless in predicting potencies of a series of derivatives (*Berking, Toxicol In Vitro 5: 109-117, 1991*), and one should not continue to waste time and money on these systems.

Using the axolotl, Drs. Oberemm and Kirschbaum in our laboratory have, e.g., induced abnormal neurulation with valproic acid which has some resemblance with defects induced in mice (*NS Arch Pharmacol 344 [suppl]: R45, 1991*). Again, we are convinced that these species do not give any advantage for "prescreening". Metabolism as well as basic development are too different from that of mammals to allow far-reaching conclusions, and there is no means of extrapolating concentrations in the medium to human exposures.

An advantage may lie in the possibility to study mechanisms of action (including molecular biology), if a simpler system is found to mimic, to some extent, processes occurring in mammals. Furthermore, it should not be underestimated that the value of such species may lie in themselves, especially for studying problems of ecotoxicology. This could be the direction to proceed in the future. For the purpose of risk assessment with respect to man the BMFT (Ministery for Research and Technology) in Germany has abandoned the funding, probably rightly.

Kavlock: You mentioned that one application of culture methods was to evaluate the possible role of maternal toxicity in embryotoxicity. Do you have an example where this approach has suggested or identified a maternal influence?

Neubert: A combined *in vitro/in vivo* approach has been used for a number of agents (*Environ Health Persp 70: 89, 1986*), such as in the case of aciclovir which first revealed a teratogenic potential in *in vitro* studies. The palate culture has been used in collaboration with some pharmaceutical companies to reveal whether a substance exhibits a direct effect on palate closure in a given concentration range. The prerequisite is that you must have a variety of systems available in order to choose the most appropriate one for a special problem.

Anderson: You have very attractive *in vitro* systems for screening effects and for evaluating pharmacokinetics, have you used these systems to define or explore mechanisms of action for use in biologically based quantitative dose-response modelling?

Neubert: We have (Prof. Merker) made a grant application for doing exactly that, including studies on oncogenes (*in situ* hybridization). We have not done any dose-response modeling.

Morphological Endpoints in In Vitro Testing

Hans-Joachim Merker, Jamaledin Ghaida and Dieter Blottner

Introduction

Before starting the *in vitro* testing of potentially toxic or teratogenic substances goal and purpose of the investigations must be known. Only then are an exact planning of the experiments and the choice of appropriate methods possible. This also includes the definition of *endpoints*, i.e. parameters and functions that are to be measured in dependence on time. These endpoints may comprise special individual aspects or steps of cell metabolism or the more general behaviour of the cells, in other words the sum of individual steps. In the first case we are confronted with a vast amount of individual parameters, i.e. endpoints, which explains why the right choice is of such importance. In the second case we investigate the sums of individual functions which are, therefore, less specific and allow fewer conclusions as to the elucidation of the modes of action. This group includes, among others, the investigation of cytotoxicity or plating efficiency. - The study of an influence on or lesion of the genome requires other techniques that will not be discussed in this connection.

Methods of the In Vitro Technique

Before discussing the problem of the *in vitro* testing of substances, the technical possibilities of the *in vitro* methodology must be known. In the last few years an enormous number of culture techniques have been described in the literature. Therefore, it is difficult to discuss all these methods in a brief and comprehensive manner. One possibility to do so is based on the complexity which is characterized by the three-dimensional structure and the number of cells (cell density) and cell types in culture.

A. *Isolated cells*

1) Monolayer cultures

different cell types on different substrates with different media (e.g. with growth factors);

combined cell types (feeder layer, separation through a filter, bicameral chamber systems);

B. *Without isolation*

1) Organ, piece or slice cultures,

2) Whole-embryo cultures.

The choice of one of these techniques has obviously a great influence on the definition of the endpoints: The simpler the *in vitro* system - the easier the definition of the endpoints. Adhesion, migration and cell-cell communication, for example, can easily be measured in monolayer cultures. The more complex cultures, however, involve technical difficulties in the definition of endpoints, because numerous individual steps proceed successively or parallel in a three-dimensional structure, in a morphologically interlocked manner. Hence, the chosen endpoints or the endpoints to be investigated determine the selection of the *in vitro* method.

Selection of the Test Strategy

Any consideration of endpoints to be measured for the planning of *in vitro* testings may also be formulated in a slightly different manner: Are "horizontal" or "vertical" strategies to be used? "Horizontal" means the use of as many *in vitro* models and tissues as possible to be able to demonstrate many parameters or endpoints. This does not always necessitate a variety of material. One cell type grown in monolayer culture already allows the measurement of numerous parameters or endpoints. Such a culture system, for example, enables us to measure the following functions: adhesion, migration, proliferation, cell-cell communication, protein biosynthesis, cell respiration, cytotoxicity, plating efficiency. A "vertical" testing, however, starts from *one in vitro* model only. This is to serve the demonstration of as many steps of one metabolic or morphogenetic sequence as possible. It is obvious that the more complex culture techniques are better suited for this kind of testing. An example is the organoid or high-density culture of limb bud cells which serves to investigate blastema formation, differentiation of blastemal cells to cartilage cells, cartilage-specific matrix synthesis and further maturation or cartilage.

Monolayer Cultures

The selection of an appropriate *in vitro* method depends, of course, on the technical facilities of the researcher. The easiest way is the use of commercially available permanent cell strains in monolayer culture for the determination of cytotoxicity. This is an often applied technique which, however, appears to be somewhat restricted in its expressiveness and relevance. Another disadvantage is the relatively unspecific response of these cells in addition to the limited number of measurable endpoints (only the number of damaged or still vital cells is measured). However, there are pronounced differences among the individual cell strains.

The demonstration of these differences in the various cell strains must not necessarily be a disturbing factor. Comparison of the behaviour of different cell strains *in vitro* may yield interesting and valuable indications. In view of such differences, several cell strains or cell types should be used also for the testing by means of "simple" monolayer culture experiments. Since we know of differences among fibrocytes from different regions, even the use of several fibroblast strains is advisable. These considerations are above all true of experiments with epithelial or tumour cells. Hence, in monolayer cultures also of commercially available cells, the specificity of the effects in the different strains can be investigated as an endpoint.

Given an improved technology, such simple monolayer cultures can also serve for the study of other parameters. One example is the so-called stain uptake assay. This technique meets three fundamental requirements of *in vitro* testing: (1) easy performance, (2) the possibility to evaluate much material in a short period, and (3) its suitability to employ statistical methods. For this purpose cells are grown on multiwell plates and a stain is added for a certain period. The stain which is bound in various ways (see below), is then released due to disaggregation of cellular membranes and measured quickly, accurately and with statistical relevance in large amounts of material via determination of extinction. Using this technique, three parameters can be measured:

(1) Adhesion (measurement during the first few hours after seeding)
(2) proliferation (measurement during the first few days)
(3) cytotoxicity (addition of the test substance and measurement after confluence has been reached).

Additional information is obtained by the type of stain chosen. Neutral red, for example, is only taken up by vital cells and stored in lysosomes. Therefore, the storage of neutral red is an indication of the amount and the activity of lysosomes in a cell population. Moreover, this technique can also be used for the measurement of cell numbers. Firstly, the great number of measured cells masks the individual differences and secondly, according to experience, only very few substances to be tested affect lysosomal systems. Nevertheless, the morphology of these organelles should always be considered.

Other stains are taken up by vital cells and transformed in mitochondria through oxidation into a blue derivative. Consequently, transformation into a blue stain depends on the number and activity of the mitochondria. However, if a sufficient number of cells is measured and an effect on the mitochondria excluded, this technique can also be used for the measurement of the cell number. - A third example for the various staining techniques is the staining of proteins.

Another staining technique is the Trypan blue method. This stain only penetrates into necrotic cells. Therefore, a rise in the measured values after Trypan blue can be taken as an indication of cytotoxicity.

Consequently, these relatively simple methods are not only suited for the measurement of the cell number as endpoints, but it also gives indications of the sites and modes of action by means of the different binding sites of the stain within the cell.

When choosing appropriate endpoints or *in vitro* models, the knowledge of the substance to be tested as well as purpose and aim of the study play a decisive role in addition to the technical possibilities. In the case of an *unknown substance*, horizontal testing should be preferred, since only this test method yields extensive information. In prenatal toxicology, so-called highly complex systems, such as "whole-embryo", organ and organoid (i.e. high-density) cultures are employed for this purpose, for the complexer the culture model, the more parameters can be determined and the more likely is the recognition of a toxic effect.

In the case of a substance with a *known mode and site of action* the spectrum of test models can be reduced. This situation is given when, for example, derivatives or metabolites of a substance with known effects are tested, or when dose-response curves are calculated. This allows a selective procedure, in other words, the use of one or only few *in vitro* systems is sufficient.

An intermediate position take those substances whose *site of action is known not*, however, *their mode of action*. A substance which attacks at the cell membrane is to serve as an example. If we neglect biochemical-physiological techniques, for we want to discuss the morphological aspect, a number of techniques with different endpoints are at our disposal:

Morphology of the cell membrane:

LM: Immunofluorescence microscopic demonstration
Membrane proteins
Other immuno techniques (demonstration of receptors)
Demonstration of lectin for the characterization of the surface coat
Demonstration of enzyme activities

EM: Descriptive morphology (shape, formation of blebs)
Immunomorphology (demonstration of receptors, binding studies)

Other techniques

Adhesion and migration tests
Chemoluminiscence demonstration of ATP release
Calcium influx
Cell-cell communication

Depending on the localization of the structure or function that are to be demonstrated, such test methods can be grouped and selected according to the purpose of the investigation. However, some reference points as to the mode and site of action should already be known for this special procedure.

In the definition of endpoints we have so far started from the available techniques, from our knowledge of the substance to be tested and from the aim of our *in vitro* investigation. - Independent of these contemplations endpoints can also be distinguished according to the type of their demonstration: There are endpoints that, from the morphological point of view, are of *qualitative* and others that are of *quantitative* nature.

These properties do not exclude one another. Any testing for the demonstration of the effects of a substance should be guided by the quantitative aspect. Only on the basis of quantitative findings is it possible to calculate dose-response curves. Nevertheless, the occurrence and consideration of qualitative effects are very useful. They facilitate data collection and contribute to a higher accuracy. In addition, conclusions can often be drawn from such qualitative findings to the mode and site of action of the substance to be tested.

The action of 6-aminonicotinamide is to serve as an example of such a purely *qualitative effect* that cannot be evaluated quantitatively. This substance induces pronounced dilatations of the perinuclear cisterna. This effect appears to be specific and in the case of the rat only relates to cells of ectodermal origin. This leads to an "all or nothing" effect, in other words, either all cells are affected or none are. That means the effect is not dose-dependent. Neither the number of affected cells nor, for example, the areas of the perinuclear cisternae are of any statistical or morphometric use.

Another example of a *qualitative effect* that can be evaluated quantitatively, is the effect of cytotoxic substances which can be demonstrated by scanning electron microscopy. In simple monolayer cultures of spread fibroblasts or epithelial cells the cells become round under the influence of cytotoxic substances at the initial stage of necrosis. The rounding of the cells which can easily be demonstrated in the SEM, represents a qualitative change. However, the number of rounded cells is clearly dose-dependent and can therefore well be used quantitatively.

In contrast, *quantitative changes* in the narrower sense are not accompanied by "qualitative" deviations in structure. The morphologically demonstrable cell and tissue components exhibit only changes in their diameter, size or area. Therefore, these parameters are often not recognizable in initial evaluations. Their distinct demonstration requires statistical, especially morphometric methods. Thus, the question arises whether all material of all experiments should be evaluated morphometrically. Two objections can be raised:

(1) Morphometric methods are rather time-consuming and costly and require large amounts of material. Despite contrary rumours, the problems of an automatic discrimination and recognition of structures has not yet been solved satisfactorily. Therefore, not all material can be evaluated quantitatively.

(2) These difficulties become even more obvious if the vast amounts of structures to be measured are taken into consideration. Theoretically, all cell organelles should be measured morphometrically, which is, however, not feasible.

Hence, there is only a pragmatic solution to these problems. That means the knowledge of the researcher or of the mode/site of action should play a decisive role in the selection of the material, in other words, it should be his responsibility to decide which material should be evaluated.

In the following some examples are given of a purely statistical-morphometric evaluation of drug effects on various morphological levels. This list is opened by the quantitative evaluation of skeletal elements. Some teratogenic substances reduce the amount of blastema via various mechanisms. This effect is, in most cases, compensated. If there is no compensation, the amount of blastema is reduced which leads to hypoplasia or even aplasia. Consequently, a slight hypoplasia may well be a minisymptom of a teratogenic effect. The more important is the demonstration and evaluation even of slight changes in growth. Such effects can be induced, for instance, by cyclophosphamide.

Another example of purely quantitative changes is the behaviour of mitochondria. Gestagens lead to an increase in the size of mitochondria in the uterus epithelium, i.e. circumference, area, diameter and volume increase. In thin sections (50 nm) for electron microscopic work it is more difficult to determine number and length of mitochondria. Other techniques, such as the fluorescence microscopic demonstration on the light microscopic level must be employed. This technique is well suited to demonstrate mitochondria *in toto* in monolayer cultures, i.e. in flat, spread cells.

Definition of Endpoints

Having discussed the techniques of the demonstration and evaluation of endpoints, e.g. qualitative and quantitative, light or electron microscopic, the endpoints themselves should now be considered more closely.

Defining the term "endpoints" two types should be distinguished. Their selection is more or less arbitrary. We can differentiate between (1) endpoints in the course of cell activities or metabolic processes and (2) endpoints during morphogenesis.

The term "endpoint" clearly indicates that it comes at the end of a series or sequence. As already said, this point may be chosen arbitrarily. Formation of a matrix component, e.g. collagen type II, may serve as an example. This sequence starts with the activation of chromosomal segments, followed by RNA synthesis in the nucleus, release into the cytoplasm, synthesis of collagen at the rough endoplasmic reticulum, different processing steps up to secretion. Here, in the extracellular space, the substances are deposited in typical areas. In this sequence, RNA synthesis, for example, can be measured autoradiographically on a morphological level. However, this measurement is rather unspecific, unless the technique of *in situ* hybridisation is employed. Moreover, the obtained findings do not allow any conclusions as to possible disturbances after the step of RNA synthesis, e.g. inhibition of protein synthesis or secretion. It is possible to demonstrate the formation of protein in general and also specifically by the incorporation of amino acids, e.g. prolin. In addition, the cytoplasm can be evaluated with morphological, electron microscopic means by the behaviour of the rough ER, the Golgi apparatus, the secretion granules, etc. The features are, however, unspecific and do not say anything about the type of protein that is synthesized or secreted. This problem can only be solved with immunomorphological techniques. They allow the intracellular and extracellular demonstration of the special product, in this case collagen type II and its localization. Disturbances in the sequence of individual steps do, of course, finally lead to a reduced secretion of collagen. However, this finding does not give any information about the site of disturbance in the sequence.

Therefore, to demonstrate disturbances in this sequence in general it is advisable to define the endpoint as far in the final part of the sequence as possible. However, a change in deposition of the matrix component, to recall the mentioned example, be it qualitative or quantitative, would only indicate that some step of these processes is disturbed. The closer one gets to the site of attack of the disturbance, the more accurate is the demonstration and evaluation of the effect and mode of action. When prescreening, the endpoints should be defined far at the end of a sequence, for the elucidation of the mode of action, however, they should come immediately after the site of attack.

Somewhat different is the determination of the endpoints in the case of *morphogenetic processes*. These endpoints play a decisive role in teratology. A morphogenetic step also consists of numerous metabolic processes. But more important is the question when and where they occur. Hence, the time factor plays an important role rather than the abstract process. In the case of collagen type II, for example, the absence of the collagen type is not necessarily due to an interruption of synthesis and secretion, but may also be attributed to the absence of differentiation of the cell type responsible for this synthesis. Thus, it is likely that in this case cartilage cells *in vitro* show a normal synthesis and secretion of collagen type II, but *in vivo* the corresponding producers are missing. The effect is identical, i.e. cartilage is missing.

Consequently, the prerequisites of the selection of endpoints *in vitro* for the investigation of embryotoxic and teratogenic effects must differ. As far as the general toxicological aspect is concerned, metabolic sequences can be investigated, in the case of the teratological aspect time and place are important, i.e. morphology plays a fundamental role. Choosing a certain method one must always bear in mind that the "simpler" methods, e.g. monolayer cultures, are less suited for these purposes. More complex techniques, such as the organoid or high-density culture, the organ culture or whole-embryo culture should be preferred for the simulation of morphogenetic processes and for the assessment of site and time of attack.

However, all these *in vitro* models have one fundamental disadvantage: They are not able to metabolize. Substances with a known mode of action are not affected by this disadvantage. Substances with an unknown mode of action, however, involve a factor of uncertainty that makes all efforts appear useless. The inability to metabolize can be compensated by the use of microsome preparations or co-cultures with functionally active hepatocytes.

Acknowledgements: This work was supported by grants from the Deutsche Forschungsgemeinschaft awarded to Sfb 174.

Selected References

Barrach HJ, Neubert D (1980) Significance of organ culture techniques for evaluation of prenatal toxicology. Arch Toxicol 45: 161-187

Ekwall B, Gómez-Lechón MJ, Hellberg S, Bondesson I, Castell JV, Jover R, Högberg J, Ponsoda X, Romert L, Stenberg K, Walum E (1990) Preliminary results from the Scandinavian Multicentre Evaluation of In Vitro Cytotoxicity (MEIC). Toxic in Vitro 4: 688-691

Faustman EM, Kirby Z, Gage D, Varnum M (1989). *In vitro* developmental toxicity of five direct-acting alkylating agents in rodent embryos: Structure-activity patterns. Teratology 40: 199-210

Flint OP (1986) An *in vitro* test for teratogens: Its practical application. Fd Chem Toxicol 24: 627-631

Flint OP (1988) *In vitro* toxicology: a commercial proposition? Xenobiotica 18: 707-714

Flynn TJ (1987). Teratological research using *in vitro* systems. Environm Health Perspect 72: 203-211

Golberg L (1986) Charting a course for cell culture alternatives to animal testing. Fund Appl Toxicol 6: 607-615

Goldberg AM (1985) Integration of fundamental knowledge and *in vitro* testing strategies. In: Homburger F (ed) Concepts in Toxicology (Vol. 3), Karger, Basel, pp 1-5

Guntakatta M, Matthews EJ, Rundell JO (1984) Development of a mouse embryo limb bud cell culture system for the estimation of chemical teratogenic potential. Teratogen Carcinogen Mutagen 4: 349-364

Hassell JR, Horigan EA (1982) Chondrogenesis: A model developmental system for measuring teratogenic potential of compounds. Teratogen Carcinogen Mutagen 2: 325-331

Kimmel GL, Smith MK, Kocchar DM, Pratt RM (1982) Proceedings of the concensus workshop on *in vitro* teratogenicity testing. Teratogen Carcinogen Mutagen 2: 374-382

Klug S (1991) Whole-embryo culture - interpretation of abnormal development *in vitro*. Reprod Toxicol 5: 237-244

Knutson JC, Poland A (1980) 2,3,7,8-Tetrachlorodibenzo-*p*-dioxin: Failure to demonstrate toxicity in twenty-three cultured cell types. Toxicol Appl Pharmacol 54: 377-383

Kochhar DM (1982) Embryonic limb bud organ culture in assessment of teratogenicity of environmental agents. Teratogen Carcinogen Mutagen 2: 303-312

Kochhar DM, Hickey T (1985) Goals and potential value of alternative to teratogenicity tests. In: Homburger F (ed) Concepts in Toxicology (Vol. 3), Karger, Basel, pp 6-15

McCormick DA (1990) Refinements in the *in vitro* slice technique and human neuropharmacology. Trends Pharmacol Sci 11: 53

Merker H-J, Barrach H-J, Kochhar DM, Neubert D (1986) Die Bedeutung der sog. "limb bud"-Kultur für die experimentelle Embryologie. Verh Anat Ges 80: 137-149

Neubert D (1982) The use of culture techniques in studies on prenatal toxicity. In: Papp JG (ed) Pharmacology and Therapeutics (Vol. 18), Pergamon Press, Oxford, pp 397-434

Neubert D (1985) Benefits and limits of model systems in developmental biology and toxicology (*in vitro* techniques). In: Marois M (ed) Prevention and Physical and Mental Congenital Defects. Part A, Alan R Liss Inc, New York, pp 91-96

Neubert D, Barrach H-J (1983) Effect of environmental agents on embryonic development and the applicability of *in vitro* techniques for teratological testing. In: Kolber A, Wong TK, Grant LD, DeWoskin RS, Hughes TJ (eds) In Vitro Toxicity Testing of Environmental Agents. Part B, Plenum Publ Co, New York, pp 147-172

Neubert D, Merker H-J (eds) (1981) Culture Techniques. Walter de Gruyter Verlag, Berlin - Heidelberg - New York

Neubert D, Blankenburg G, Lewandowski C, Klug S (1985) Misinterpretation of results and creation of "artifacts" in studies on developmental toxicity using systems simpler than *in vivo* systems. Developmental Mechanisms: Normal and Abnormal. Alan R Liss Inc, New York, pp 241-266

Neubert D, Blankenburg G, Chahoud I, Franz G, Herken R, Kastner M, Klug S, Kröger J, Krowke R, Lewandowski C, Merker H-J, Schulz T, Stahlmann R (1986) Results of *in vivo* and *in vitro* studies for assessing prenatal toxicity. Environm Health Perspect 70: 89-103

Oesch F, Glatt H, Utesch D (1988) Metabolic perspectives on *in vitro* toxicity tests. Xenobiotica 18: 35-44

Oglesby LA, Ebron MT, Beyer PE, Carver BD, Kavlock RJ (1986) Co-culture of rat embryos and hepatocytes: *In vitro* detection of a proteoglycan. Teratogen Carcinogen Mutagen 6: 129-138

Robbana-Barnat S, Lafarge-Frayssinet C, Frayssinet C (1989) Use of cell cultures for predicting the biological effects of mycotoxins. Cell Biol Toxicol 5: 217-226

Safe S, Mason G, Sawyer T, Zacharewski T, Harris M, Yao C, Keys B, Farrell K, Holcomb M, Davis D, Safe L, Piskorska-Pliszczynska J, Leece B, Denomme MA, Hutzinger O, Thoma H, Chittim B, Madge J (1989) Development and validation of *in vitro* induction assays for toxic halogenated aromatic mixtures: a review. Toxicol Industr Health 5: 757-775

Saxén L (1983) *In vitro* model systems for chemical teratogenesis. In: Kober A, Wong TK, Grant LD, DeWoskin RS, Hughes TJ (eds) In Vitro Toxicity Testing of Environmental Agents. Part B, Plenum Press Corp, New York, pp 173-190

Uphill PF, Wilkins SR, Allen JA (1990) *In vitro* micromass teratogen test: results from a blind trial of 25 compounds. Toxic in Vitro 4: 623-626

Wilk AL, Greenberg JH, Horigan EA, Pratt RM, Martin GR (1980) Detection of teratogenic compounds using differentiating embryonic cells in culture. In vitro 16: 269-276

Discussion of the Presentation

Kavlock: Could you please comment further on the effect of 6-AN on the perinuclear membrane - is the effect related to the embryotoxicity, and if so, would that suggest a threshold type phenomena in the embryo? Do you know the mechanism for the swelling, and is there any hypothesis on why this would be seen only in the rat?

Merker: A long time ago, we found that after being converted into the 6-AN-NADP analogue, this antimetabolite inhibits 6-phosphogluconate dehydrogenase, leading to a dramatic accumulation of 6-P-gluconate (*Curr Top Pathol 69: 311, 1980*). Whether the dilatation of the perinuclear cisternae is the direct result of this accumulation we do not know. We also have no explanation why this very typical and pronounced effect does not occur in the mouse, the species showing the pronounced malformations (in contrast to the rat).

Clinical Aspects of Risk Assessment

Significance of Epidemiological Studies for Risk Assessment in Prenatal Toxicity

Janine Goujard

Introduction

It is rather widely accepted that results of human experience is ultimately necessary to confirm or refute the results of predictions derived from animal tests. But if human studies are closer to reality, they are obviously not amenable to the rigourous control that is possible in animal studies and, consequently, are exposed to criticisms, being judged too often as not relevant enough to come to a conclusion.

Therefore, studies should be conducted taking a number of methodological constraints into account. But one should bear in mind the limitations and inherent error sources, either known or unknown, that cannot always be avoided or eliminated.

Design of human studies as well as procedures of evaluation of their results have to take into account the main following points.

Type of Study Conducted

Large cohort studies have often been viewed as the best answer to hypotheses, closest to experimental studies, avoiding both memory misclassification and selected memory bias, easily interpreted. Their usefulness in the evaluation of the teratogenic risk of pharmaceutical drugs, after the thalidomide tragedy, has been undisputed, (Crombie et al. 1970; Degenhardt et al. 1972; Heinonen et al. 1977; Rumeau-Rouquette et al. 1978). They also offered the possibility of following up groups with different levels of exposure.

However, these studies carry heavy constraints related to the active and necessary complete follow-up of a large population in order to reach sufficient statistical power. That is why other epidemiological approaches should be considered (Table 1). They have to be carefully chosen on account of the question to be studied.

Correlation studies present the major advantage of making use of available health statistics and indicators of industrial exposure when these data exist. The main problems in these studies are the lack of possibility to check if, indeed, the exposed population had the highest risk, and to separate the effects of the relevant agent from other potential risk factors. That is why these studies should only be considered as studies generating hypotheses.

Cross-sectional studies in one area are most relevant when the prevalence of the reproductive failure is high: spontaneous abortions are the best example of this. They are easily implemented and can quickly provide a first hint. However, we have to be aware of a possible selection bias; the population present at one time might represent a selected target population. We must also bear in mind that the time sequence between exposure and effect is difficult to establish.

Case-control studies are suitable for the study of rare potential adverse outcomes. But retrospective exposure assessment and confounding factors might be sources of bias.

Selection of Comparison Groups

This step is probably one of the most difficult ones in epidemiological studies, and especially in the selection of the referent groups. An example is the assessment of the risk of spontaneous abortions using case-control studies. Theoretically, controls should be selected among all intrauterine pregnancies still alive at the gestational stage when the abortion occurred. The use of a normal birth as a control, easier in practice, is not completely adequate when induced abortions are common in the population studied.

Table 1. Methodology: Recommended human studies according to type of exposure

	Type of Exposure		
	Drug	Environmental chemical or exposure	Occupational exposure
Rare outcome (birth defect, cancer)	- Case-control study within a population where the drug is prescribed (e.g.: DES and cancer) (Herbst et al. 1971)	- Comparison of rates in several geographical areas: correlation studies (e.g. herbicides and neural tube defects) (Hanify et al. 1981) - Case-control in the area where exposure is present (Problem: low contrast of exposure) (e.g. ambient vinyl chloride and birth defects) (Theriault et al. 1983)	- Retrospective cohort studies - Case-control study in the area around industry or inside industry (Problem: low frequency of exposure)
Frequent outcome (e.g. low birth weight, spontaneous abortion)	- Prospective study comparing drug users (disease) to non-drug users (with or without disease) (Problem: prescription of the drug must be frequent enough)	- Cross-sectional study in the area (Problem: low contrast between exposed and non-exposed)	- Cross-sectional study within the industry (e.g. spontaneous abortion in hospital staff (Hemminki et al. 1982), DBCP and sperm anomalies (Whorton et al. 1977) - Cohort study of exposed and non-exposed

Source: Guidelines for the evaluation of reproductive toxicity. Commission of the European Communities. Luxembourg 1988

Considerable care should be taken in the assessment of studies in which there are selective non-responders. This situation - a possible source of bias, whatever the type of study conducted - should be evaluated quantitatively and qualitatively: how many are they, and who are they? Such information is imperative for the interpretation and extrapolation of results.

Studies based on a voluntary population - which is most often observed in cross-sectional or retrospective cohort studies - are also possible sources of errors. The acceptance to participate might be closely related with the notion of exposure *and* the existence of the adverse outcome. For instance, who are the workers having chosen to participate in surveys on Dibromochloropropane (DBCP) exposure and evaluation of the spermatogenesis (Lipshultz et al. 1980)?

A possibility to limit bias in the referent groups consists in establishing two controls for each case enrolled, as far as possible matched on one or more potentially confounding factors, and to check that these control groups are not statistically different in the rate of the reproductive outcome studied and in the confounding factors that were not taken into account in the matching procedure.

Sample Size

Too many epidemiological studies on prenatal toxicity have limited results on account of their too small sample size, leading to false-negative or false-positive results.

In cohort studies, the number of cases required in both exposed and unexposed groups to identify a significant increase of the risk of adverse outcome could be considerable (Table 2). For example, to identify a doubling of the risk of spina bifida after valproic acid or low chemical exposure, the prevalence rate of this anomaly being 5 per 10.000 births, we would need to study about 2 million pregnancies, a formidable task.

Table 2. Cohort survey and minimum number of cases required in each group for different relative risk (RR) (α = 5%, ß = 20%)

RR	Incidence of the adverse outcome		
	1 p 10.000	1 p 1.000	1 p 100
2	100.000	10.000	1.000
3	70.000	7.000	700
10	10.000	1.000	100

In case-control surveys, calculations of minimum detectable relative risk (with α = 5% and ß = 20%) for various frequencies of exposure and various numbers of cases, lead to a smaller sample size (Table 3). In comparison with the animal experiments where very high doses of chemical substances are often used during the most sensitive period of pregnancy, human data consist mainly of relatively low exposure (for a rather undefined period of time). Then, the reduction of the group size used means that an increase in the number of cases of malformations, for example, is seldom observed at low doses.

Table 3. Human studies. Calculations of minimum detectable relative risk (RR) (α = 5%, ß = 20%) for different frequencies of exposure and number of cases*

Frequency of exposure	Number of cases	
	50	100
p = 5%	RR min = 5,2	RR min = 3,5
p = 10%	RR min = 3,7	RR min = 2,7
p = 25%	RR min = 2,8	RR min = 2,1

* Using a procedure described by S.D. Walter (1977)

Evaluation of the Exposure

Data to be used to assess a health risk requires a qualitative assessment, and also a quantitative one (notions of levels, time, duration) for the making-out of dose-response relationships and possibly the definition of a safe level of exposure.

In practice, beside specific problems related to memory bias, the situation is relatively easy when the aim is to evaluate the risk related to drug exposure, particularly for drugs not used temporarily: an embryo is regarded as exposed when the pregnant woman used the drug for all the time of the pregnancy or at least during the sensitive time period of the pregnancy. Usually, investigations are recorded during the pregnancy or at birth. Very few studies are performed at distance to the outcome of the pregnancy. The experience shows that it is generally possible to control the interview of the women through their prescriptions. Moreover, the analysis of a dose-relationship is feasible.

Measurement of exposure to chemical agents is not so clear, the ideal situation being rarely available. The optimal situation is found when the study analyses a specific exposure in a specific occupation, for example "DBCP exposure in a pesticide factory and reproductive outcome". A less precise situation lies in studies analysing a specific work place situation to an average exposure: for example, "operating room personnel and reproductive outcomes". Most of the time, the situation is loosely defined, with a mixture of products, as it is the case with "work in laboratories and reproductive outcomes". Furthermore, even when it is possible to conduct studies where a qualitative assessment is performed, the quantitative assessment is generally difficult to achieve because estimates are retrospective, sometimes many years after the studied outcome, or the range of exposure is not recorded or recorded for a rather undefined period of time or indirectly recorded. For example, even when the work place is clearly defined, only the length of time when the mother or the father had worked could be used as a measure of exposure under investigation, since she or he was assigned interchangeably to different tasks. It is also difficult to achieve a quantitative assessment when the range of exposure is not sufficient to establish a dose-response or because several toxic compounds are present at the same time.

Obviously, suitable methods should be developed in the future and employed to monitor human exposures and reproductive effects after a drug, some agrochemical or household chemical has entered the market or when significant populations may be exposed to industrial chemicals (occupational or environmental pollutants). This should be done, even though we are aware that studies performing multiple comparisons are to be conducted as hypothesis-generating. Of course, a choice has to be made, partly for economic reasons and partly for practical ones. The role of linked registries, routinely collecting information recorded among large populations shall be essential.

An attempt to improve the methodology for exposures to which the population of pregnant women has already been subjected is being conducted in a sub-project of the EEC Concerted Action on Congenital Malformations (EUROCAT) (Cordier et al. 1989). The general purpose is the assessment of the risk of congenital malformations in the offspring of women exposed to chemical agents at their workplace and at home, prior to conception and during early pregnancy, excluding cases of known environmental origin or associated with chromosomal or monogenic syndromes. The methodology used is a multicentric case-control design. All information are obtained by means of an interview of the mother in the maternity unit, using a standard questionnaire, always following a similar procedure for each case-control pair. Exposure assessment is done according to a procedure used in some on-going case-control studies on cancer. The interviews are reviewed and analysed by experts on industrial hygiene for validation and classification without knowledge of the mothers' status. The chemical sheet for each worker takes into account the route of exposure (inhalation, skin contact or both), the level of exposure (high, medium, or low), the frequency of exposure (50 to 100% of time spent at work, 5 to 50%, less than 5%), the reliability of assessment (certain, probable, possible). All the questionnaires are reviewed on a central level to ensure reproducibility in exposure assessment.

Definition and Sources of Outcome Data

Among the outcomes being now most frequently studied on humans, some are easily and objectively measured (birth weight or perinatal deaths), some require specific investigations (perturbation of gametogenesis, gestational age) or careful ascertainment (spontaneous abortions and birth defects).

Early miscarriages seem to be very sensitive indicators of environmental hazards. But a high number of human conceptions die very early during their development, and even before or shortly after their implantation. Women do not notice such pregnancies, and consequently they cannot be recorded with the usual epidemiological investigations that use interviews or questionnaires. The timing of observation is essential to evaluate false-negative conclusions. It is also likely that other factors may influence the memory or the interpretation of early pregnancy loss versus delayed menstruation: education, socioeconomic factors, reproductive history, all confounding factors that should be taken into account in the evaluation of data.

Consequently, the data-source used should be carefully evaluated when miscarriage rates are compared, whether inside the study or between studies: for instance, cases recorded through women's interviews should not be compared with those recorded in computerized discharge registries available from women hospitalized for miscarriage, nor when sophisticated biochemical evidence of the pregnancy was performed.

It is also impossible to evaluate rates when the variables used in the denominator are not specified and we may notice a great deal of imagination in some studies.

As the concept of congenital malformation is not strictly defined, rates of congenital malformations unaccompanied by definition or enumeration are rather meaningless. The inclusion or exclusion of minor anomalies may considerably change the recorded rate. These minor anomalies are undoubtedly important indicators of a disturbed morphogenesis and of teratogenic agent, but they are the most difficult ones to delimit. If strict definitions are not applied, rates are difficult or impossible to evaluate. Internal malformations are only selected after special investigations. The access to these investigational techniques, the advancements of such techniques, the autopsy rate, will influence the recorded rate.

Physicians are more inclined to diagnose an anomaly when they know about a risk factor (a drug as an occupational exposure) and they have the notion of the mother's exposure for one of these risk factors. The risk of false-positive and the solutions to eliminate or, at least, to minimize this risk might carefully be considered in the protocol.

The infant's age at the time of the diagnosis will also influence the recorded rate.

These notions play an important role in the interpretation of the data: how long have the infants been followed? how was the recording made? when was the diagnosis done? to what extent were minor and variable anomalies included? have the data been collected in the same way everywhere? Looking at the notion of a strict methodological comparability between the compared groups is an important step in the evaluation.

Control for Confounding Factors

Confounder bias caused by factors linked to both the outcome and the risk factor studied might be very important in all studies, except when the prospective approach is chosen. If confounding factors (maternal age, previous pregnancies, the mother's place of residence) cannot be eliminated by adequate matching, they should be taken into account in the statistical analysis (Mantel Haenszel Adjustment method and multivariate analysis).

Statistical Analysis

No positive association is found.

Then, is it possible to conclude in favour of the absence of risks? Since we are in the situation of observation, the academical answer is No. In practice, this negative result could be informative if the study clearly showed that:

- the material is suitable for testing the hypothesis of the study;
- the study groups are large enough for the purpose;

- specific power calculations are available to evaluate the size of the excess frequency of a specific kind of embryo-fetal damage, which the study should be able to demonstrate with a high degree of probability (e.g. 80%).

Meta-analysis is an interesting statistical approach to consider in the future, when reliable studies do not have the statistical power to conclude. With meta-analysis, the results of isolated studies are added. The combined estimates of the relative risk and of the 95% confidence interval are calculated and, due to an increase in the statistical power, the meta-analysis could lead to a global significant result (Cuckle and Wald 1989).

There is a positive association. Does it mean chance, bias or causality?

The risk of chance correlations always exists especially in screening programs without a definite hypothesis, when a great number of variables are compared with several groups of defects for example. Naturally, determination of statistical significance indicates the magnitude of this risk in a single association, but it is infrequently corrected for multiple testing situations, and even if corrected, it cannot pinpoint the "nonsense" findings. No doubt many such chance correlations have found their way into teratologic literature, from which they are being slowly removed when subsequent studies fail to confirm them.

The necessary criteria for evidence for human studies could be summarized as follows:

- there is an increased risk of embryo-fetal damage in a group of exposed persons; positive bias as confounding factors can be excluded;
- the association between the risk of embryo-fetal damage and exposure is strong and unlikely to be attributable to chance variations;
- a dose-response connection can be seen between exposure and the risk of embryo-fetal damage;
- the same results can be found in several, completely separate, studies.

The fact that the same results have been observed in animal studies is a supplementary support. The subsequent constitution that elimination of the presumed cause has led to a decrease in the risk of occurrence of the embryo-fetal damage under investigation (attributable risk) is the ultimate argument.

Overall Conclusions

A number of "ad hoc" epidemiologic surveys were conducted in order to assess the risk of birth defects associated with specific environmental factors or specific occupational categories. For a majority of positive associations, however, there was no conclusive evidence of a real risk because either the number of cases was too small, or the effect of confounding factors could not be dismissed.

The scarcity of observations on humans clearly justifies the need for carefully designed studies aimed at assessing more precisely the perinatal risk associated with environmental exposure at the workplace and at home.

A weakness in epidemiological reproductive studies so far is that, in a large number of studies performed on pregnant women or at birth, the general population is used as the research material. A more efficient way is the use of selected segments of a population such as those who work in a specific chemical industry. This opens up the way for large multicentric studies, even though we are fully aware that they are always difficult to organise and finalise.

But selecting specific working conditions, and consequently specific work places is an open door to ethical problems that we also have to consider and to solve.

Another drawback of past studies was the difficulty of identifying women in very early susceptible stages of pregnancy. As home pregnancy tests become widely used, we shall be able to study more early embryos in the future.

A better knowledge of the necessity of large population studies and consequently of the necessity for people, either exposed or non-exposed, to collaborate in epidemiological studies, their sensibilisation to the environmental factors and to the quality of life, the notion of "precious child", are favourable material for the development of human perinatal studies.

Behaviour teratology is certainly the most difficult endpoint to achieve in human studies. It requires large cohort surveys, a long follow-up; the role of confounding factors such as the socioeconomic status, are so important that the success of these studies is very compromised.

While waiting for the development of relevant human studies, these sentences shall remain significant:

> *"Not to be used by pregnant women until more is known about its safety".*
>
> *"Whenever possible, pregnant women should be removed from workplaces where chemical exposure of significance can possibly occur".*

References

Cordier S, Goujard J, De Wals Ph (1989) Collaborative Study on Environment and Pregnancy. EUROCAT Guide 4, Catholic University of Louvain, Brussels

Crombie DL, Pinsent RJF, Slater BC, Fleming D, Cross KW (1970) Teratogenic drugs - RCGP Survey. Br Med J 4: 178-179

Cuckle H, Wald N (1989) Ovulation induction and neural tube defects. Lancet ii: 1281

Degenhardt KH, Kerken H, Knorr K, Koller S, and Wiedemann HR (1972) Drug usage and fetal development. Preliminary evaluations of a prospective investigation. In: M Klingberg, A Abramovici, J Chemke (eds), Drugs and Fetal Development, Plenum Press, New York, pp 467-479

Hanify JA, Metcalf P, Nobbs CC et al. (1981) Aerial spraying of 2,4,5-T and human birth malformations: an epidemiological investigation. Science 212: 359-370

Heinonen OP, Slone D, and Shapiro S (1977) Birth Defects and Drugs in Pregnancy. PSG Publishing Co., Littleton, MA

Hemminki K, Mutanen P, Saloniemie I et al (1982) Spontaneous abortions in hospital staff engaged in sterilising instruments with chemical agents. Br Med J 185: 1461-1463

Herbst AL, Ulfelder H, Poskanzer DC. (1971) Adenocarcinoma of the vagina-association of maternal stilbestrol therapy with tumor appearance in young women. New Engl J Med 284: 878-881

Lipshultz LI, Ross CE, Whorton D, Milby T, Smith R, Joyner RE (1980) Dibromochloropropane and its effects on testicular function in man. J Urol 124: 464-468

Rumeau-Rouquette C, Goujard J, Huel G, Kaminski M (1978) Malformations congénitales. Risques périnatals. Enquête prospective. Publication INSERM. Série Santé Publique, Paris

Theriault G, Ittura H, Gingras S (1983) Evaluation of the association between birth defects and exposure to ambient vinyl chloride. Teratology 27: 359-370

Walter SD (1977) Determination of significant relative risks and optimal sampling procedures in prospective and retrospective comparative studies of various sizes. Am J Epidemiol 105: 387-397

Whorton D, Krauss RM, Marshall S et al. (1977) Infertility in male pesticide workers. Lancet ii: 1259-1261

Discussion of the Presentation

Manson: Dr. Goujard, do you think data from Teratogen Information Services described by Dr. Peters can be used in classical epidemiology studies? Are there confounding factors?

Goujard: My position is that they should be considered on the same line as case reports: a hypothesis generating system. The risk of selection bias is too large to consider data from these services on the same levels as those from classic epidemiological studies. The probability to call on the Teratogen Information Services is surely related to a high probability of acquiring congenital anomalies, similarly as previous malformation histories. Therefore, a false relationship could be established. In order to minimize this bias, it is important that for each woman concerned a record of some selected characteristics (potential confounding factors) should be provided, and this distribution should be studied in comparison to pregnant women of the general population. Then, my position might be reevaluated.

Maternal Thyroid Autoantibodies and Fetal Thyroid Growth

Annette Grüters, Ulrich Bogner, Petra Schumm-Draeger and Hans Helge

Introduction

Congenital hypothyroidism due to a malformation of the thyroid gland is a disease which has been well known since the 17th century (Cranefield 1962). The pathophysiological mechanisms leading to these malformations are still unknown today in the majority of cases. The frequency of congenital hypothyroidism, as assessed by neonatal screening programs, is 1 in 3000 newborns in most of the countries (Toublanc 1990). Only 25% of the newborns have a gland of a normal size and in the normal position, and in these patients hypothyroidism is most probably caused by an inherited defect of thyroid hormone synthesis. 75% of the patients with congenital hypothyroidism have a malformed thyroid gland. In 40% the thyroid cannot be detected by radionuclear imaging studies, but modern ultrasound techniques demonstrated small remnants of thyroid tissue in those athyrotic patients (Hassan et al. 1989). In the remaining 35% of the patients, the glands are hypoplastic or located in an ectopic position. As pathogenic mechanisms for the defective migration infectious or genetic causes were discussed (Aagaard and Melchior 1959; Mihai et al. 1979).

Since autoimmunity is a frequent cause for thyroid diseases in older children and adults, as early as 1960 it was proposed that congenital hypothyroidism might be caused by autoantibodies (Blizzard et al. 1960), especially because dysgenetic congenital hypothyroidism - such as, e.g., Graves disease and Hashimotos' thyroiditis - is two to three times more common in females (Goujard et al. 1981). We therefore investigated possible teratogenic effects of maternal thyroid autoantibodies.

Patients

Investigations were carried out in 52 newborns with congenital hypothyroidism detected by a neonatal screening program and their mothers. In 32 patients a dysgenesis of the thyroid gland was diagnosed. 15 patients with athyrosis, 14 with thyroid ectopy and three patients with a hypoplastic gland. In ten patients a normal gland was detected and in ten patients no definitive diagnosis had yet been made. Antibodies were also determined in the serum of 28 older patients with congenital hypothyroidism (5 to 26 years of age). In the newborns and their mothers blood was drawn within the first three weeks after birth.

Methods

Several thyroid autoantibodies have been described. In suspected autoimmune thyroid disease today routinely microsomal (anti-thyroidperoxidase) and anti-thyroglobulin antibodies are determined. Furthermore, in most patients with Graves disease antibodies that bind competitively to the TSH-receptor have been identified. In this study with a special interest in possible teratogenic effects, the investigations were also expanded to cytotoxic and growth blocking effects of immunoglobulins in the sera of newborns with congenital hypothyroidism and their mothers.

Preliminary results of the effect of thyroid growth blocking antibodies identified *in vitro* on thyroid xenotransplants in nude mice will be described.

TSH-Binding Inhibiting Antibodies (TBIAb)

TSH-receptor antibodies were analysed by a radio-receptor assay. In brief, 50 μl serum samples were preincubated at room temperature with 50 μl of a solubilized TSH-receptor preparation for 15 min, 100 μl of ^{125}I-TSH were then added and incubations were carried out at 37°C for 60 minutes. Subsequently, polyethyleneglycol was added to precipitate the TSH-receptor preparation and the added IGG, the samples were then centrifuged and the radioactivity of the pellets counted. The upper normal limit was 10% binding.

Antibody Dependent Cell Mediated Cytotoxicity (ADCC)

ADCC was determined as described previously (Bogner et al. 1989). Human thyroid cells were obtained from surgical material. After isolation and enzymatic processing they were kept in primary culture for three days and then transferred in suspension, washed and diluted to 10^5 cells/ml and incubated with 100 μCi Na_2 $^{51}CrO_4$ for one hour. Then they were preincubated with 100 μl heat inactivated and 1:10 diluted serum samples. After one hour 100 μl mononuclear cells of a normal, healthy donor were added in an effector: target ratio of 25:1. After 18 hours of incubation at 37°C an aliquot was measured in a gammacounter (exp). The nonspecific chromium release(nonspec) and the 100% value (max) were determined from an aliquot of the supernatant of samples without effector cells.The specific lysis was calculated as follows:

$$\text{\% specific lysis:} \quad \frac{\text{cpm(exp)-cpm(nonspec)}}{\text{cpm (max)-cpm(nonspec)}}$$

Cytotoxicity was considered positive if the specific lysis of the patients serum was above the 95th percentile of the control serum samples determined in the same assay.

Thyroid Growth Blocking Antibodies (TGBAb)

For the measurement of the growth blocking activity human thyroid cells were incubated with 100 μl serum or IGG in the presence of 10^{-9} m TSH and 1 μCi ^{3}H-thymidine for two days. Then the cells were processed with 0.5/0.2% EDTA and then the DNA was precipitated with 20% TCA. After harvesting of the cells the ^{3}H-activity was measured in a betacounter. The values were expressed as % of the ^{3}H-thymidine uptake in the presence of a TSH-concentration of 10^{-9} M (100%) without the addition of normal serum. A positive blocking activity was defined with a value below two standard deviations of the normal controls (n=10), which were run in each assay.

Xenotransplantation of Nude Mice

Normal human thyroid tissue was transplanted to the back of nude mice. After four weeks 0.3 ml of serum or NaCl 0.9% (controls) were injected i.p. into the animals, so that the following groups were formed:

Group 1 (n =10): Control

Group 2 (n =10): Serum of patients with Graves disease

Group 3 (n =10): Serum of newborns with congenital hypothyroidism

After three days ^{131}Iodine (5 mCi, 0.1 ml i.v.) were injected 24 hours before sacrifice. Thyroid growth was assessed morphologically by measurement of the nuclear volume and by the ^{131}Iodine uptake.

Results

The frequency of TSH-receptor antibodies in newborns with congenital hypothyroidism was low. Only three newborns and their mothers and none of the older patients had clearly detectable antibodies. One of these mothers had primary myxedema and was therefore substituted with l-thyroxine. In this mother and her newborn with a hypoplastic gland, these IGG blocked the cAMP release from human and rat FRTL-5 thyroid cells. In the two other patients normal thyroid tissue was detected by ultrasound studies, therefore the disturbed thyroid development in most of the cases is most likely not caused by a defective TSH action (Fig. 1).

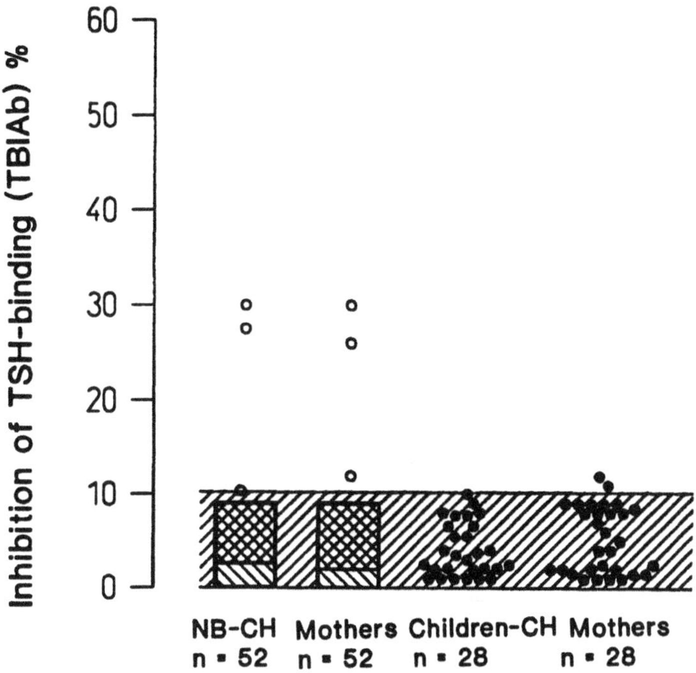

Fig. 1. TSH-binding inhibitory antibodies (TBIAb) in newborns with congenital hypothyroidism (NB-CH), their mothers, children with congenital hypothyroidism (Children-CH) and their mothers. Hatched area = range of normal values

Cytotoxic antibodies were present in 13 of the 52 (25%) newborns and their mothers. Interestingly cytotoxic antibodies were also present in 11 of 36 (30%) older children with congenital hypothyroidism and their mothers (Fig. 2). In 60 normal newborns (serum, which was left for the determination of bilirubin and hematocrite and twenty adult normal volunteers) no positive result was obtained. In contrast, of 28 children and adolescents with Graves disease and Hashimotos' thyroiditis, only in three the determination of cytotoxic antibodies was negative.

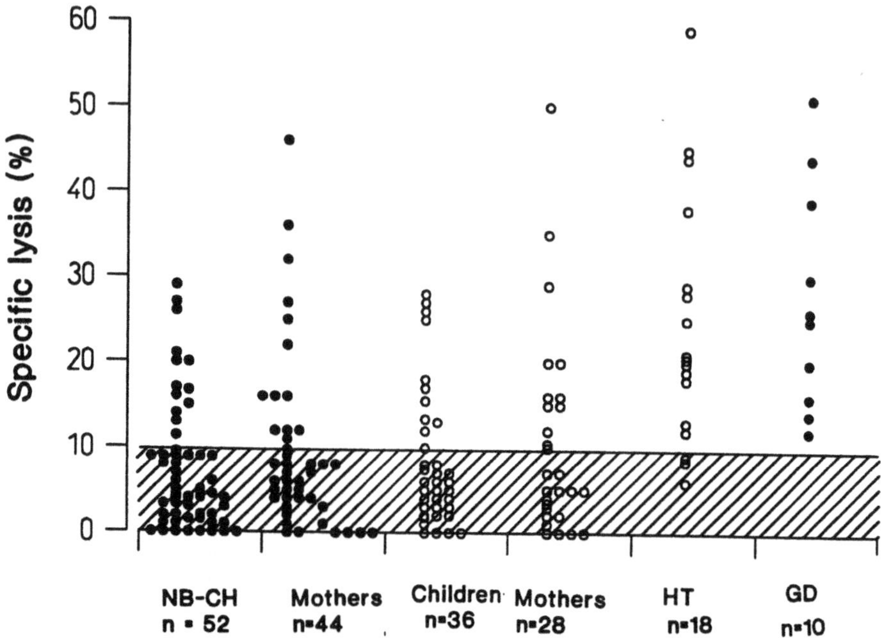

Fig. 2. Antibody dependent cell mediated cytotoxicity (ADCC) in newborns with congenital hypothyroidism (NB-CH), their mothers, children with congenital hypothyroidism (Children-CH) and their mothers and in children with Hashimotos thyroiditis (HT) and Graves disease (GD). Hatched area: Specific lysis in 60 normal newborns and adults (± 2 SD)

Thyroid growth blocking antibodies, as determined by a decreased ^{3}H-thymidine uptake into human thyroid cells, were present in 14 of the 52 newborns (27%) and in 7 of 33 of their mothers (21%) (Fig. 3). The results of the experiments in the xenotransplanted nude mice are given in Table 1.

A further reduction of the nuclear volume in group 3 might be expected after a prolonged period with repeated administration of serum or IGG.

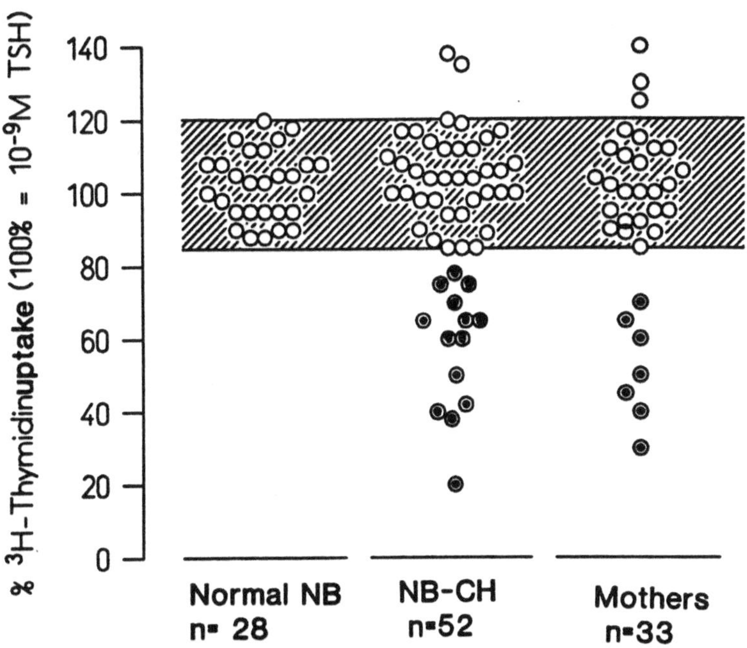

Fig. 3. ^{3}H-Thymidine uptake into human thyroid cells in the presence of serum of normal newborns (normal NB), newborns with congenital hypothyroidism (NB-CH) and their mothers. Hatched area: range of normal values in normal newborns and control adults.

Table 1. Nuclear volume and ^{131}Iodine uptake into human xenotransplants in nude mice after administration of 0.9% saline (Group 1), serum of patients with Graves disease (Group 2) and patients with congenital hypothyroidism (Group 3).

	Nuclear volume (μm^3)	^{131}Iodine uptake (%)
Group 1 (Control)	89.0 ± 5.0	16.5 ± 0.5
Group 2 (Graves disease)	143.0 ± 6.1	39.3 ± 1.5
Group 3 (CH)	76.0 ± 5.5	9.1 ± 0.7

Discussion

From the results of this study it can be concluded that cytotoxic and thyroid growth blocking antibodies can be frequently detected in newborns with congenital hypothyroidism and their mothers. Their organ specific effects can be clearly demonstrated in *in vitro* and *in vivo* experiments.

Like others we were not able to show an increased frequency of TSH-receptor antibodies in patients with congenital hypothyroidism (Brown et al. 1990). However, transient neonatal hypothyroidism has been identified in infants of mothers with autoimmune thyroiditis and could be related to the transplacental passage of antibodies that blocked the TSH-binding to the thyroid cell (Matsuura et al. 1980; Iseki et al. 1983). In most of the cases the hypothyroidism disappeared after a couple of weeks, but in some patients whose mothers presented with primary myxedema, the condition seemed to be long-standing. Since these conditions are very rare, a routine anti-TSH-receptor antibody determination in every pregnant woman to predict the possible occurrence of transient hypothyroidism is not suggested, but in pregnancies of women with atrophic primary myxedema it is strongly recommended.

Old (Blizzard et al. 1960; Sutherland et al. 1960) and recent studies (van der Gaag et al. 1985, 1986; Bogner et al. 1989) have suggested a relationship between maternal autoimmune thyroid disease, although mothers with autoimmune thyroid disease usually give birth to unaffected infants (Parker et al. 1961). Cytotoxic and growth blocking mechanisms of the autoantibodies have been studied, because a study of newborns in a mass screening program had shown that the common antibodies against thyroid peroxidase and thyroglobulin are not frequent in newborns with thyroid failure (Dussault et al. 1980). Using a cytochemical assay it has been shown (van der Gaag et al. 1985) that thyroid growth blocking antibodies are present in the serum of more than 50% of mothers of infants with congenital hypothyroidism. The significance of these findings was debated because of the limitations due to the methodological difficulties of the assay used. The ^{3}H-thymidine uptake used in our study is also subject to be discussed in its relevance for the interpretation of cell growth (Dumont et al. 1987). Since in our study ^{3}H-thymidine uptake was controlled by counting the cell number and the mitosis index, and since other pitfalls such as high serum iodine concentrations were excluded (Grüters et al. 1990), the reported incidence of 27% of positive antibodies is supported by the preliminary results of the *in vivo* experiments in nude mice, which also showed a growth blocking effect. This effect seems not to be mediated through a TSH-receptor dependent mechanism, although TSH is one of the growth factors involved in thyroid growth (Roger and Taton 1987), since all patients with thyroid growth blocking antibodies were negative for the TSH-binding inhibition.

Our results of thyroid antibody dependent cell mediated cytotoxicity in patients with congenital hypothyroidism have been recently confirmed by others, showing a similar frequency with porcine thyroid cells (Rodien et al. 1991). These results can be interpreted as evidence for the presence of antibodies which specifically bind to thyroid surface antigens, targeting the cytotoxic action of natural killer cells. These antibodies have been detected in the maternal and patients' sera several years after birth. The antigen-antibody reaction involved in ADCC does not appear to be correlated with the thyroperoxidase system (Bogner et al. 1991).

The results of our study are still the subject of further research, since they are not yet defined on a molecular basis. Although the results are in favour of the responsibility of maternal factors in the dysembryogenesis of the fetal thyroid gland, further studies are necessary before this concept can be completely accepted. Then maternal markers might be available to predict fetal hypothyroidism already before or during pregnancy and interactions of environmental and nutritional factors, e.g. iodine supply, will have to be assessed in their effects on the prevalence of these immunological disturbances.

References

Aagaard K, Melchior J (1959) The simultaneous occurrence of congenital toxoplasmosis and congenital hypothyroidism. Acta paediat 48: 164-168

Blizzard RM, Chandler RW, Landing BH, Petit MD, West CD (1960) Maternal autoimmunization to thyroid as a probable cause of athyrotic cretinism. New Engl J Med 263: 327- 336

Bogner U, Grüters A, Sigle B, Helge H, Schleusener H (1989) Cytotoxic antibodies in congenital hypothyroidism. J Clin Endocrinol Metab 68: 671-675

Bogner U, Kotulla P, Peters H, Schleusener H (1990) Thyroid peroxidase/microsomal antibodies are not identical with thyroid cytotoxic antibodies in autoimmune thyroid disease. Acta Endocrinol 123: 431-437

Brown RS, Keating P, Mitchell E (1990) Maternal thyroid-blocking immunoglobulins in congenital hypothyroidism. J Clin Endocrinol Metab 70: 1341-1344

Cranefield PF (1962) The discovery of cretinism. Bull Hist Med 36: 489-511

Dumont JE, Roger PP, Ludgate M (1987) Assays for thyroid growth immunoglobulins and their clinical implications: Methods, concepts and misconceptions. Endocrin Rev 8: 448-454

Dussault JH, Letarte J, Guyda H, Laberge C (1980) Lack of influence of thyroid antibodies on thyroid function in the newborn infant and on a mass screening program for congenital hypothyroidism. J Pediatr 96: 385-389

van der Gaag RD, Drexhage HA, Dussault JH (1985) Role of maternal immunoglobulins blocking TSH-induced thyroid growth in sporadic forms of congenital hypothyroidism. Lancet 1: 246-249

van der Gaag RD, Frisch H, Weissel M, Wick G, Drexhage H (1986) Congenital hypothyroidism in a Turkish family: the role of immunoglobulins blocking the trophic effects of TSH *in vitro*. Acta Endocrinol 111: 44-49

Goujard J, Safar A, Rolland A, Job JC (1981) Epidemiologie des hypothyreoidies congenitales malformatives. Arch fr. Pediat 38: 875-879

Grüters A, Bogner U, Biebermann H, Peters H, Helge H (1990) Thyroid growth blocking antibodies in congenital hypothyroidism: pitfalls in their determination. Horm Res 35 (S3): 41

Hassan M, Garel C, Leger J, Czernichow P (1989) Cervical ultrasound in congenital hypothyroidism. In: Delange F, Fisher DA, Glinoer D (eds) Research in Congenital Hypothyroidism. Plenum Press, New York, pp 193-198

Iseki M, Shizume YM, Oikawa T, Hojo H, Arikawa K, Ichikawa Y, Momotami N, Ito K (1983) Sequential serum measurements of thyrotropin binding inhibition immunoglobulin G in transient neonatal hypothyroidism. J Clin Endocrinol Metab 57: 384-387

Mihai K, Ichihara K, Amino M, Nose O, Yabuuchi H, Tsuruhara T, Oura T, Kurimura T (1979) Seasonality of birth in sporadic cretinism. Early Hum Dev 3: 85-88

Matsuura N, Yamada Y, Nohara Y et al. (1980) Familial neonatal transient hypothyroidism due to maternal TSH-binding inhibitory immunoglobulins. New Engl J Med 303: 738 741

Parker RH, Beierwaltes WH, Elzinga KE et al. (1961) Thyroid antibodies during pregnancy and in the newborn. J Clin Endocrinol Metab 21: 792-798

Roger PP, Taton M (1987) TSH is a direct growth factor for normal human thyrocytes. Ann Endocrinol 48: 163

Rodien P, Madec AM, Bornet B, Stefanutti A, Orgiazzi J (1991) Assessment of antibody dependent cell mediated cytotoxicity in autoimmune thyroid disease using porcine thyroid cells. Abstract 10th Int. Thyroid Conference, The Hague, p 168

Sutherland JM, Esselborn VM, Burket RL et al. (1960) Familial non-goitrous cretinism apparently due to maternal antithyroid antibody. N Engl J Med 263: 336-341

Toublanc JE and the working group on congenital hypothyroidism of the ESPE (1990) Epidemiological inquiry on congenital hypothyroidism in Europe. Horm Res 34: 1-3

Pre- and Postnatal Risk Factors for the Development of Atopy in Childhood

Ulrich Wahn, Renate Bergmann, Maria-Elisabeth Herrmann, Annette Grüters, Werner Luck and Markos Schmitt

Introduction

Atopic diseases (bronchial asthma, allergic rhinitis and conjunctivitis, atopic eczema, allergic urticaria and gastro-intestinal allergy) represent important health problems in infancy and childhood. They are associated with the genetically mediated capacity to mount specific IgE-antibody responses to minute amounts of environmental allergens. The cumulative prevalence of these disorders in Germany is not exactly known, from other western countries it is reported to range between 25 and 35% (Croner and Kjellman 1990).

Genetic Aspects

It has been known for generations that atopy runs in families. Studies performed by Marsh (1982) indicate that there are at least two different genetic control mechanisms regulating the human IgE response. From family studies it has become obvious that the high IgE phenotype is controlled by one recessive gene, whereas specific IgE responses are closely linked to certain HLA-DR phenotypes. Thus, a high correlation between serum-IgE to the antigen V of ambrosia pollen (ragweed) and the HLA-DR 2 phenotype could be demonstrated. Other associations between specific IgE-antibody responsiveness towards highly purified allergens and HLA phenotypes have been described.

Recently it was claimed by another group (Cookson et al. 1989) that one gene on the short arm of the 11th chromosome segregates with the atopic phenotype. However, there are serious questions of the definition of atopy in this study, which included clinical symptoms, specific IgE antibody responses as well as elevated serum-IgE levels. Therefore this hypothesis of the atopy gene needs further evaluation and confirmation.

Prediction of Atopy

The risk of a neonate to develop atopic symptoms is primarily determined by the atopic state of his parents and siblings.

6.5% out of 7000 newborns recruited from a population based study had at least two first degree relatives with a convincing history of atopy. As it has been shown by Croner and Kjellman (1990) the risk of these children to develop atopic symptoms by the age of 10 is approximately 60 - 80%, compared to 10% in the general population.

During the embryo-fetal period the IgE production begins around week 11. Neonates with cord blood IgE concentrations higher than 0.9 ng/ml have been found to be at risk to develop atopic symptoms early in life. According to our own German multicenter study approximately 8.9% of all neonates are born with cord blood IgE concentrations higher than 0.9 ng/ml. The risk populations defined by family history and cord blood IgE are two different groups of infants. Only a small subgroup (0.7%) presenting with both risk markers (double positive family history and high cord blood IgE) could be identified by us. Preliminary data indicate that this subgroup runs the highest risk to develop serious atopic symptoms already in infancy.

The sensitivity as well as specificity of both predictive parameters, family history and cord blood IgE is not sufficient to use these criteria as general *screening* tests. Therefore, there is a need to study other parameters such as lymphocyte surface markers as the low affinity receptor for IgE (CD 23) including its soluble fragments in serum (sCD 23 or IgE binding factors), interleukins, especially those which are involved in the regulation of IgE (IL 4), the spectrum of fatty acids in cord blood, or the capacity of cultivated lymphocytes to produce immunoglobulin E *in vitro*.

Recent studies in our laboratory have shown that under certain experimental conditions (in the presence of hydrocortison and interleukin 4) cord blood lymphocytes are able to produce immunoglobulin E *in vitro*, however, we were unable to detect a correlation of *in vitro* IgE production with cord blood IgE levels.

Prenatal Risk Factors

There are a number of studies addressing the question of prenatal sensitization to food and inhalant allergens (Chandra et al. 1986; Fälth-Magnusson and Kjellman 1987; Lilja et al. 1988). From these investigations it can be concluded that, if intrauterine sensitization occurs, it seems to be very infrequent and without major clinical relevance. Assessment of the *in utero* exposure to maternal smoking by cotinine measurements in cord blood have also been performed in a few studies. It has been suggested that maternal smoking during pregnancy is causally related to cord blood IgE, and possibly also to a later development of atopic symptoms. Since high cord blood IgE level is predictive of infant allergy, further studies are urgently needed. Also prenatal sensitization to food has been reported to be rather rare. During the last year a number of prospective controlled studies have been undertaken to determine the effect of maternal diet, void of milk and egg or other potential allergens during the third trimester of pregnancy. While a Canadian study (Chandra et al. 1986) suggested a beneficial effect of diet during pregnancy, two Scandinavian studies (Fälth-Magnusson and Kjellman 1987; Lilja et al. 1988) were unable to detect differences between prenatally randomized groups of mothers with or without diet.

Postnatal Sensitization to Foods

For several decades it has been debated, whether breast feeding can influence the risk of atopic diseases, and numerous studies have been conducted to test this hypothesis. Many studies lack strict diagnostic criteria. It appears that if high risk children are followed in prospective studies, exclusive breast feeding for a sufficient time (4 to 6 months) obviously is protective. As it was demonstrated by Høst (1989) and colleagues (Høst et al. 1988), early inadvertent and occasional exposure to cow's milk proteins, which is rather common in obstetric wards, may initiate sensitization in predisposed neonates. Subsequent exposure to minute amounts of bovine milk proteins in human milk may act as booster dose eliciting allergic reactions. Recent studies have indeed reported the presence of minute quantities of food proteins, such as ß-lactoglobulin, ovalbumin and ovomucoid in human milk. Consequently, studies on the effect of maternal diet during the lactation period, devoid of eggs, cow's milk and fish during the first three months of breast feeding were untertaken. They indicate that a "hypoallergenic" diet of the mother during the lactation period leads to a decreased incidence of atopic dermatitis during the first six months of life in infants with a positive family history for atopy (Zeiger et al. 1989).

The early introduction of solid foods into the diet of infants at risk for atopy seems to increase the risk of early manifestation of atopic eczema by the factor two to three (Fergusson et al. 1990).

In general, hypersensitivity reactions to inhalant allergens do not become manifest before the end of the first year of life.

Beside pollen allergens, allergenic proteins from fecal particles of house dust mites as well as from pets play a major role. In a cross-sectional study our own group (Lau et al. 1989) could demonstrate that there is a dose-response relationship between allergenic exposure to house dust mite allergens and the risk of sensitization (production of specific IgE antibodies). Sporik et al. (1990) recently suggested that the degree of exposure to indoor allergens, produced by house dust mites, not only determines the risk of sensitization, but also the time of primary manifestation of asthma in infancy and early childhood.

Conclusions

The development of atopic allergy is controlled by genetic factors and depends on a number of environmental factors. Up to now, there is little evidence for a major role of environmental factors during pregnancy, however, further data on the influence of maternal smoking are needed. Different environmental factors are clearly important in the neonatal period and early infancy. Among those the early exposure to food allergens and probably also indoor inhalant allergens play a key role. Other adjuvant factors, such as passive smoking, pollution and respiratory infections may increase the risk of sensitization and influence the natural course of disease.

A general screening program for atopy should not be recommended before preventive protocols have been shown to be effective.

References

Chandra RK, Puri S, Suraiya C, Cheema PS (1986) Influence of maternal food antigen avoidance during pregnancy and lactation on incidence of atopic eczema in infants. Clin Allergy 16: 563-71

Cookson WOCM, Sharp PA, Faux JA, Hopkins JM (1989) Linkage between Immunoglobulin E responses underlying asthma and rhinitis and chromosome 11q. Lancet 1: 1292-1295

Croner S, Kjellman N-IM (1990) Development of atopic diseases in relation to family history and cord blood IgE levels. Ped Allergy Immunol 1: 14-20

Fälth-Magnusson K, Kjellman N-IM (1987) Development of atopic disease in babies whose mothers were receiving exclusion diet during pregnancy - a randomized study. J Allergy Clin Immunol 80: 868-75

Fergusson DM, Horwood J, Shannon FT (1990) Early solid feeding and recurrent childhood eczema: A 10-year longitudinal study. Pediatrics 86, No 4:541-546

Høst A (1989) The influence of early allergen contact on the development of atopy in childhood. Allergologie Jg 12, Kongreßausgabe: 186-191

Høst A, Husby S, Oesterballe O (1988) A prospective study of cow's milk allergy in excessively breast-fed infants. Acta Pædiatr Scand 77: 663-670

Lau S, Falkenhorst G, Weber A, Werthmann I, Lind P, Buettner-Goetz E, Wahn U (1989) High mite allergen exposure increases the risk of sensitization in atopic children. J Allergy Clin Immunol 90: 718-725

Lilja G, Dannaeus A, Magnusson KF (1988) Immune response of the atopic woman and fetus: effects of high- and low-dose food allergen intake during late pregnancy. Clin Allergy 18: 131-42

Marsh DG, Hsu SH, Roebber M (1982) HLA-DW2: a genetic marker for human immune response to short ragweed pollen allergen Ra 5. J Exp Med 135: 1439-1451

Sporik R, Holgate ST, Platts-Mills TA, Cogswell JJ (1990) Exposure to house dust mite allergen (Der p I) and the development of asthma in childhood. N Eng J Med 323: 502-507

Zeiger RS, Heller S, Mellon MH, Forsythe AB, O'Connor RD, Hamburger RN, Schatz M (1989) Effect of combined maternal and infant food-allergen avoidance on development of atopy in early infancy: A randomized study. J Allergy Clin Immunol 84: 72-89

Discussion of the Presentation

Neubert: Is there a risk of a child developing an allergy if the mother experiences an allergic reaction during pregnancy? Do you think that the antigens are transferred to the fetus in order to trigger the IgE response in the fetus? Or could this be endogenous antigens?

Wahn: As far as we know intrauterine sensitization to allergens, to which the mother is reacting, occurs, if at all, only very infrequently.

Segerer: As to the preventive measures during infancy: is it really possible to prevent allergic symptoms or are they just postponed?

Wahn: So far we only have hard data on the first 18 months of life. In this period atopic symptoms can be significantly reduced by preventive measures. It may very well be that later in childhood prevention has to focus on factors other than diet.

Schleusener: Do IgE-antibodies cross the placenta like IgG?

Wahn: No, they don't. IgE-antibodies in cord blood are produced by the fetus. However, avoiding of contamination by mother's blood during sampling of cord blood is crucial.

Does Smoking During Pregnancy Alter Brain Perfusion in the Neonate? A Doppler Study

Hashem Abdul-Khaliq, Hugo Segerer and Michael Obladen

Abstract

Cerebral blood flow velocities (CBFVs) were significantly higher in 41 infants of smoking mothers compared to 59 control infants although gestational age, birth weight, and systolic blood pressure were lower in infants exposed to smoke prenatally. We speculate that increased CBFVs in these neonates are due to disturbed prostanoid synthesis.

Abbreviations. CBFV = Cerebral blood flow velocity; BP = Arterial blood pressure

Introduction

In Western countries, 25 - 30% of pregnant women smoke (Sokol et al. 1990; Cnattingius et al. 1988). Effects of maternal smoking on fetal and neonatal morbidity and mortality are well established (Merritt 1981; Sokol et al. 1990; Cnattingius et al 1988; Nash and Persaud 1988).

Maternal smoking has been shown to increase fetal blood flow velocities in the descending thoracic aorta and in umbilical vessels (Morrow et al. 1988; Lindblad et al. 1988). Ahlsten et al. (1987) showed an impaired vascular reactivity of the skin in newborn infants of smoking mothers 24 - 48 h after birth. Assuming that postnatal CBFVs could be influenced by prenatal smoke exposure, we measured CBFVs by Doppler sonography in infants who were born to smoking mothers, and in healthy newborns of non-smoking mothers.

Methods

Between January and May 1991, forty-one subsequent infants born to mothers who smoked 10 or more cigarettes per day during their pregnancy were studied. In each case, the mother was asked about smoking habits, drug abuse, and medications. We excluded infants who presented with factors that are known or suspected to influence CBFVs: immaturity with less than 35 completed weeks of gestation, birth weight below 10th percentile, mean arterial blood pressure < 30 mmHg during the first day of life, perinatal asphyxia with an umbilical artery pH < 7.15, intra- or periventricular hemorrhage, signs of persistent ductus arteriosus, venous hematocrit > 65%, infants of mothers with diabetes, gestosis, or hypertension with blood pressures > 160/100 mmHg during pregnancy, or drug abuse. Informed parental consent was obtained for the Doppler examination and urine collection.

Fifty-nine healthy newborn infants born to non-smoking mothers in the same time period served as controls after informed parental consent had been obtained. The same exclusion criteria were applied to these infants.

At examination, all infants were between 20 and 42 hours old. The Doppler examination was postponed if the infant was not quiet. BP was measured at the time of the Doppler examination using a Dinamap[R] (Criticon Inc., Florida, U.S.A.). A pulsed-Doppler, two-dimensional ultrasound scanner (Sonoline SL-2, Siemens, Erlangen, Germany) with a 7.5-MHz Doppler probe was used to exclude structural abnormalities and to measure CBFVs. All Doppler examinations were carried out by the same investigator (H. A-K.). We examined CBFVs in the internal carotid artery immediately beneath the lateral edge of the sella turcica, in the anterior cerebral artery directly anterior to the third ventricle beneath the lower edge of the corpus callosum, and in the basilar artery between pons and the skull base (Deeg and Rupprecht 1989). Doppler frequencies were recorded when sharpest characteristically visual and highest audible signals were obtained (Perlman 1985). A tracing of at least 10 heart cycles was printed out (Mitsubishi Video Copy Processor K 70S, Mitsubishi Electric Corporation, Japan). From these recordings, maximal systolic, mean, and end-diastolic velocities could be read (Fig. 1).

Differences between variables of smoke exposed and control infants were compared by means of the Mann-Whitney test.

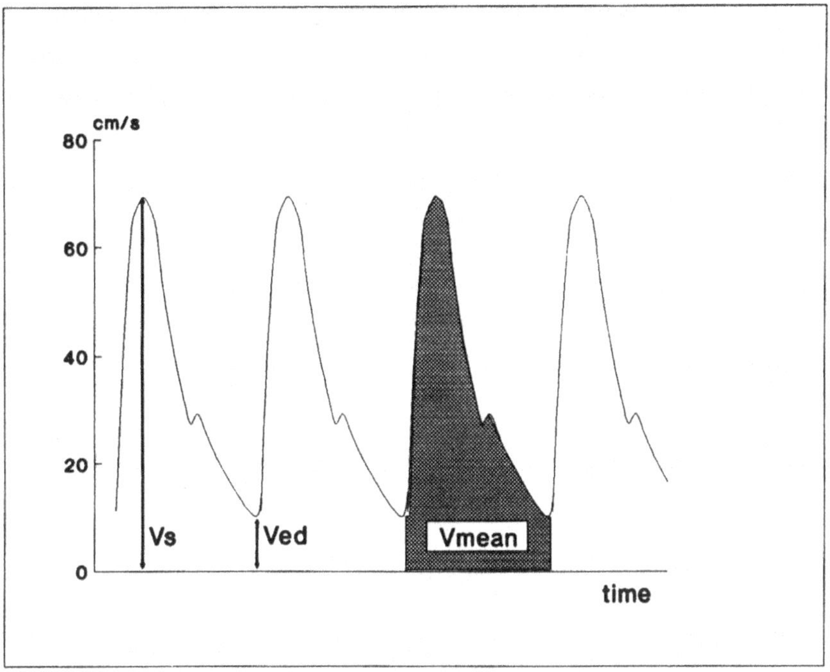

Fig. 1. Example of a blood flow velocity tracing (Doppler spectrum). V_s: systolic (maximal) flow velocity, V_{ed}: end-diastolic velocity. The area under the Doppler curve represents the mean flow velocity

Analysis of variance (ANOVA) was used to assess a possible influence of birth weight, gestational age, postnatal age, heart rate, and systolic BP on CBFVs. All calculations were carried out with the SPSS-PC+ statistical package (SPSS Inc., Chicago, ILL). A statistically significant difference was assumed if p was lower than 5%.

Results

Mean gestational age and mean birth weight were significantly lower in infants exposed to cigarette smoke before birth (Table 1). The mean heart rate of these babies was slightly but not significantly higher as compared to control infants. Systolic BP was lower in infants of smoking mothers than in the control infants. Between smokers' and control infants, there were no significant differences in umbilical artery pH, Apgar scores at 1, 5 or 10 min., diastolic or mean BP at examination and postnatal age at examination. None of the 100 infants studied had to be referred to a special care nursery.

Table 1. Clinical data (mean ± SD) of control and smoke exposed infants

	Controls n = 59	Smoke exposed n = 41	Statistics
Gestational age (w)	39.6 ± 1.0	38.8 ± 1.6	$p < 0.01$
Birth weight (g)	3402 ± 375	3147 ± 469	$p < 0.01$
Umbilical art.-pH	7.3 ± 0.06	7.3 ± 0.06	NS
Apgar score			
1 min	9.1 ± 0.5	9.0 ± 0.5	
5 min	9.9 ± 0.4	9.9 ± 0.3	NS
Age at Doppler examination (h)	31.7 ± 6.4	29.5 ± 5.7	NS
Syst. blood pressure	72.1 ± 11.3	67.7 ± 10.8	$p < 0.05$
Mean art. pressure	55.8 ± 9.7	53.8 ± 8.8	NS
Heart rate (/min)	127 ± 16	134 ± 17	$p = 0.06$

In each of the three cerebral arteries examined, maximal systolic flow velocity and mean flow velocity were significantly higher in infants of smoking mothers than in control infants. Figure 2 gives an example; it depicts maximal systolic flow velocities in the anterior cerebral artery of both groups of infants. Analysis of variance showed that birth weight, gestational age, age at examination, systolic or mean BP, and heart rate did not contribute to this observation (data not shown).

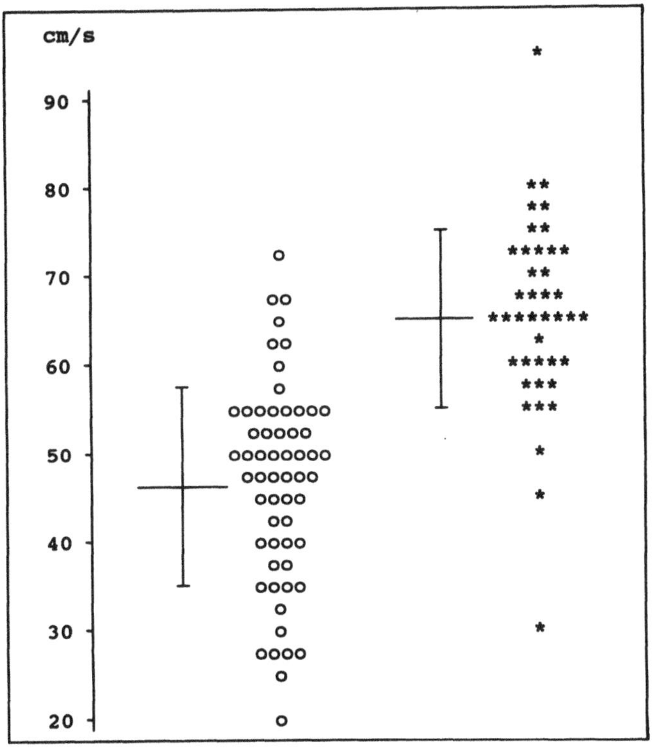

Fig. 2. Systolic blood flow velocities measured in the anterior cerebral arteries of infants of smoking mothers (*) and of control (*o*) infants. Horizontal bars represent means, vertical bars represent standard deviations in each group

Discussion

CBFVs in newborn infants are influenced by various factors: birth weight, gestational age, postnatal age, drug administration and intracranial or cardio-pulmonary abnormalities (Perlman 1985; Raju 1991). In our study, according to the lower birth weight and gestational age of babies born to smoking mothers compared to control infants, lower CBFVs were to be expected in the former group (Deeg and Rupprecht 1989). However, we found significantly higher CBFVs in smokers' babies.

We did not find factors other than prenatal exposure to tobacco smoke to explain higher CBFVs in the study group. The CBFVs in our control infants were in the same range as those observed by others (Deeg and Rupprecht 1989).

Prenatal Doppler studies have demonstrated an acute increase in fetal blood flow velocities associated with elevated maternal BP and heart rate as well as an increase in fetal heart rate immediately after maternal smoking (Morrow et al. 1988; Lindblad et al. 1988). We do not think that the differences in CBFVs observed in our study between infants of smoking mothers and controls can be explained by such acute effects of nicotine, catecholamines, or carbon monoxide, both nicotine and carbon monoxide have a half-life too short to explain a prolonged influence in infants 20 hours or more after birth. Also, if nicotine were directly responsible for the increased CBFVs, concomitant catecholamine-mediated changes in BP and heart rate should be expected but could not be observed in our study.

Our data support observations of Ahlsten et al. (1987) who demonstrated that the reactivity of skin vessels in newborn infants of smoking mothers was impaired 24 - 42 hours after birth. Similar changes may occur in other vessels. The factors causing vascular tone alterations after smoking are still unclear. Prostanoids play an important role in the regulation of the fetoplacental as well as of the cerebral circulation (Chaudhuri et al 1991; Leffler and Busija 1987). Reduced synthesis of prostacycline, for example, has been found in umbilical arteries from newborns of smoking mothers, and might well play a role also in other vessels (Ahlsten et al. 1987; Dadak et al. 1981).

Further studies are necessary to elucidate whether these changes are related to long-term effects of maternal smoking during pregnancy such as SIDS (Bulterys et al. 1990) or impaired mental development (Naeye and Peters 1984).

Acknowledgements: These studies were supported by the Deutsche Forschungsgemeinschaft, Sfb 174-A9.

References

Ahlsten G, Ewald U, Tuvemu T (1987) Impaired vascular reactivity in newborns of smoking mothers. Acta Paediatr Scand 76: 248-253

Bulterys MG, Greenland S, Kraus JF (1990) Chronic fetal hypoxia and sudden infant death syndrome: Interaction between maternal smoking and low hematocrit during pregnancy. Pediatrics 86: 535-540

Chaudhuri G, Heglund B, Meirik O (1991) Endothelium-derived vasoactive substances in fetal placental vessels. Seminar Perinatol 15: 4-10

Cnattingius S, Haglund B, Meirik O (1988) Cigarette smoke as risk factor for late fetal and early neonatal death. Br Med J 279: 258-261

Dadak C, Leithner C, Sinzinger H (1981) Diminished prostacycline formation in umbilical arteries of babies born to women who smoke. Lancet i: 94

Deeg KH, Rupprecht Th (1989) Pulsed Doppler sonographic measurement of normal values for the flow velocities in the intracranial arteries of healthy newborns. Pediatr Radiol 19: 71-78

Leffler CB, Busija DW (1987) Arachidonic acid metabolites and perinatal cerebral hemodynamics. Seminar Perinatol 11: 31-42

Lindblad A, Marsal K, Andersson K-E (1988) Effect of nicotine on human fetal blood flow. Obstet Gynecol 72: 371-382

Merritt TA (1981). Smoking mothers affect little lives. Am J Dis Child 135: 501-502

Morrow RJ, Ritchie JWK, Bull SB (1988) Maternal cigarette smoking: The effects on umbilical and uterine blood flow velocity. Am J Obstet Gynecol 159: 1069-1071

Naeye RL. Peters EC (1984) Mental development of children whose mothers smoked during pregnancy and effects on the fetus. Obstet Gynecol 64: 601-607

Nash JE, Persaud TVN (1988) Embryopathic risks of cigarette smoking. Exp Pathol 33: 65-73

Perlman JM (1985) Neonatal cerebral blood flow velocity measurement. Clin Perinatol 15: 179-193

Raju TNK (1991) Cerebral Doppler studies in the fetus and newborn infant. J Pediatr 119: 165-174

Sokol JR, Drugan A, Evans M (1990) Substance abuse, smoking in pregnancy. In: Eden RD, Boehm FH (eds) Assessment and care of the fetus, physiological, clinical and medicolegal principles. Prentice-Hall International Inc., pp 687-691

Discussion of the Presentation

Weissinger: You presented a list of "exclusion criteria" for neonates of smoking parents including gestation period <35 weeks, patent ductus arteriosus, etc. Then you presented that "no exclusion criteria" were used for non-smoking mothers. How can you draw comparative conclusions across the two sets of data?

Segerer: "No exclusion criteria" was meant to mean "No *additional* exclusion criteria". In actuality identical exclusion criteria were used for both groups, however no control neonates fulfilled the exclusion criteria, anyway.

Neubert: Do you know anything about the smoking behaviour of the women in the study? When did they smoke their last cigarette? It will be very important to assess whether the effect you observed is very persistent. Do you plan a follow-up on the children?

Segerer: We don't have data on the time interval between the last cigarette and birth or time of examination, respectively. I agree that a further study is necessary that looks into such data, that includes repeated measurements of CBFV's, and that allows to relate these findings and socioeconomic data to later outcome.

Scott: Did the infants have the opportunity to nurse after birth and before your measurement of cerebral blood flow therefore perhaps taking in nicotine or metabolites via mothers' milk?

Segerer: Yes, some of the infants were nursed by their mothers, but only very small amounts of milk can be expected at 20 to 44 hours after birth, which was the time of examination.

Scott: Is cerebral blood flow different in male vs. female infants?

Segerer: No.

Risk Assessment of Tocolytic Therapy in Pregnancy

Ruth Hildebrandt

Introduction

The risk of drug treatment exists throughout the entire course of pregnancy. The induction of malformations is but a part of the whole risk. Effects on organ function might be induced in late pregnancy and their intensity might be dependent on the ability of the fetus and the newborn child to compensate. It is evident that gestational age at the time of treatment, the time interval between the end of treatment and birth as well as additional factors such as - for example - perinatal asphyxia or concomitant diseases of the child have an influence on what can be observed. Furthermore, drugs are prescribed to treat a disease of the pregnant woman or a complication of pregnancy which in itself might be harmful to the fetus. Therefore, the benefit of the treatment has to be taken into account. In addition, data on the transplacental passage of drugs in the human being are rare due to ethical and experimental problems.

For these reasons it is evident that there is a need for clinical studies which are designed to describe what happens in the human being under the condition of drug treatment of a disorder in pregnancy. The design of the study should enable us to test hypotheses on suspected effects of the drug as well as supply us with data for an exploratory data analysis in search of any other effect.

Premature labour is one of the main complications in pregnancy requiring drug treatment with about 10% of pregnant women experiencing premature labour. Prematurity is one of the major causes of perinatal morbidity and mortality in our times. Treatment is given between gestational age 20 to 37 weeks by use of β_2-adrenergic drugs. In German speaking countries, Fenoterol is mostly used.

The pharmacological properties of β_2-adrenergic drugs - relaxation of smooth muscles leading to a stop of uterine contractions, tachycardia, lowering of diastolic blood pressure and serum potassium, hyperglycaemia - are well known. Studies on the effects of β_2-adrenergic drugs on the fetus and child revealed the risk of impairment of myocardial function (Vogtmann et al. 1982; Plieth et al. 1980; Löser et al. 1981) and the risk of hypoglycaemia (Epstein et al. 1979; Procianoy and Pinheiro 1982) in the postnatal period. In addition, pediatricians reported a decrease in muscle tone in children exposed to Fenoterol *in utero* which could not be explained otherwise (personal communication).

From February 1986 onwards we conducted a clinical study on tocolytic therapy with Fenoterol. The aim of the study was to assess the risk of Fenoterol treatment for the fetus and on child development. The hypotheses were that Fenoterol will impair myocardial performance, will decrease glucose levels in the newborn and will decrease muscle tone dose-dependently. It was assumed that the probability for a drug-effect relationship is high if the effects can be shown to be dose-dependent.

Methods

Between February 1986 and June 1990, 298 pregnant women were enrolled who met the following inclusion criteria: singleton pregnancy, no premature rupture of membranes, absence of severe maternal disease. Women were not enrolled if the treatment was begun more than a fortnight before enrollment or if any other β_2-adrenergic drug had been used in this pregnancy. Informed written consent was obtained. For ethical reasons an untreated control group of pregnant women with premature labour could not be introduced. Due to the considerable variation of the disease "premature labour and/or cervical dilatation", no fixed dosage regimen was chosen.

The study protocol included a careful record of the Fenoterol dose administered. Serial determinations of Fenoterol plasma levels and pharmacodynamic measurements were carried out simultaneously in the pregnant women throughout the entire treatment period. The area under the plasma concentration time curve

(AUC) and the overall mean concentration were estimated in each individual. At delivery a sample of cord blood was collected for the determination of Fenoterol, insulin and glucose.

In a subgroup of fifty children neonatal behaviour was assessed by use of the Brazelton Scale. In all children myocardial performance was measured within the first week of postnatal life by systolic time intervals (from ECG, phonocardiogram and arterial pulse curve) and echocardiography (left ventricular shortening fraction). Thereafter the children were examined by a pediatrician at regular intervals until the age of two including neurologic examinations. Information on the development and any disease of the children was collected carefully. Psychomotor development was measured using the Bayley Scales of Infant Development at the age of 12 and 24 months.

Data analysis was performed using the SAS Program package. The main approach to analyse the data was to prove the hypothesis of a dose-effect relationship by showing a correlation between dose and effect. In a first step "dose" or exposure was defined as the AUC, the peak plasma concentration and the duration of exposure. The main endpoints were myocardial performance, insulin and glucose levels in cord blood and muscle tone of the children seen up to two years. Data are presented as median, Q_{25} and Q_{75}.

Results

The retrospective analysis of the enrollment and follow-up is shown in Table 1. 64.3% of the eligible subjects were enrolled in the study and 17% were lost to enrollment despite daily visits of two trained nurses on all wards. Up till now, 193 woman/child pairs could have completed the study protocol; of these, 82% completed the entire course.

Table 1. Retrospective analysis of enrollment and follow-up

Enrollment

Between 86-2-15 and 90-6-30

799/6511 (12.3%)	Women treated with Fenoterol delivered in our institution
434/ 799 (54.3%)	Met the inclusion criteria
73/ 434 (16.8%)	Were lost to enrollment
82/ 434 (18.9%)	did not consent to participate
279/ 434 (64.3%)	were enrolled in the study

In addition, 19 enrolled subjects were lost to follow-up before delivery.

Follow-up (No. 1-193)

	N	(%)
Enrolled	193	100.0
Lost to follow-up before delivery	16	8.3
Delivery seen	177	91.7
Stillborn (< 25 w.gest.)	3	
Died within 3 months	2	
Living children	172	100.0
Drop outs (0-2 yrs)	24	13.9
Lost to follow-up (0-2 yrs)	7	4.1
Investigations completed	141	82.0

The treatment records are summed up in Table 2 and show a great variability with respect to dose, time of treatment and route of administration. The same holds true for the values of AUC, peak plasma concentration and mean plasma concentration. The median peak plasma concentration was found to be 870.0 pg/ml (Q_{25} : 353.0 pg/ml, Q_{75} : 1360.0 pg/ml). Fenoterol concentrations in cord blood were below the level of detection in about 50% of the children. However, in 10% of the children the cord blood concentrations were found to be in the same range as in the pregnant women undergoing treatment.

Table 2. Treatment records and pharmacokinetic parameters

	Q_{25}	Med	Q_{75}	
Start of treatment	23.6	26.8	30.7	w.gest
End of treatment	31.1	33.9	35.1	w.gest
Duration of treatment	10	24	52	days
Treatment free interval	4	17	34	days
Cumulated dose				
i.v.	0	2.16	7.28	mg
p.o.*	240	650	1425	mg
	i.v.	p.o.	i.v. + p.o.	
Route of administration	3.5	36.4	60.1	%
	Q_{25}	Med	Q_{75}	
AUC**	2953.7	7019.7	17112.6	(pg/ml)days
Peak plasma concentration	353.0	870.0	1360.0	pg/ml
Mean plasma concentration	203.0	321.0	521.0	pg/ml
Fenoterol concentration in cord blood	n.d.+	< 200	$\geq$ 200	pg/ml
	52.0	36.5	11.5	%

* Bioavailability of Fenoterol is < 5 %
** Area under the plasma concentration time curve
\+ Not detected

On the whole it can be stated that the majority of the children studied here presented normal values with respect to all endpoints. Within the wide range observed no correlation between dose - expressed as AUC, peak plasma concentration or length of the treatment period - and effects could be found.

The data on myocardial function, thickness of the interventricular septum, insulin and glucose levels in the newborn are presented in Table 3. The median values are all in the normal range (Lange et al. 1983; Lundell and Wallgren 1982).

Muscle tone was found to be normal throughout year one and two in 52% of the children. 27.5% were found to be slightly hypotone on one or more occasions. However, the majority of these children were found to be slightly hypotone once at the age of 3 months and then presented with normal muscle tone later on. 1.8% of the children were found to be hypotone at least once. Values of the Psychomotor and the Mental Development Index of the Bayley Scales are presented in Table 4.

Table 3. Results

	Q_{25}	Med	Q_{75}	
Myocardial Performance				
PEP/LVET*	0.304	0.335	0.372	
SF**	29	34	39	%
Interventricular septum	3	4	5	mm
Metabolism				
$Insulin_{cb}$+	7.6	9.6	11.6	uU/l
$Glucose_{cb}$	47.0	65.0	81.0	mg%
$Glucose_{2\text{-}4h\ pp}$	40.0	40.0	60.0	mg%
Muscle Tone				
No abnormal findings			52.7	%
Slightly hypotone at least once			27.5	%
Hypotone at least once			1.8	%

* Preejection period over left ventricular ejection time
*** Left ventricular shortening fraction
\+ Cord blood

Table 4. Bayley Scales of Infant Development

	Q_{25}	Med	Q_{75}
Mental Development Index			
12 months	100	109	118 %
24 months	98	114	127 %
Psychomotor Development Index			
12 months	92	99	115 %
24 months	96	100	105 %

Discussion

From our study any risk of a tocolytic treatment for the fetus and child with respect to myocardial performance, blood glucose in the postnatal period and the psychomotor development could not be identified.

However, there are more points to be considered:

1. There might be subgroups of children (for example, those with low gestational age, short treatment free interval, high cord blood levels, etc.) who exhibit the suspected effects such as impairment of myocardial function and others. Our data will be analysed accordingly.
2. Unwanted adverse effects in the fetus or the child may occur via maternal side-effects. The well-known symptoms of β_2-adrenergic therapy such as tachycardia, palpitations, tremor, hypokalaemia and hyperglycaemia may influence pregnancy outcome and hence the child. As stated above, we collected data on the pharmacodynamics of Fenoterol in the pregnant women which will analysed with respect to the relationship between maternal side-effects and pregnancy outcome.
3. From our data we could not prove the hypothesis of adverse effects of an intrauterine exposure to Fenoterol in the neonates and the developing child. On the other hand, we were able to demonstrate the transplacental passage of Fenoterol - plasma levels of Fenoterol in cord blood after i.v. bolus injection during second stage labour amounted to 70% of maternal plasma levels - and, in addition, a concentration-dependent rise in fetal heart rate under constant rate intravenous infusion. Therefore, we may conclude that Fenoterol exerts an effect in the fetus.
4. When the risk of drug treatment is estimated, the benefit of the treatment has to be taken into account. In contrast to the undoubted effects of Fenoterol on uterine contractions when treatment is started by the intravenous route, the efficacy of a long-term and an oral treatment has not yet been proven. Oral treatment was thought not to be effective due to low plasma levels. In our study, plasma levels under long-term multiple dose oral treatment (recommended dose 8 x 5 mg/d) were about 50% of those under the lowest recommended infusion rate of 0.06 mg/hr with a high interindi-

vidual variability in mean trough plasma levels. The latter might be due to differences in bioavailability (less than 5% as shown by pharmacokinetic data collected in healthy volunteers by our group). However, there is no reason why Fenoterol given by the oral route should not exert an effect on the uterus if the appropriate concentration can be measured. On the other hand, tachyphylaxia is known to occur with β_2-adrenergic drugs. We could demonstrate this with respect to tachycardia and hypokalaemia in the pregnant women and it has to be assumed to exist with the tocolytic effect as well. In conclusion, the dosage regimen has to be individualized to avoid the development of tolerance and to avoid underdosing.

The clinical study presented here provided us with data on the effects of a drug on the fetus and on child development under the condition of drug treatment. The quality of the data may be less than those one can obtain from studies in laboratory animals or even in healthy volunteers. The number of data points which can be collected are limited and variability is great. However, a clinical trial may be an instrument to collect data on all aspects of pharmacokinetics and pharmacodynamics, including prenatally induced adverse effects on the offspring.

References

Epstein MF, Nicholls E, Stubblefield PG (1979) Neonatal hypoglycemia after beta-sympathomimetic tocolytic therapy. J Pediatr 94: 449-453

Lange L, Fabecic-Sabadi V, Beni G (1983) Vergleichende Übersicht echokardiographischer Normalwerte vom Frühgeborenen bis zum Adoleszenten. Herz 8: 105-121

Löser H, Steinkamp U, Müller KM, Pfefferkorn JR, Dame WR, Hilgenberg F (1981) Kardiotoxische Wirkung des Tokolytikums Fenoterol (Partusisten[R]). Münch med Wschr 123: 49-52

Lundell BPW, Wallgren CG (1982) Assessment of left ventricular adaption to extrauterine circulation. Systolic time intervals in the newborn infant. Acta Paediatr Scand 71: 745-752

Plieth M, Ache-Ebelt H, Fitzner R, Bein G, Karkut G (1980) Untersuchung der Herzfunktion nach Langzeittokolyse mit Fenoterol und Isoxsuprin bei Neugeborenen. Z Geburtsh u Perinat 184: 275-282

Procianoy RS, Pinheiro CEA (1982) Neonatal hyperinsulinism after short-term maternal beta sympathomimetic therapy. J Pediatr 101: 612-614

Vogtmann Ch, Kögler B, Ruckhäberle KE, Richter Th (1982) Myokardiale Kontraktilität Neugeborener nach Tokolyse mit Beta-Sympathikomimetika. Z Geburtsh u Perinat 186: 136-140

Discussion of the Presentation

Peters: First of all I would like to congratulate you on your presentation, with the careful study you have done, and the learned way in which you presented the results and your risk assessment. Such a contribution is what we need on the use of necessary drugs in pregnancy. Difficult to avoid is a bias with respect to the selection of who should be treated and who not?

Hildebrandt: The decision to treat the pregnant woman with premature labor was done by the physician at the clinic and not by the medical persons involved in the study. On the whole, the number of pregnant women treated with fenoterol remained constant throughout the study period.

Weissinger: How do these human data presented compare to data on effect of administration of fenoterol in animals and offspring?

Hildebrandt: I don't have an overview of the data in animals.

Dudenhausen: You mentioned the increase of the fetal heart rate during treatment with fenoterol. Did you see a difference in the increase between oral or intravenous route of treatment?

Hildebrandt: We haven't looked for this difference up till now, but we do know that the question has to be answered.

Neubert: If only considering the bioavailability and the pharmacokinetic data, one could come to the conclusion that fenoterol is not a substance to be administered orally. Since you have good comparative data on oral versus intravenous treatment, would you still recommend oral treatment on the basis of your findings? Is it possible to assess the therapeutic effect achieved?

Hildebrandt: The bioavailability of fenoterol after oral administration is below 5% and displays a great variability. After oral administration we found plasma levels which were in the range of those after intravenous treatment. Therefore, I think oral treatment can be effective and may be recommended if the woman is monitored carefully. In the literature a few studies are reported which attempted to assess therapeutic efficacy. Betamimetic drugs are thought to be effective in preventing premature delivery. It is difficult to answer this question with respect to oral treatment since an untreated control group cannot be introduced in a study design nowadays for ethical reasons.

Histochemical and Immunocytochemical Investigations of the Fetal Extravascular and Vascular Contractile System in the Normal Placenta and During Preeclampsia

Renate Graf, Hans-Georg Frank and Taylan Öney

Introduction

In the normal human placenta, angiotensin II is generated in the fetoplacental vascular bed and is thought to be important in the local modulation of vascular resistance (Maguire et al. 1988). The angiotensin II-degrading peptidase aminopeptidase A (L-α-aspartyl (L-α-glutamyl)-peptide hydrolase, EC 3.4.11.7, formerly angiotensinase A, now glutamyl aminopeptidase, EAP) has been separated biochemically and its angiotensinase activity confirmed (Mizutani et al. 1981). Therefore, it may be assumed that in analogy to the kidney renin-angiotensin system, placental angiotensin may also be cleaved by EAP. Histochemically, low activities of this enzyme have been demonstrated in the media of fetal placental blood vessels (Gossrau et al. 1987).

In the fetal placenta a second blood pressure-regulating system may exist in addition to the angiotensin system: In the stroma of term placental cotyledon evidence has been given for the presence of a blood pressure-regulating system in fetal blood vessels and an extravascular contractile myofibroblastic system. Together they may lead to a rhythmic self-movement of the placenta and their function is thought to be based on the cleavage of substance P (Heymann and Mentlein 1984). Substance P has been demonstrated in umbilical cord blood and biochemically shown to be cleaved by dipeptidyl peptidase IV (EC 3.4.14.5, DPP IV, Skrabanek et al. 1980; Heymann and Mentlein 1984). DPP IV activity has been localized histochemically in endothelial cells of placental fetal blood vessels (Gossrau et al. 1987). Thus, obviously two different fetal blood pressure-regulating systems are present in the placenta and the occurrence of both enzymes, DPP IV and EAP, has already been demonstrated in the fetal blood vessel system of the term placenta. Therefore, the aim of the present study was to investigate whether alterations of these peptidases would occur in preeclampsia. The studies were carried out using histochemical and biochemical means.

Material and Methods

Placentae of uncomplicated pregnancies and from patients with preeclampsia were delivered by elective caesarian section between the 39th and 41st week of gestation. Preeclampsia was defined as the presence of blood pressure values of at least 140/90 mmHg in a previously normotensive woman measured at least twice with 6 or more hours between the individual measurements. In some patients hypertension was concomitant with significant proteinuria (> 0.3 g/24 h).

Histochemistry: EAP and DPP IV were investigated using 10 μm thick acetone/chloroform-pretreated (100% 1:1 v:v, 5 min at -25°C) cryostat sections. Detection of EAP activity was performed with 1.5 mM L-α-glutamyl- or L-aspartyl-2-naphthylamide (α-Glu- or Asp-NA, Bachem, Heidelberg, FRG) or α-Glu- or Asp-4-methoxy-2-naphthylamide (α-Glu- or Asp-MNA) as substrates in the presence of 1 mM $CaCl_2$ or 1 mM EDTA, or without addition of $CaCl_2$ or EDTA, and DPP IV activity using 1.5 mM glycyl-prolyl-MNA (Bachem). Media were prepared with 0.1 M cacodylate or Tris-maleate buffer, pH 7.2. Simultaneous azo-coupling for both enzyme detections was performed using 1 mg Fast Blue B salt pure (Serva, Heidelberg, FRG) per ml incubation medium (Lojda and Gossrau 1980).

Kinetic fluorometry of EAP: 20% (w/v) homogenates of whole placentae in 0.05 M imidazole buffer, pH 7.4, with 2% Triton X-100 were prepared as described previously (Graf et al. 1990). For the measurements of EAP activity media were prepared with the same buffers as for the histochemical investigations using 4 mM to 0.004 mM α-Glu-MNA in the presence of 2 mM $CaCl_2$ or 1 mM EDTA or without addition of $CaCl_2$ or EDTA. 50 μl supernatants were added to 1 ml substrate solution. 30 seconds after thorough mixing of the supernatant and substrate solution kinetic measurements were carried out for 5 min at room temperature with a quinine sulphate calibrated Farrand Foci Ratiofluorometer (New York, USA; primary filter 350 nm, secondary filter 450 nm, sample mode measurement). Computer-aided calculations of the kinetic parameters Vmax, Km, Kis (inhibition constant of substrate), r (relative inhibition constant formed by the ratio Kis/Km) and normalized plotting of data points and calculated curves were based on the modified enzyme kinetics for the case of high substrate inhibition as given by Dixon and Webb (1979).

Immunocytochemistry: 4 μm thick acetone-pretreated (5 min at -25°C) and subsequently air-dried cryostat sections were incubated with monoclonal anti-α-actin (Boehringer, Mannheim, FRG) or polyclonal anti-desmin (Laboserv, Gießen, FRG) for 1h at room temperature. Reactivity was demonstrated using goat anti-mouse IgG/TRITC (Nordic, Tilburg, Netherlands) or swine anti-rabbit IgG/TRITC (Dakopatts, Hamburg, FRG).

Results

Glutamyl aminopeptidase. In placentae of healthy women generally little reaction product was detectable. Preeclampsia highly enhanced the hydrolysis rate of EAP in all active sites (Figs. 1 and 2). In all placentae enzyme activity was localized in cytotrophoblasts and in the media of large fetal blood vessels. In cytotrophoblasts reaction product was distributed throughout the cytoplasm (Fig. 3). Reaction product often displayed a checked pattern, which in slender cell processes assumed a striated appearance. In preeclampsia, cytotrophoblast cell processes positive for EAP activity, which seemed to be thicker than in normal placentae, were seen to underline large areas of the syncytiotrophoblast (Figs. 1 - 3). Light microscopic means did not allow any clear-cut statements as to whether these EAP-positive cell processes surrounded the whole cotyledon or were continuous to an EAP-positive thick layer, which, because of its morphological pattern, resembled the basement membrane of the trophoblast cells.

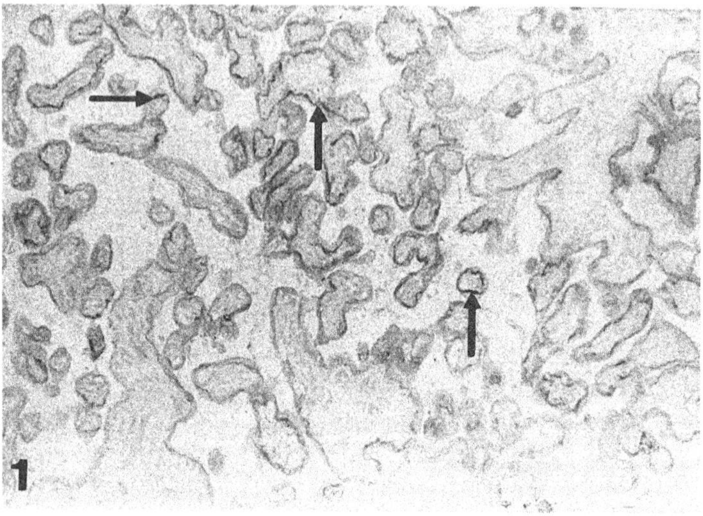

Fig. 1 - 3. Glutamyl aminopeptidase (angiotensinase A) activity in the human term placenta

Fig. 1. Uncomplicated pregnancy. Low amounts of reaction product (arrows)

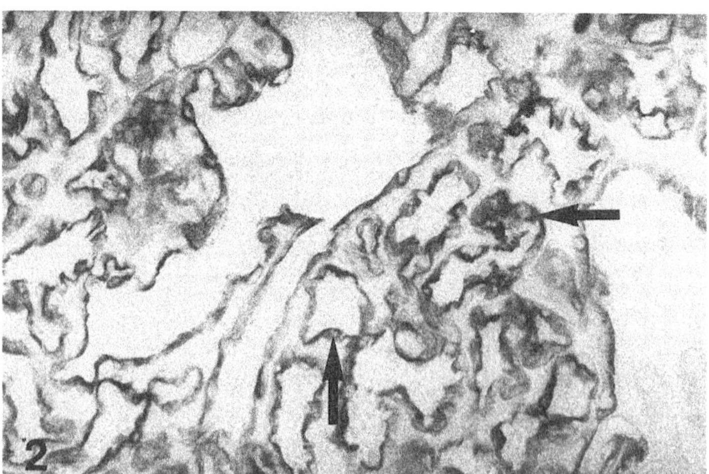

Figs. 2. Preeclampsia. Enhanced enzyme activity. Reaction product is localized in the cytotrophoblast (arrows)

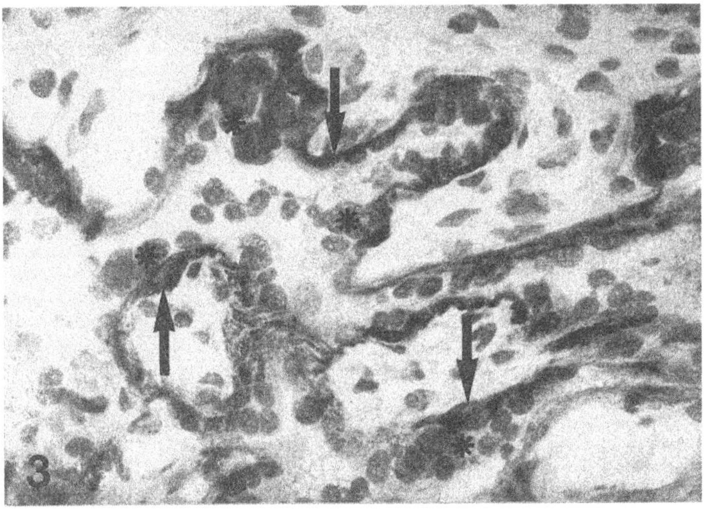

Fig. 3. Preeclampsia. Enhanced enzyme activity. Reaction product is localized in the cytotrophoblast (arrows). Counterstained with hemalum. Syncytiotrophoblast (*)

Histochemical detection of EAP activity in the human placenta essentially depended on the substrate and buffer applied. The use of α-Glu-MNA yielded higher hydrolysis rates than Asp-MNA. Only extremely diffusely distributed reaction product was found after incubation with α-Glu- or Asp-NA as substrates. In placentae of healthy women, application of cacodylate buffered media, in contrast to Tris-maleate buffer, led to a lack of reaction product in cytotrophoblasts and to a decrease in the media of fetal blood vessels. Addition of EDTA inhibited calcium-chloride activated EAP activity.

Kinetic characterization of EAP clearly indicated the presence of high substrate inhibition (Fig. 4). The maximally achievable Vr (Vrmax) increased with addition of calcium and was lowest in the presence of EDTA. The Sr-value at Vrmax was identical with the square root of the relative inhibition constant r, that was high upon addition of calcium and reduced by EDTA. The changes of Vrmax and r clearly showed that the reaction was inhibited by EDTA and activated by calcium. In contrast to the histochemical results, according to which incubation with Tris-maleate buffered media yielded higher EAP activity than cacodylate buffered media, cacodylate buffer was shown to be superior in fluorometric detection.

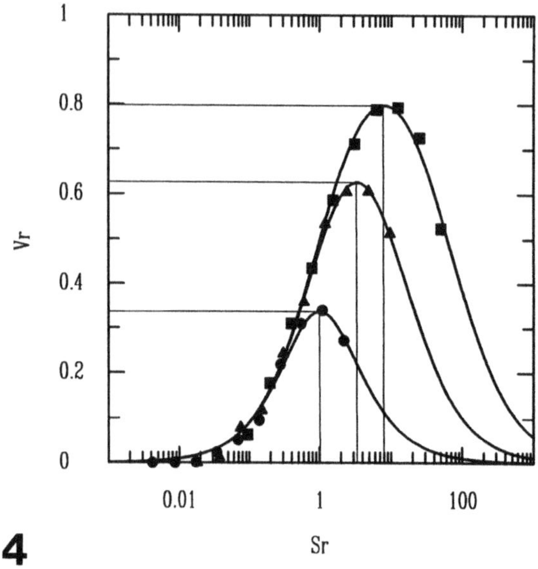

4

Fig. 4. Normalized plotting of kinetic features of glutamyl aminopeptidase. Relative velocity (Vr, defined by v/Vmax) on the y-axis and relative substrate concentration (Sr, defined as S/km) on the logarithmically scaled x-axis. X-coordinates of the maxima of curves are identical with the square root of the relative inhibition constant r. Media contained either 1 mM EDTA (●), no additional calcium (▲) or 2 mM calcium (.)

Dipeptidyl peptidase IV. In placentae of healthy women very moderate activity of DPP IV occurred in the syncytiotrophoblast, whereas high amounts of reaction product were demonstrated in cytotrophoblasts and fibroblast-like cells of the villous stroma (Fig. 5). In terminal villi DPP IV-positive elongated cell processes reached the basement membrane of the trophoblast and seemed to be attached there (Fig. 6). It was difficult to clarify whether these processes belonged to the highly DPP IV-active fibroblast-like cells. High amounts of reaction product were also seen in endothelial cells of all fetal blood vessels. Furthermore, in stem and intermediate villi conspicuous long and slender spindle-shaped cells lying in the adventitia of large fetal vessels and predominantly running in longitudinal direction in their vicinity in the villous stroma, exhibited high DPP IV activity (Fig. 7). A number of these cells extended to the basement membrane of the cytotrophoblasts and seemed to be attached there. In placentae of preeclamptic women a generalized rather moderate inhibition of DPP IV activity was demonstrated.

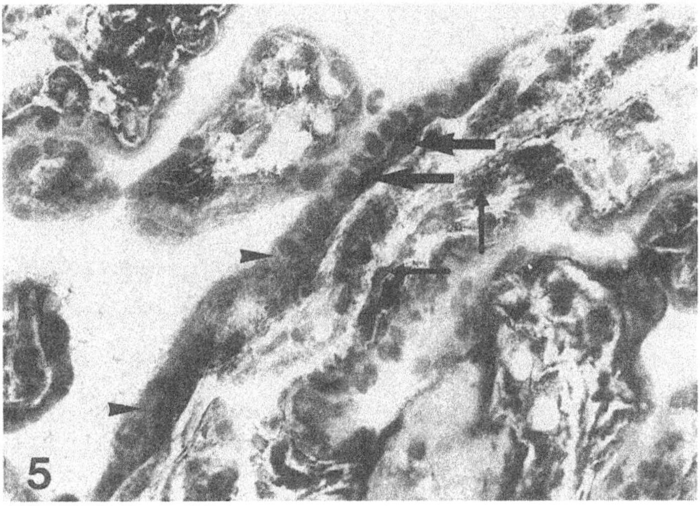

Fig. 5 - 7. Dipeptidyl peptidase IV activity in the normal term placenta. Hemalum counterstaining

Fig. 5. Low amounts of reaction product in the syncytiotrophoblast (arrowheads), high amounts in cytotrophoblasts (arrows) and fibroblasts (thin arrows)

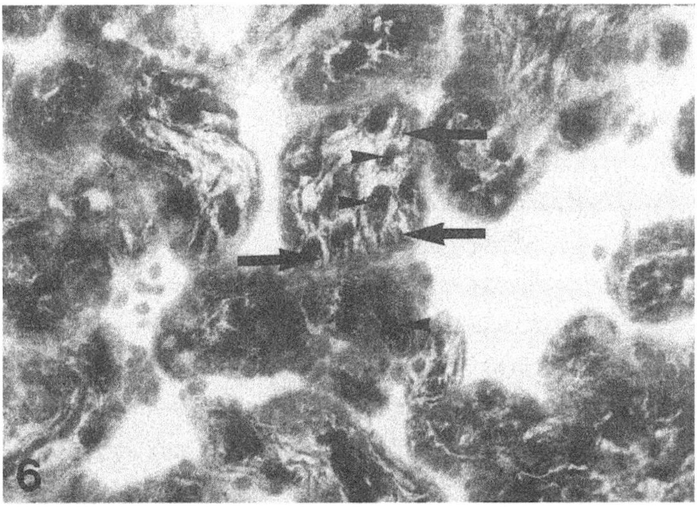

Fig. 6. DPP IV reaction in cell processes extending to the trophoblast (arrows). (Arrowheads = capillary endothelial cells)

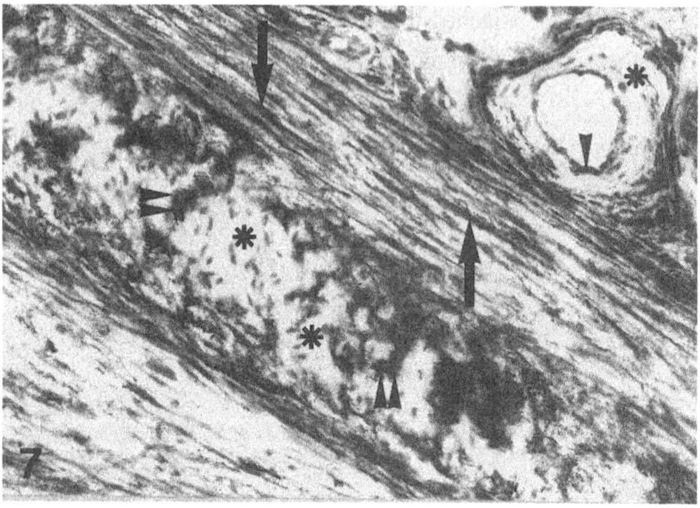

Fig. 7. Stem villous. DPP IV reaction in extravascular spindle-shaped cells (arrows), adventitia (double arrowheads) and endothelial cells (arrowheads). Note absence of reaction product in media muscle cells (*)

Anti-actin reactivity. In normal placentae, in all villi fluorescence could be localized in media muscle cells of large and small fetal blood vessels. Furthermore, in stem and intermediate villi reactivity was seen in long and slender spindle-shaped cells lying in the adventitia of large vessels and running in longitudinal direction in their vicinity in the stroma (Fig. 8). No reactivity was present in stromal cells of terminal villi. In placentae of women with preeclampsia less fluorescence and fewer fluorescent structures could be detected in most small vessels of intermediate and terminal villi. Some vessels were unaffected.

Anti-desmin reactivity. In normal placentae, in all large fetal blood vessels of stem and intermediate villi immunofluorescence was present at the same sites as anti-actin reactivity. It occurred also in long and slender spindle-shaped cells lying in the adventitia of the vessels and in their vicinity. No fluorescence was found in small blood vessels. In intermediate and terminal villi anti-desmin reactivity was detected in a delicate network formed by cells with long processes. No alterations were found in placentae from preeclamptic patients.

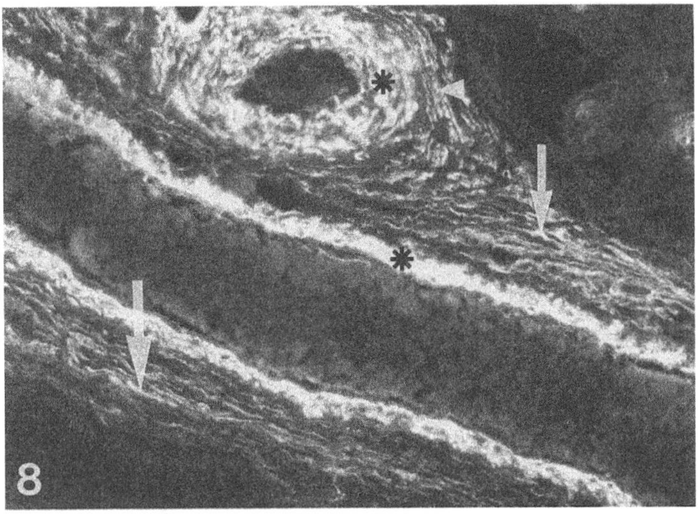

Fig. 8. Stem villous, normal placenta. Anti-actin reactivity in spindle-shaped extravascular cells (arrows), adventitia (double arrowheads) and media muscle cells (*)

Discussion

From our study it is evident that in the fetal part of the term placenta EAP and DPP IV, which are involved in the regulation of two different blood pressure systems, are both localized in cytotrophoblasts, but are localized at different sites in the fetal blood vessels and in the stroma of the villi. Only DPP IV-active cells seemed to be attached to the basement membrane of the trophoblast and only EAP showed clear alterations in preeclampsia, which concerned cytotrophoblasts and fetal blood vessels.

The histochemical demonstration of EAP in the human placenta raises the question whether this enzyme activity is the same as the one demonstrated by biochemical measurements and which was confirmed to be angiotensinase A (Mizutani et al. 1981). Our histochemical detection method is based on the method used in previous extensive histochemical and biochemical investigations of angiotensin-degrading aminopeptidases in the rat kidney (Kugler 1982) and of aminopeptidase A (glutamyl aminopeptidase) in various rat organs (Lojda and Gossrau 1980). EAP catalyses the removal of N-terminal Glu- and Asp-residues from natural peptides including angiotensin II and from the synthetic peptides of NA and MNA which therefore are suitable substrates in histochemistry (Glenner et al. 1962; Lojda and Gossrau 1980; McDonald and Barrett 1986). Our results for EAP activity are in agreement with studies which have demonstrated the general superiority of MNA-substrates in qualitative protease histochemistry (Nachlas et al. 1960) and of α-Glu-MNA compared to Asp-MNA in the case of EAP in rat organs (Lojda and Gossrau 1980; Kugler 1982).

In a previous histochemical study using α-Glu-MNA as substrate and cacodylate buffer, no EAP activity was detected in cytotrophoblast cells of the normal human placenta (Gossrau et al. 1987). In the present study, positive results were obtained in cytotrophoblasts by the use of Tris-maleate buffered media. In fluorometry, however, the use of cacodylate buffer was proven to be superior to Tris-maleate buffer. Media used for fluorometric measurements differ from those used in histochemical enzyme detection only by the lack of the coupling agent Fast Blue B. Therefore, interactions of either cacodylate or Tris-maleate buffer with Fast Blue B may be responsible for the different results in histochemistry and fluorometry (Frank 1990; Graf et al. 1990).

Activation of EAP by $CaCl_2$ and inhibition by EDTA, as demonstrated in our study by histochemistry and fluorometry, are proposed as additional criteria for angiotensinase A (McDonald and Barrett 1986) and therefore confirm that the histochemically detected EAP is identical with the angiotensin II-degrading enzyme. High substrate inhibition seems to be a kinetic feature also for other exopeptidases (Hartmann and Gossrau 1983).

High activities of EAP and DPP IV were localized in cytotrophoblasts. These cells possess a function in metabolic processes, but mostly serve as a kind of stem cells for the regeneration of the syncytiotrophoblast (Kaufmann 1972). Cytotrophoblasts could aid in the general cleavage of proteins or peptides providing the fetus with amino acids. However, they may also cleave their specific substrates angiotensin II and substance P, passing the cells from the maternal side and thus protecting fetal vessels from an overflow of these additional blood pressure-regulating substances and/or they may mediate vaso- and/or stromal constriction in co-operation with the myofibroblasts.

A striking feature in placentae from preeclamptic patients was that most villi were almost completely surrounded by a thick layer of reaction product generated by EAP. Reaction product was located in cell processes of cytotrophoblasts but may also be present in the basement membrane. This differentiation, however, is difficult to perform by light microscopy. In preeclampsia or under conditions of hypoxia it is well established that cytotrophoblasts proliferate, differentiate and become hyperplastic and then cover up to 80% of the whole placental cotyledon (Schiebler and Kaufmann 1981; Kaufmann and Stark 1972; Kaufmann 1972). Our findings of enhanced EAP activity in preeclampsia may be connected to these alterations. It cannot be ruled out, however, that EAP is also secreted into the basement membrane, since this protease can also be measured in body fluids (Lalu et al. 1984; McDonald and Barrett 1986). The detection of EAP in cytotrophoblasts leads to the assumption that these cells are, additionally, involved in the local modulation of the placental renin-angiotensin system, an assumption which is underlined by the immunocytochemical localization of renin in cytotrophoblasts (Maguire et al. 1988).

The long and slender spindle-shaped cells in the stem and intermediate villi exhibiting DDP IV activity showed a striking similarity to those with anti-actin and anti-desmin reactivity. It is generally accepted that actin and desmin are present in smooth muscle cells and myofibroblasts and both antibodies utilized in this study are declared as markers for both cell types by the manufacturer. Feller et al. (1985) and Schmidt et al. (1986) have described anti-actin and DPP IV positive cells in placental villous stroma that fulfill all criteria known for myofibroblasts. They point out that these cells differ from smooth muscle cells which lack DPP IV activity. Because of their studies it is tempting to speculate that at least the extravascular spindle-shaped cells in the stem and intermediate villi, which were altogether DPP IV-, anti-actin- and anti-desmin-positive represent the same cell type and may be considered as myofibroblasts. However, morphologically these cells differ from the myofibroblasts described by Feller et al. (1985) and Schmidt et al. (1986). They rather resemble the spindle-shaped smooth muscle cells in the stem villi described by Krantz and Parker (1963) and the vimentin-, desmin- and actin-positive cells surrounding large vessels as described by Beham et al. (1988). In the intermediate and terminal villi a second fibroblast-like DPP IV-positive cell type was localized, which, however, showed no reactivity with the monoclonal anti-actin antibody used in this study. According to its morphology, this cell type resembles the myofibroblasts described by Feller et al. (1985) as well as the second branching smooth muscle cell type described by Krantz and Parker (1963). Further studies have to clarify whether these cells are also identical with the anti-desmin-positive cells as shown in our study. Thus, we assume that two DPP IV-positive cell types are present in the human term placenta: The spindle-shaped myofibroblast which is mainly restricted to the stem and intermediate villi and the myofibroblast-like type, which is localized in the intermediate and terminal villi. Both cell types obviously reach the basement membrane of the trophoblast. In modification of the hypothesis of Heymann and Mentlein (1984) and in accordance with Krantz and Parker (1963) we suppose that the spindle-shaped myofibroblasts of the stem and intermediate villi may provide a suction and pressure pump for the fetal blood flow. Whether this function is due to the cleavage of substance P or may be affected by other modulating systems must be the subject of further investigations.

In conclusion, this study shows that in preeclampsia only one of the two presumably fetal blood flow-regulating systems, the renin-angiotensin system, is altered. This concerned cytotrophoblasts and fetal blood vessels. Since the action of both investigated proteases is located in fetal placental cells and tissues, this implies the possibility of an induction of blood pressure regulation by the fetus and therefore also the possibility that the fetus provides reasons for a dearrangement of regulatory mechanisms, which in turn may encroach on the maternal organism.

Acknowledgements. The authors are indebted to the staff of the Dept. of Obstetrics and Gynecology for excellent cooperation, M. Gutsmann, M. Khakpour, U. Sauerbier and H. Tersch for technical assistance, and B. Steyn for grammatical corrections. This work was supported by the German Research Foundation (Deutsche Forschungsgemeinschaft) (Sfb 174).

References

Beham A, Denk H, Desoye G (1988) The distribution of intermediate filament proteins, actin and desmoplakins in human placental tissue as revealed by polyclonal and monoclonal antibodies. Placenta 9: 479-492

Dixon W, Webb EC (1979) Enzymes. Longman Group Ltd, London, 3rd ed, pp 126-137

Feller AC, Schneider H, Schmidt D, Parwaresch MR (1985) Myofibroblasts as a major cellular constituent of villous stroma in human placenta. Placenta 6: 405-415

Frank HG (1990) Interactions of Ala-4-methoxy-2-napthylamine in enzyme-free incubation media. Transact Roy Microsc Soc 1: 581-584

Glenner GG, McMillan PJ, Folk JE (1962) A mammalian peptidase specific for the hydrolysis of N-terminal α-L-glutamyl and aspartyl residues. Nature 194: 867

Gossrau R, Graf R, Ruhnke M, Hanski C (1987) Proteases in the human full-term placenta. Histochemistry 86: 405-413

Graf R, Frank H-G, Szabó A (1990) Histochemical and kinetic fluorometric investigations with Ala- and Leu-4-methoxy-2-naphthylamide (MNA) and Ala- and Leu-2-naphthylamide (NA) as substrates for proteases in the normal rat placenta and after application of steroid hormones. Transact Roy Microsc Soc 1: 585-588

Hartmann K, Gossrau R (1983) Zur Eignung von Aminomethylcoumarin- und Naphthylaminsubstraten für mikrochemische Peptidasenmessungen. Acta histochem Suppl 28: 295-301

Heymann E, Mentlein R (1984) Beeinflußt Dipeptidylpeptidase IV Blutdruck und Gerinnung? Klin Wochenschr 62: 2-10

Krantz KE, Parker JC (1963) Contractile properties of the smooth muscle in the human placenta. Clin Obstet Gynecol 6: 26-38

Kaufmann P (1972) Untersuchungen über die Langhanszellen in der menschlichen Plazenta. Z Zellforsch 128: 283-302

Kaufmann P, Stark J (1972) Enzymhistochemische Untersuchungen an reifen menschlichen Plazentazotten. I. Reifungs- und Alterungsvorgänge. Histochemie 29: 65-82

Kugler P (1982) On angiotensin-degrading aminopeptidases in the rat kidney. Adv Anat Embryol Cell Biol 76: 1-86

Lalu K, Lampelo S, Nummelin-Kortelainen M, Vanha-Perttula T (1984) Purification and characterization of aminopeptidase A from the serum of pregnant and non-pregnant women. Biochim Biophys Acta 789: 324-333

Lojda Z, Gossrau R (1980) Study on aminopeptidase A. Histochemistry 67: 267-290

Maguire MH, Howard RB, Hosokawa T, Poisner AM (1988) Effects of some autacoids and humoral agents on human fetoplacental vascular resistance: Candidates for local regulation of fetoplacental blood flow. In: P Kaufmann, RK Miller (eds), Trophoblast Research Vol. 3, Plenum Medical Book Company, New York, London, pp 203-214

McDonald JK, Barrett AJ (1986) Mammalian Proteases. A Glossary and Bibliography. Exopeptidases, Vol 2, Academic Press, London, Orlando, San Diego

Mizutani S, Okano K, Hasegawa E, Sakura H, Yamada M (1981) Aminopeptidase A in human placenta. Biochim Biophys Acta 678: 168-170

Nachlas MM, Monis B, Rosenblatt DH, Seligman AM (1960) Improvement in the histochemical localization of leucine aminopeptidase with a new substrate, L-leucyl-4-methoxy-2-naphthylamide. J Biophys Biochem Cytol 7: 261-264

Schiebler TH, Kaufmann P (1981) Reife Plazenta. In: V Becker, TH Schiebler, F Kubli (eds), Die Plazenta des Menschen. Thieme, Stuttgart, New York, pp 51-100

Schmidt D, Feller AC, Parwaresch MR (1986) Villous myofibroblasts and their possible implication in blood flow. 2nd Meeting of the European Placenta Group, September 24-27, Rolduc Monastery, Netherlands

Skrabanek P, Balfe A, McDonald D, McKaigney J, Powell D (1980) Substance P in human cord blood and its degeneration by placenta *in vitro*. Eur J Obstet Gynecol Reprod Biol 11: 157-161

Discussion of the Presentation

Dudenhausen: Did you see any correlation between the activity of the mentioned enzymes and the degree of the pre-eclampsia?

Graf: At the moment, we don't have enough placentae to form groups of different degrees of pre-eclampsia. We are still collecting samples and will look into this problem in the near future.

Neubert: Would you like to speculate on a possible cause for these changes? And, since you now have a much better understanding of the ongoing pathological changes, could you see any therapeutic consequences in the light of your findings? I realize that this is a rather difficult question.

Graf: I think that from these results it is not possible to draw any conclusions about therapeutic consequences. The speculation about possible causes for the changes is quite difficult. It is usually thought that changes in the fetal part of the placenta during pre-eclampsia are secondary responses to the hypoxic situation. Our study shows that only one of the two potentially blood flow-regulating systems may be affected (via angiotensinase A), while the regulation of substance P seems to be unaffected. These results may underline the autonomy of the placenta. I think that more attention should be drawn on the fact that the fetus, or even the placenta itself, might be involved in the induction of pre-eclampsia.

Manson: At what stage of pregnancy did you obtain placenta for measurement of dipeptidyl peptidase IV and angiotensinase A ?

Graf: We generally used placentae of the 39th to 41st week of gestation.

Rohde: Are there any risks for the offspring?

Graf: From 13 born babies, one baby had a very bad Apgar score. Others obviously had difficulties because of their mothers anaesthesia. However, we did not directly look at the babies and did not follow their development, but concentrated on the placentae.

Some Data on the Pattern of Lymphocyte Subsets in Blood During the Perinatal Period

Reinhard Neubert, Isabella Delgado, Ursula Jacob-Müller,
Joachim Wolfram Dudenhausen and Diether Neubert

Introduction

As a prerequisite for studies of possible effects of prenatal exposure on lymphocytes in newborn, the knowledge of peculiarities of lymphocyte patterns in the blood of normal, unexposed *newborn* is essential. In our special case, information on both the human blood and the experimental model to be used is necessary.

It is worth mentioning that the lymphocyte pattern in the *mother* also changes in the course of a spontaneous delivery (Neubert et al. 1991e; Delgado et al. 1991), probably due to a "stress" reaction, and that some of these changes are also reflected in the newborn.

The information available, up till now, indicates that the lymphocyte pattern in the newborn deviates from that in the adult. As may be expected, the pattern of the newborn suggests a more immature state. This certainly has to be taken into account when attempting a risk assessment.

Within the last year we have obtained first information on both lymphocytes from human newborn (umbilical cord blood) and from marmosets. The discussion here is confined to the T cells, and our results are summarized in the following.

Lymphocyte Pattern in Maternal Blood

Although studying umbilical cord blood is the most convenient way of assessing a situation existing in the newborn, the status evaluated in cord blood may not represent the normal situation during the late fetal stage nor that of early infancy. Through the process of delivery, which represents a "stress" situation for the mother and the child, deviations may be expected to be superimposed on the normal variables.

In our studies we observed clear-cut alterations in hematological variables in maternal blood during spontaneous delivery (Neubert et al. 1991b; Delgado et al. 1991). Some of these alterations do not occur when a primary section is performed.

The alterations revealed during spontaneous delivery include (Table 1):

- a leukocytosis, and
- a lymphopenia.

These changes are consistent with the assumption of a "stress"-effect mediated by substances, such as e.g. glucocorticoids.

Table 1. Differences in maternal white blood count after spontaneous delivery and after primary C-section

	Cells/μl blood (Median and Range)		
	Non-pregnant (n = 8)	Spontaneous delivery (n = 12)	Primary sectio (n = 6)
Leukocytes	7,000 (5900-9,600)	*18,850* *(9,500-29,900)*	8,100 (6,700-9,700)
Lymphocytes	2,650 (1,800-3,800)	*1,250* *(480-1,500)*	1,550 (1,300-2,500)
CD4+	1,150 (670-1,680)	*480* *(130-600)*	850 (460-1,100)
CD8+	790 (470-1,430)	*220* *(90-570)*	360 (230-820)

Numbers in italics are significantly different from non-pregnant values ($p < 0.05$)

An important finding was that when the mothers exhibited a pronounced effect with respect to leukocytosis, apparently due to a long and painful delivery, a similar change was also observed in cord blood (Table 2). Leukocyte levels in cord blood were also drastically increased after spontaneous delivery.

When a regression analysis was made, there was a good correlation (Fig. 1) between the increase in leukocyte count in the mothers and a corresponding effect in the newborn (cord blood). These alterations have to be taken into consideration when evaluating possible effects of, e.g., drug treatment during the perinatal period.

Lymphocyte Pattern in Human Cord Blood

About twenty years ago, Xanthou (1970) and later Manroe (1979) discovered that the blood of newborn children contains a rather high number of *total* white blood cells (predominantly neutrophiles), which decline within a few days after birth. The data presented here show that the very high values only occur subsequent to spontaneous delivery (*see:* Table 2 and Fig. 1).

Table 2. Leukocyte count in maternal blood and in cord blood after spontaneous delivery or after primary C-section (*median* and [range])

	Cells/μl blood (Median values and Range)	
	Maternal* blood	Cord* blood
Spontaneous delivery** (n = 13)	*19,700* [9,500-29,900]	*17,400* [10,900-25,600]
C-section** (n = 5)	*7,800* [6,700-8,800]	*9,000* [6,500-9,300]

* These are all mother/child-pairs
** No significant difference (Mann-Whitney test) between mother and child

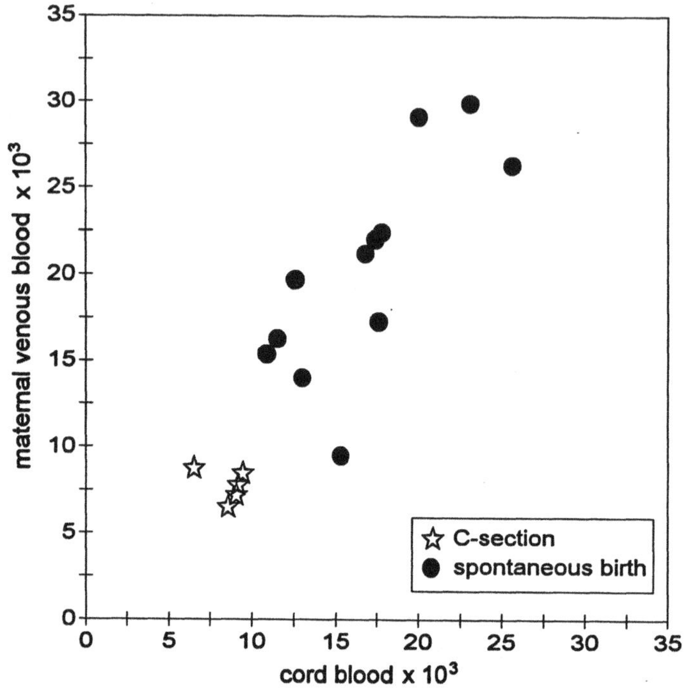

Fig. 1. Regression analysis of the number of total white blood cells in umbilical cord and in maternal blood. Each point represents a mother/child-pair. When the maternal white blood count is high, this is also the case in the cord blood. The regression equation is (n=17): Cord blood = 4140 + 0.62 x maternal blood (R-sq[adj] = 76.1%)

Data published on the pattern of lymphocyte *subpopulations* in cord blood are rather controversial. This may be explained by some intrinsic difficulties in performing such studies and the fact that multicolour flow cytometric analyses have only been used recently.

From the data of our studies it is obvious that virtually all of the $CD4^+$ cells ("helper T cells") carry the CDw29 marker (Neubert et al. 1991f; Delgado et al. 1991). However, in contrast to cells from adults the vast majority (an average of 93%) of the lymphocytes of newborn express this surface marker at a *low epitope density* (5-30 x 10^3 epitopes per cell, = $CDw29(ld)^+$ cells). In adult blood an average of 60% of the CD4$^+$CDw29$^+$ cells express this marker at a high epitope density (3-70 x 10^4 epitopes per cell, CDw29(hd)$^+$ cells), and only 40% at low density. This difference is demonstrated in the original FACScan-dotplot shown in Figure 2, and in the data compiled in Table 3.

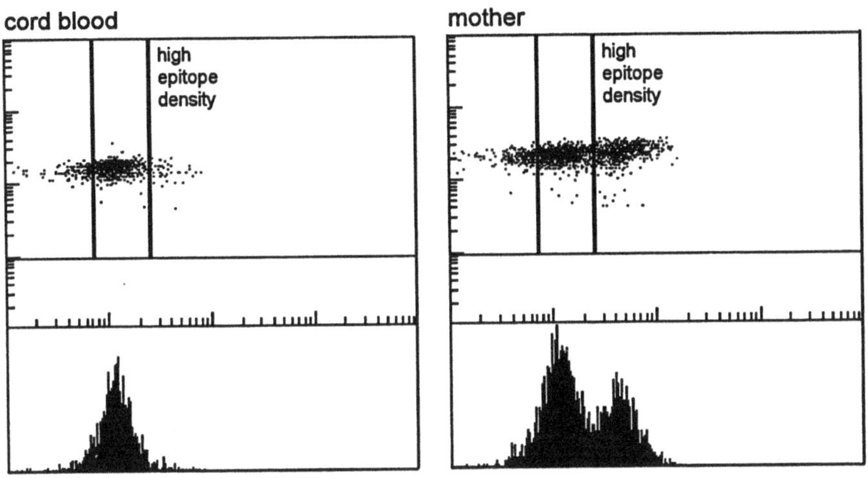

Fig. 2. Original FACScan dotplots and histograms showing the difference in the epitope density on the surface of CD4$^+$CDw29$^+$ cells in maternal and cord blood. A shift to the left denotes a decrease in epitope density (logarithmic scale). It is obvious that there is virtually no CD4$^+$ subset with a high CDw29$^+$ epitope density (>5 x 10^4 epitopes/cell) in cord blood, in contrast to the situation (two peaks) in adult blood

On the other hand, only very few of the CD4$^+$ cells in cord blood (< 3% of the total lymphocytes) were found to carry the CD45R0 marker (Table 3), which are believed to identify "memory cells". Furthermore the bearing of the CD45RA receptor is generally in conjunction with a low epitope density of the CDw29 receptor. Thus, as a result of the triple-colour analyses, more than 90% of the CD4$^+$ cells in cord blood were:

- CDw29(ld)$^+$CD45RA$^-$, or
- CDw29(ld)$^+$CD45R0$^-$, or
- CD45RA$^-$CD45R0$^-$.

Table 3. Differences in T cell subsets in blood of adults and in cord blood

Lymphocyte subtype	Cells/μl blood (Median values) Adult* blood (n = 10)		Cord blood (n = 18)	
Total lymphocytes	2650		4600	
$CD4^+$	1170	(44.2%)	2250	(48.9%)
$CD4^+CDw29(hd)^+$	*630*	*(23.8%)*	*130*	*(2.8%)*
$CD4^+CDw29(ld)^+$	420	(15.9%)	1820	(39.6%)
$CD4^+CD45R0^+$	*190*	*(7.2%)*	*40*	*(0.9%)*
$CD4^+CD7^+(hd)$	420	(15.9%)	2280	(49.6)
$CD4^+TQ1^+$	870	(23.8%)	2040	(44.3%)
$CD3^+HLA\text{-}DR^+$	72	(2.7%)	27	(0.6%)
$CD5^+CD20^+$	89	(3.4%)	340	(7.4%)
Ratio: *$CD4^+CDw29(hd)^+/CD4^+CDw29(ld)^+$*	*1.50*		*0.07*	

Italics: especially pronounced differences
* Blood from non-pregnant women
(hd) and (ld): high and low epitope density on the cell surface
(%): % of total lymphocytes

Our results are in accord with remarks published by Sanders et al. (1988), but they are in clear contrast to results published by Pirruccello et al. (1989) and of Clement et al. (1990), who reported that >95% of the $CD4^+$ lymphocytes in cord blood were $CD4^+CD45RA^+$ cells, and only few (<5%) belonged to the $CD4^+CDw29^+$ subpopulation. Apparently these authors did not recognize the $CDw29(ld)^+$ cells, because no closer two- or tri-colour analysis was performed.

It is noteworthy that the $CD4^+CDw29(ld)^+CD45RA^-$ cells of human cord blood can be converted *in vitro* into $CD4^+CDw29(ld)^+CD45RA^+$, $CD4^+CDw29(ld)^+$ $CD45R0^+$ and cells bearing the $CD4^+CDw29(hd)^+CD45R0^+$ receptors by culturing with various mitogens (e.g. pokeweed mitogen, Concanavalin A) and antigen (Neubert et al. 1991b).

The lack of these $CD45R0^+$ cells in the original cord blood provides the basis for our suggestion that lymphocytes in umbilical cord blood are less mature than cells found in the blood of adults.

Lymphocyte Pattern in Newborn Marmosets

Virtually no data are available, until now, on the pattern of lymphocytes of newborn rats, but we have recently obtained some data on the pattern of lymphocyte subpopulations in newborn of a non-human primate (*Callithrix jacchus*).

Comparative investigations of lymphocyte surface receptors in man and marmosets indicate numerous *similarities*. However, there are also some pronounced species *differences*, which have to be taken into account when a comparative risk assessment is attempted.

We have compared the pattern of lymphocyte subsets in adult human and marmoset blood, and in newborn of the two species. Some of the major findings are compiled in Figure 3. The following differences between man and marmoset are obvious:

- In the adults, the percentage of total $CD4^+$ cells is somewhat higher in man; there is no difference in the newborn.
- In both species the $CD4^+CDw29(low)^+$ population is higher in the newborn; in contrast the corresponding cells with a high epitope density on the surface, $CD4^+CDw29(high)^+$ cells (helper-inducer cells), are much lower on the first postnatal day than in adults.
- With respect to the $CD4^+CD45RA^+$ subsets, cells exhibiting a low epitope density are exceptionally frequent in the newborn marmoset; furthermore the marmoset contains more cells with a high epitope density than man, and this difference is especially pronounced in the newborn.
- There is a comparatively high percentage of "double positive" thymocytes ($CD4^+CD8^+$) in the blood of newborn marmosets.

- The newborn marmoset contains a higher percentage of cytotoxic T cells ($CD56^+CD8^+$) than the other three groups studied.
- The marmoset (adult and newborn) has more pan-B-cells ($CD20^+$) in it's blood than man.
- Human blood contains more non-activated T cells ($CD2^+HLADR^-$) than the marmoset, or the marmoset (adult as well as newborn) has more activated T cells ($CD2^+HLADR^+$) than man.

These data now form a solid basis for comparative risk assessments between marmosets and man on possible effects on lymphocytes after prenatal drug exposures. It has still to be investigated at which developmental stage the changes observed in the newborn as compared with adult marmosets disappear.

Conclusions

Too few data are available on the normal characteristics of the pattern of lymphocyte subsets in the blood of fetuses to allow meaningful studies on the susceptibility to xenobiotics of earlier developmental stages when compared with the adult organism. However, the technology to perform such studies, e.g. in the marmoset, has been worked out in adult organisms, and comparative studies can now be extended to the perinatal period.

Comparative studies on leukocytes and lymphocytes in human maternal and newborn blood have revealed drastic changes induced during spontaneous delivery, in both the mother and the newborn. Since many of these changes (e.g. leukocytosis) are not seen after primary C-section, they are probably due to the physiological "stress" situation, e.g. induced by glucocorticoid discharge.

It is not known whether these alterations are confined to man, or whether they also occur in experimental animals. The normal lymphocyte count of the adult marmoset is already much higher than in humans.

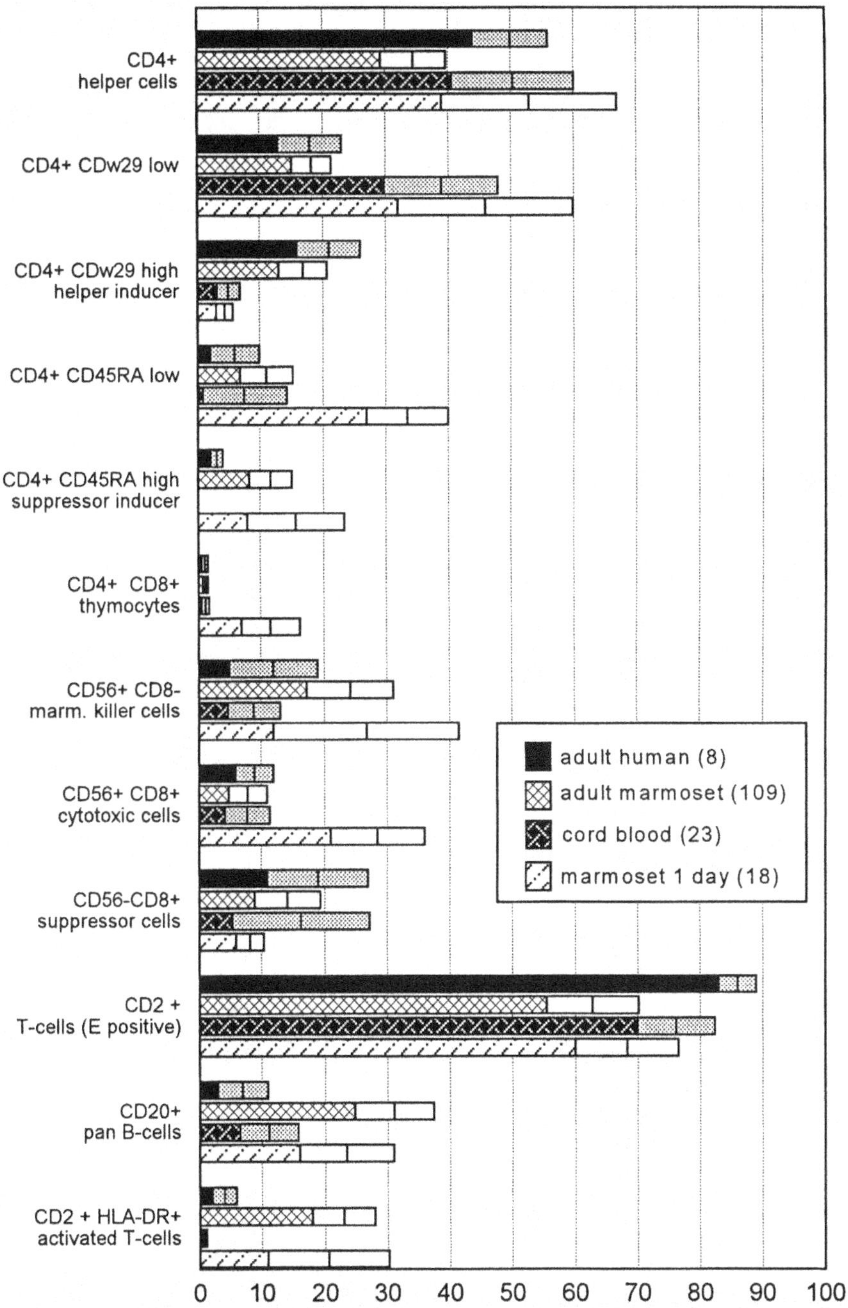

Fig. 3. Similarities and differences observed in the pattern of lymphocyte subsets in adult and newborn marmosets and humans. ( ): number of blood samples analysed. The lymphocyte pattern in newborn seems to represent a less mature state.

Nevertheless, these pronounced changes occurring during spontaneous delivery may partially obscure the typical features in the lymphocyte pattern during this developmental stage, and have to be taken into account when assessing possible additional effects of medicinal products given for therapeutic reasons, or when considering possible actions of environmental exposures or of stimulants on reactions of the immune system.

Although there are many similarities in the pattern of lymphocyte subsets in blood of marmosets and man, we have found also clear-cut differences which can now, and should be taken into account when attempting a risk assessment.

References

Clement LT, Vink PE, Bradley GE (1990) Novel immunoregulatory functions of phenotypically distinct subpopulations of $CD4^+$ cells in the human neonate. J Immunol 145: 102-108

Delgado I, Neubert R, Dudenhausen JW (1991) Lymphozyten-Subpopulationen im peripheren Blut bei Mutter und Kind während der Geburt. Perinatal Med 3: R153(21) [abstract]

Manroe BL, Weinberg AG, Rosenfold CR, Browne R (1979) The neonatal blood count in health and disease. I. Reference values for neutrophilic cells. J Pediat 95: 89-98

Neubert R, Delgado I, Dudenhausen JW, Neubert D (1991a) Conversion of $CD4^+CDw29^+CD45RA^-$ lymphocytes of human cord blood to $CD4^+CDw29^+CD45RA^+$ cells. Naunyn-Schmiedeberg's Arch Pharmacol 344 [suppl 2]: R123(78) [abstract]

Neubert R, Pegg S, Delgado I, Dudenhausen JW, Neubert D (1991b) Lymphopenia occurring in women during parturition. Naunyn-Schmiedeberg's Arch Pharmacol 344 [suppl]: R98(258) [abstract]

Neubert R, Pegg S, Delgado I, Dudenhausen JW, Neubert D (1991c) T cell subpopulations in human cord blood. Naunyn-Schmiedeberg's Arch Pharmacol 344 [suppl]: R98(259) [abstract]

Pirruccello SJ, Collins M, Wilson JE, McManus BM (1989) Age-related changes in naive and memory CD4+ T cells in healthy human children. Clin Immunol Immunopathol 52: 341-345

Sanders ME, Makgoba MW, Sharrow SO, Stephany D, Springer TA, Young HA, Shaw S (1988) Human memory T lymphocytes express increased levels of three cell adhesion molecules (LFA-3, CD2, and LFA-1) and three other molecules (UCHL1, CDw29, and Pgp-1) and have enhanced IFN-g production. J Immunol 140: 1401-1407

Xanthou M (1970) Leukocyte blood picture in healthy full-term and premature babies during the neonatal period. Arch Dis Childhood 45: 242-249

Subject Index